Big Data: Techniques and Technologies in Geoscience

Big Data: Techniques and Technologies in Geoscience

Edited by Nicolas Chapman

Larsen & Keller
www.larsen-keller.com

Big Data: Techniques and Technologies in Geoscience
Edited by Nicolas Chapman
ISBN: 979-8-88836-072-9 (Hardback)

Larsen & Keller
Published by Larsen and Keller Education,
5 Penn Plaza,
19th Floor,
New York, NY 10001, USA

Cataloging-in-Publication Data

Big data : techniques and technologies in geoscience / edited by Nicolas Chapman.
p. cm.
Includes bibliographical references and index.
ISBN 979-8-88836-072-9
1. Geology--Data processing. 2. Geology--Technological innovations. 3. Big data. 4. Geographic information systems.
I. Chapman, Nicolas.
QE48.8 .B54 2023
550.285--dc23

For more information regarding Larsen and Keller Education and its products, please visit the publisher's website www.larsen-keller.com

Table of Contents

Preface

Big data refers to a collection of data that is enormous in volume and is expanding at exponential rates with the passage of time. It is a dataset that is too large and complex for traditional data management tools to process or store it effectively. There are various types of big data including semi-structured, structured and unstructured. Big data technologies are used in a number of industries and fields including banking, communication, healthcare, geosciences, astrophysics, and trade. The geographic information systems (GISs) are an important application of big data in geosciences, which aid in creating, managing, analyzing and mapping all types of geospatial data. GISs generate huge amounts of data, which are analyzed using big data technologies. Big data computing for GIS data enables analysis and decision making from massive datasets by using spatiotemporal data mining, algorithms and query processing. This book elucidates the concepts and innovative models around prospective developments with respect to big data techniques and technologies used in geoscience. It will serve as a reference to a broad spectrum of readers.

The information shared in this book is based on empirical researches made by veterans in this field of study. The elaborative information provided in this book will help the readers further their scope of knowledge leading to advancements in this field.

Finally, I would like to thank my fellow researchers who gave constructive feedback and my family members who supported me at every step of my research.

Editor

1

Using High-Performance Computing to Address the Challenge of Land Use/Land Cover Change Analysis on Spatial Big Data

Xiaochen Kang [1,*], Jiping Liu [1,2,*], Chun Dong [1] and Shenghua Xu [1]

1 Research Center of Government Geographic Information System, Chinese Academy of Surveying and Mapping, Beijing 100830, China; dongchun@casm.ac.cn (C.D.); xushh@casm.ac.cn (S.X.)

2 Institute of Geographical Sciences, Henan Academy of Sciences, Zhengzhou 450052, China

* Correspondence: kangxc@casm.ac.cn (X.K.); liujp@casm.ac.cn (J.L.)

Abstract: Land use/land cover change (LUCC) analysis is a fundamental issue in regional and global geography that can accurately reflect the diversity of landscapes and detect the differences or changes on the earth's surface. However, a very heavy computational load is often unavoidable, especially when processing multi-temporal land cover data with fine spatial resolution using more complicated procedures, which often takes a long time when performing the LUCC analysis over large areas. This paper employs a graph-based spatial decomposition that represents the computational loads as graph vertices and edges and then uses a balanced graph partitioning to decompose the LUCC analysis on spatial big data. For the decomposing tasks, a stream scheduling method is developed to exploit the parallelism in data moving, clipping, overlay analysis, area calculation and transition matrix building. Finally, a change analysis is performed on the land cover data from 2015 to 2016 in China, with each piece of temporal data containing approximately 260 million complex polygons. It took less than 6 h in a cluster with 15 workstations, which was an indispensable task that may surpass two weeks without any optimization.

Keywords: LUCC analysis; spatial big data; high-performance computing; spatial decomposition

1. Introduction

Land surface changes result from numerous driving forces, including tropical deforestation, rangeland modification, agricultural intensification, urbanization, social-political factors and the worldwide interconnectedness of places and people [1–5]. Telecoupling provides a new avenue of research that enables natural and social scientists to understand and generate information for managing how humans and nature sustainably coexist [6–10]. With global issues such as climate change, surging population growth and continued energy shortages, more focus is being directed toward global environmental changes [10–12]. As landscapes are enormously complex systems, most methods to study the driving forces decompose the system into different functional components and then describe the state of the landscape, the processes within the landscape, and the reactions of the landscape over time [3,13,14]. In addition, monitoring the locations and distributions of land cover change is important for establishing links between policy decisions, regulatory actions and the resulting land-use activities [15,16]. Land use/land cover change (LUCC) analysis has become a fundamental component of environmental change and sustainability research [17,18]. As stated by Li [19], except for the analysis of the driving mechanism of LUCC, other existing studies are currently dedicated to designing reasonable methods to detect the changes more accurately [11,20–22] or predicting the change trends of the spatial-temporal progress of LUCC [23–26], which can be performed at a fine scale [27–32] or a global scale [33–37].

Over the last few decades, dramatic changes have occurred throughout the world. LUCC is fundamentally a spatial process, resulting from the complex interaction of social, ecological and geophysical processes [38]. Land use maps indicating the status are more detailed for agricultural and urban land use classes than for others. Satellite data has always been the primary source for obtaining information about the earth's land surface, and a wide variety of techniques ranging from unsupervised algorithms to parametric supervised algorithms to machine learning algorithms (MLA) have been used to classify the land cover over large areas from satellite [39,40]. Generally, post-classification comparison and multi-date composite image change detection are the two most commonly used methods in change detection [41–43], and the techniques of satellite remote sensing and GIS can also be used to examine the spatial and temporal patterns of the LUCC [44]. Moreover, after classification, overlaying two layers in different time series, in raster or vector format, can directly reflect the changes in pixels or polygons [45–47]. In addition, a hybrid classification by combining different types of methods or incorporating different data sources may also provide a better choice [48,49]. In recent years, the classification accuracy of these methods has been greatly enhanced with the increase of spatial resolution, such as very high-resolution (VHR) remote sensing images [50], and the optimization of knowledge-based approaches which play a role in use-specific photointerpretation and machine learning-based algorithms [51,52]. In the near future, it is likely that more and more robust and highly adjusted approaches will emerge for recognizing objects from images.

When multi-temporal images are available, how to detect the changes among these images has become a disparate, highly variable and ever-expanding area of research [53]. After classification, the land surface can be easily partitioned into different classes of land use/land cover parcels. On this basis, some research on change analysis addresses post classification methods, termed as a post-classification comparison [54,55]. For such comparisons, pixel-by-pixel [56,57] or object-by-object comparisons [58–60] are usually conducted. As suggested by Zhou et al. [59], object-based approaches provide a better means for change detection than pixel-based methods because they can provide a more effective way to incorporate spatial information and expert knowledge into the change detection process. With the resultant objects in polygon layers from the segmentation, land parcels are further split through GIS overlay functions, and then the classification can be well preserved at the object level [1,61]. Generally, two types of spatial models are widely used when representing the earth's phenomena, vector and raster, which are the data formats for performing spatial operations [62]. Therefore, to detect the changes from multi-temporal data, GIS operations can be used on raster images, vector polygon layers, or a combination of both representing the land classifications.

With the advances in sensor and computer technology, the approach for collecting, managing and analyzing the spatial data is going through a revolution. Monitoring geographical conditions can provide consistent measurements of landscape conditions, allowing detection of both abrupt changes and slow trends over time [54]. National geographical conditions are viewed as an important part of the basic national conditions, including spatial distribution and relationships among natural elements, cultural elements, and socio-economic information [63]. Routine monitoring provides a persistent data source for the LUCC analysis, a very large amount of data and the complicated computing processes require that high-performance computing (HPC) technologies should be used to conquer these challenges. In geographical information science, geocomputation refers to the adoption of a large-scale computationally intensive approach to the problems of physical and human geography in particular, and the geosciences in general [64]. In essence, HPC is used as an important component of geospatial cyberinfrastructure (GCI) and is critical to large-scale geospatial data processing and problem-solving [65,66]. In the era of spatial big data, cluster computing, cloud computing, and GPU computing play complementary, overlapping and even crucial roles in the processing of geospatial data [67]. In fact, the utilization of HPC systems in spatial big data processing has become more and more widespread in recent years [68], such as image classification [69], land use change analysis [70], urban growth simulation [71], etc. To perform land use change analysis over massive spatial data at a national scale, a very heavy computational load is often unavoidable, especially when processing the multi-temporal land cover data with finer resolution using more complicated procedures. Particularly, to solve this geospatial problem, a theoretical approach towards capturing the computational loads can be used to direct the two necessary steps: spatial decomposition and task scheduling [72,73].

This paper presents a new approach to this problem: a cluster computing paradigm for conducting the data moving, clipping, overlay operation, area calculation and the matrix building. As a validation case, we used it to quickly detect the land use changes over China from multi-temporal layers. It should be noted that the overlay operation is performed on digital maps in a vector format for different temporal land cover layers. However, we guess that the overlay operation could also be applied to raster maps with small changes.

The remainder of the paper is organized as follows. Section 2 introduces the detailed method, including the LUCC procedures, the manner of spatial decomposition and the stream scheduling mechanism. Next, two individual experiments based on the estimating factors are given in Section 3. Section 4 provides a detailed discussion of the obtained results and finally the main conclusions of the work are presented in Section 5.

2. Materials and Methods

2.1. Procedures of LUCC Analysis Based on Multi-Temporal Vector Objects

In China, the Geographical Conditions Monitoring (GeoCM) project aims to monitor all kinds of indexes for the land surfaces in a dynamic and quantitative way, thereby achieving geographical descriptions of the spatial distributions and spatiotemporal changes of natural, economic and social factors. Since 2013, China has initiated the first national survey of geographical conditions. By the end of 2015, the complete land cover/land use information had been collected through interpretation from high-resolution imagery. Since 2016, it has entered a routine monitoring stage, and the land cover data for 2016 and 2017 are now both available.

The produced land cover data are all in vector format, which can therefore clearly depict the vegetation coverage, water and wetlands, desert and bare land, traffic networks, residential sites and facilities, etc. In 2015, for example, over 260 million complex polygons were collected, and each of the produced polygons belonged to a specific class of land cover. At a rough level, ten classes of land parcels are involved, i.e., farm-lands, garden lands, wood-lands, grass-lands, buildings, roads, constructed bodies, digging lands, desert lands, and waters, which form a complete coverage of the land area. Meanwhile, at the finest level, there are more than 80 classes involved. The detailed classification is provided in Supplementary Materials.

With the high volume of land cover data, a natural workflow is described in Figure 1, which involves five procedures, i.e., data moving, clipping, overlay of multi-temporal layers, area calculation and transition matrix building.

The data moving is the first procedure (P_1), which downloads a partition of the data from the distributed file sharing system to the local storage, according to some spatial boundaries. Particularly, the volume of the partial data to be moved should be predetermined with reference to the computing power of the local CPU resources. In fact, the following procedures should also be considered for assessing the computational load. As the original data is produced and stored in physical chunks, some redundant chunks should also be moved to a local computer, which ensures data integrity when performing the LUCC analysis according to semantic partitions, such as administrative divisions. As data moving is typically an input/output (I/O)-intensive operation, the memory consumption and time usage are both closely related to the size of the dataset.

With the available local dataset, the clipping procedure (P_2) partitions them into some semantic chunks for the subsequent overlay operation (P_3) and area calculation (P_4). Herein, the semantic chunks are usually bounded by administrative divisions. As the data clipping may cause heavy pressures both on the CPU and hard disks, its time usage is often highly relevant to the number of polygon vertices. According to our experiences, when managing a large dataset, the main proportion of the consumed time in this procedure is used in computation, instead of I/O.

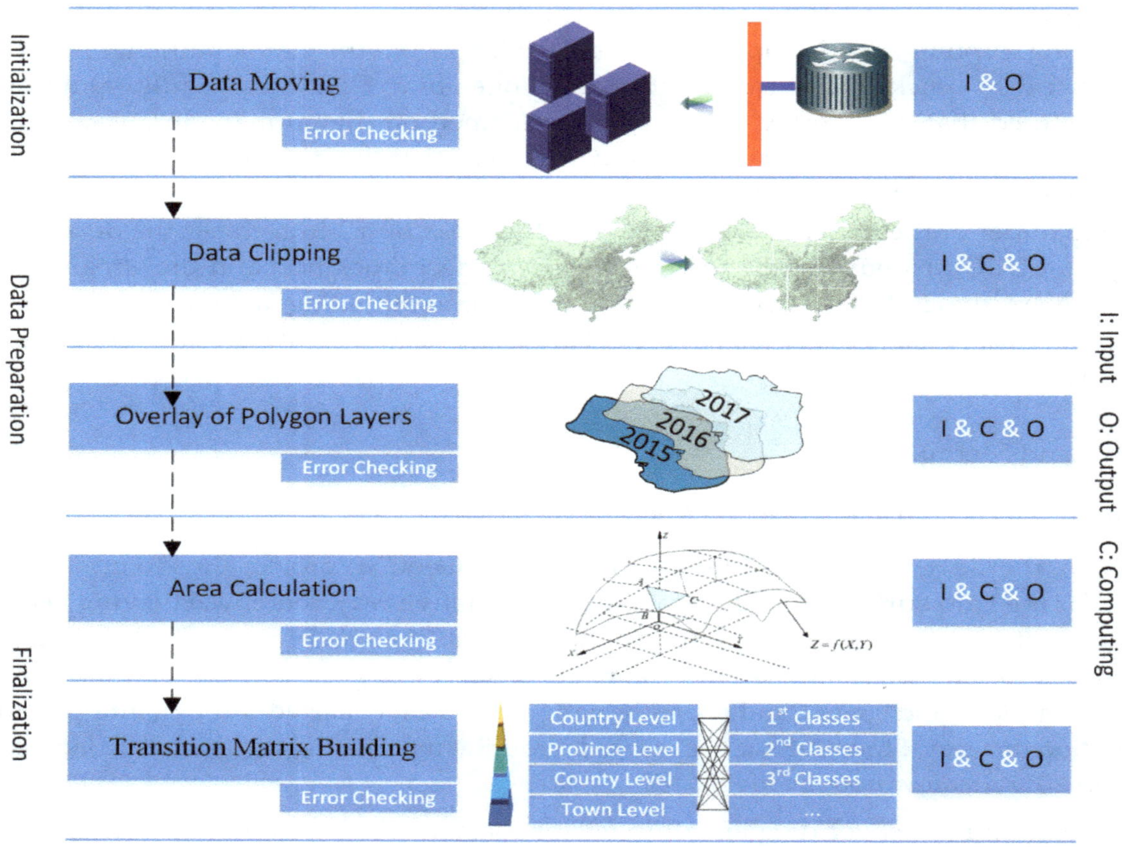

Figure 1. Workflow of the LUCC analysis.

The overlay operation creates a new layer according to two temporal layers produced in the last procedure (P_3). More importantly, the attributes of both original layers are assigned to the new layer by the intersection of the polygons. In other words, the new layer represents the LUCC information from different years. This operation is typically computationally intensive, and the time usage is determined by the number of polygon vertices to a great degree.

Area calculation is conducted to calculate the area for each of the intersected land parcels. Specifically, the area for each polygon is stored in a table row that contains the relevant information, as shown in Table 1. Naturally, area calculation is closely related to the count of the vertices in each polygon. After the above four procedures, a transition matrix is built in an MS-SQL database (P_5), which can produce the matrixes for different levels of administrative districts and different levels of land cover classes through GROUP BY and SUM operators. In comparison with the first four procedures, the time used in this procedure is negligible.

Table 1. Structure of the results for area calculation.

No.	Field Name	Notes
1	Region Code	Spatial identifier of each land parcel after intersection
2	Polygon ID	Unique identifier of each land parcel after intersection
3	Land Cover Class (from)	Land cover class before change
4	Land Cover Class (to)	Land cover class after change
5	Area	Spherical area of the land parcel after overlay operation
6	Other	A spare field

2.2. Graph-Based Spatial Decomposition

2.2.1. Administrative Level Decomposition

To perform LUCC analysis over large areas, administrative districts provide a direct and effective way for partitioning the large-scale landform, and a hierarchical structure can be drawn from the study area. Shown in Figure 2 is a 4-level hierarchy of administrative districts in China.

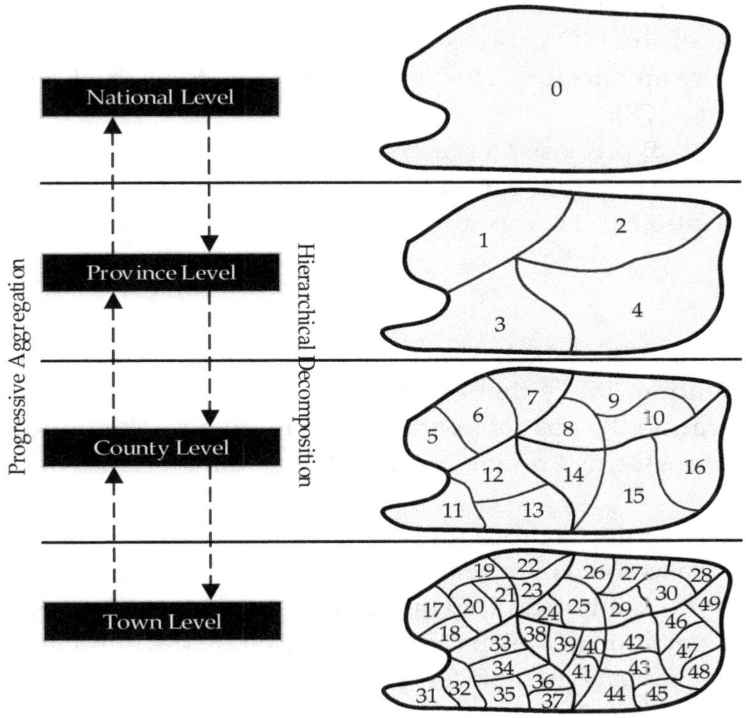

Figure 2. Administrative level decomposition in China (The numbers in these areas are unique identifiers).

From the top down, the country can be decomposed into some lower-level areas hierarchically. Meanwhile, from the bottom up, the transition matrix can be built level by level. Herein, each administrative district at the same level can be viewed as an independent partition of the country, and moreover, they can be processed simultaneously. However, different divisions usually have different areas, different numbers of land parcels, different numbers of polygon vertices, etc., which can be used to evaluate the time complexity of an algorithm or procedure and thus contribute to measuring the computational load for each of the of involved procedures. As the computational load may vary significantly in different procedures, a better strategy for load measuring requires referencing both the data and relevant operations in a combined way.

In fact, the administrative division is for the purpose of administration, and most countries in the world have three or more levels. Therefore, this initial decomposition can also be used in managing the data for other countries.

2.2.2. Computational Load Evaluation and Representation

Under the administrative level decomposition, each district in the hierarchical level involves a certain amount of computational load. How to quantitatively measure the load is a primary step for grouping the districts and scheduling the procedures. Among the five procedures in the LUCC analysis in Figure 1, the first four procedures are the most time consuming due to heavy loads in computations or I/O operations, and are thus performed in distributed workstations. Theoretically, the time usage in data moving and area calculation should present a linear correlation with the data volume to be processed, i.e., the number of polygons or the constituting vertices. Both the clipping and overlay are essentially intersection operations. According to Becker, et al. [74], the optimal time complexity of the overlay operation on two sets of polygons is $O\ (N \times \log N + K)$, where N is the number of line segments, and K is the number of line intersections. When K is significantly smaller than N, a normal selection for measuring the computational loads is the number of vertices composing the polygons. In addition, I/O consumption is another factor that cannot be ignored, which is an impediment to the scalable speedups when parallelizing the overlay operations in a shared file system [75–77]. It is noteworthy that the jitter of the I/O process is unavoidable, which is often caused by the resource contention within multicore architectures, the communication, the kernel process scheduling and the cross-application contention [78].

Wang and Armstrong [73] proposed a theoretical approach which transforms the field-based and object-based spatial dataset to a spatial computational domain and then uses a composite function to represent the I/O consumption and computing time as compu-bands.

$$f = df + of \tag{1}$$

where df is a data-centric function, which transforms knowledge about spatial data characteristics in the context of spatial operations into either memory or I/O band. of is an operation-centric function, which takes into consideration the spatial operations that are directly or indirectly related to spatial data characteristics and transforms knowledge about the characteristics of spatial operations into a computing time band.

For the LUCC analysis, multiple procedures can be identified. Herein, we concentrate on how to finish the entire workflow as soon as possible, and the first four procedures are represented as four transforming functions. $f_1(p)$ is a typical I/O operation, and its time usage is mainly determined by the data volume of a data partition p, which can be measured by the number of polygons, the number of geometry vertices, or the areas. $f_2(p)$ and $f_3(p)$ measure the computing time consumed in clipping and overlaying operations, and they have a similar measurement for the time complexity [79]. $f_4(p)$ measures the area calculation, and the number of polygon vertices is naturally used as the main influencing factor.

$$f_{1,2,3,4}(p) \propto \{L_O(p), L_V(p), L_A(p), \ldots\} \tag{2}$$

where $L_O(p)$ is the number of polygons of p, $L_V(p)$ is the number of polygon vertices of p, and $L_A(p)$ is the area of p. Moreover, the Pearson correlation coefficient (PCC) is used to measure the linear correlation between the possible variables and the time usage, and then determine which factor should be selected to measure the loads. Finally, a compressive measurement is defined based on these functions.

$$T = T_{P1}(f_1) + T_{P2}(f_2) + T_{P3}(f_3) + T_{P4}(f_4) \tag{3}$$

where T_{P1}, T_{P2}, T_{P3} and T_{P4} are fitting functions used to convert the values calculated from (2) to computing time units (e.g., seconds).

During the entire workflow, the computational loads result from both the inner part of a district and the neighboring districts. Specifically, the inner loads are closely related to the inner part of the data and the operations performed on it, while the outer loads are closely related to the neighboring divisions due to the clipping operations. According to Kang and Lin [80], each district can be represented as a graph vertex with weights (i.e., the load produced from P_1 to P_4), and the relations among the districts can be represented as graph edges with weights (i.e., the load in clipping neighboring districts). After this representation, a graph that completely illustrates the data distribution and the intensity of the computational loads in the LUCC analysis can be obtained. This graph can be formally modeled by the following formula:

$$\mathrm{Graph}(LUCC) = \mathrm{G}(\mathrm{V}[i, f_v, w_v], \mathrm{E}[j, f_w, w_e]) \tag{4}$$

where V is a graph vertex set, and E is a graph edge set. Each vertex in V contains three elements: i (the identifier for the vertex), f_v (the vertex function) and w_v (the vertex weight). f_v denotes the specific operation upon a particular object, and w_v denotes computational load of the function on that object. Similarly, each edge in E contains three elements: j (the identifier for the edge), f_e (the edge function), and w_e (the edge weight). f_e is used to merge these results from f_v, and w_e denotes the computational load of the edge function on the spatial relation between two objects. In fact, f_e is optional and can be empty if output dependence does not exist. w_v and w_e may vary greatly for different vertices and edges due to the spatial heterogeneity. For more details about the graph representation, please refer to the definition of data dependence in spatial operations [80], which also provides an operable means for generating the graph vertices and edges.

Figure 3 illustrates a sample instance by simulating seven districts, which are all represented as graph vertices. The neighboring relations are represented as graph edges. The background grids are the originally produced chunks, each of which is filled by many land class parcels. In addition, the value in each grid is the corresponding load value, and the load of a district can be obtained by summing load values from these topologically contained or intersected grids. As a matter of fact, the shapes of these grids are often irregular, and they provide complete coverage of the entire area.

In Figure 3, the weights both exist on the graph vertices and the edges. As shown in Table 2, the weight on each vertex is obtained by summing the loads of the related grids, while the weight on the edges is obtained by summing the loads of the crossing grids that intersect with the neighboring boundaries.

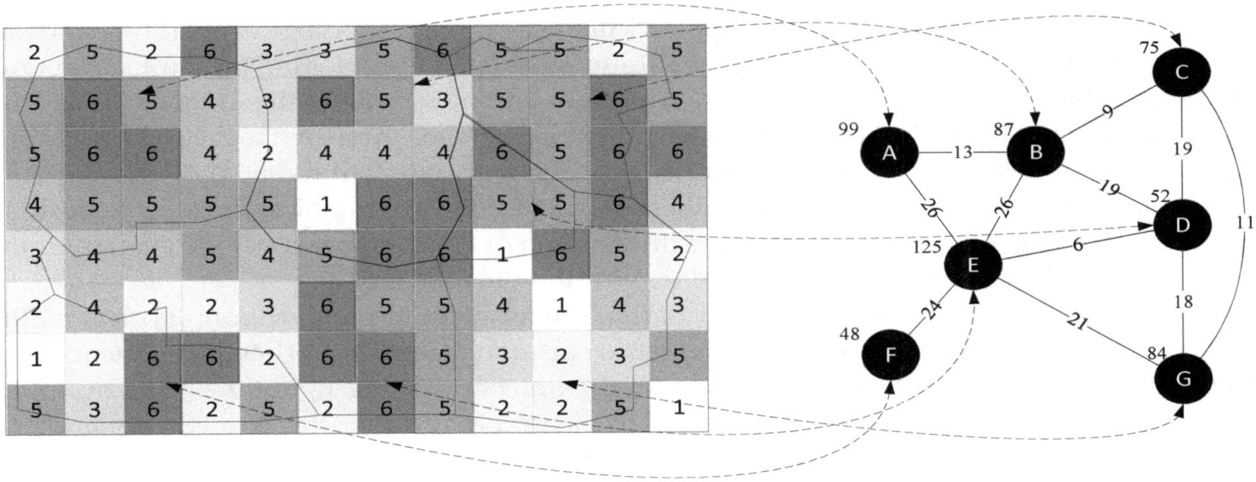

Figure 3. Graph representation of the computational loads over administrative districts (The value in each grid is the corresponding load).

Table 2. Graph weight evaluation.

Type	Vertex ID	Weights
Vertex	A	$W_A = 2+5+2+6+3+5+6+5+4+3+5+6+6+4+2+4+5+5+5+5+3+4+4 = 99$
	B	$W_B = 3+3+5+6+3+6+5+3+2+4+4+4+5+1+6+6+4+5+6+6 = 87$
	C	$W_C = 6+5+5+2+5+3+5+5+6+5+6+5+6+5+6 = 75$
	D	$W_D = 3+5+4+6+5+6+5+5+6+1+6 = 52$
	E	$W_E = 5+5+5+3+4+4+5+4+5+6+6+2+4+2+2+3+6+5+5+6+6+2+6+6+5+2+6+5 = 125$
	F	$W_F = 2+4+2+1+2+6+6+2+5+3+6+2+5+2 = 48$
	G	$W_G = 5+6+4+6+1+6+5+2+5+4+1+4+3+5+3+2+3+5+5+2+2+5 = 84$
Edges	A	$W_{AB} = W_{BA} = 3+3+2+5 = 13$; $W_{AE} = W_{EA} = 3+4+4+5+5+5 = 26$
	B	$W_{BC} = W_{CB} = 6+3 = 9$; $W_{BD} = W_{DB} = 3+4+6+6 = 19$; $W_{BE} = W_{EB} = 5+4+5+6+6 = 26$
	C	$W_{CD} = W_{DC} = 3+6+5+5 = 19$; $W_{CG} = W_{GC} = 5+6 = 11$
	D	$W_{DE} = W_{ED} = 6$; $W_{DG} = W_{GD} = 6+1+6+5 = 18$
	E	$W_{EF} = W_{FE} = 2+4+2+6+6+2+2 = 24$; $W_{EG} = W_{GE} = 6+5+5+5 = 21$

2.2.3. Graph Partitioning of the LUCC Analysis Graph

When the calculations are described as graphs in which the vertices and edges represent the computation and data dependence, they can be divided among the CPU cores of a parallel computer by partitioning the vertices of a graph [81]. In mathematics, the graph partitioning problem is defined in the form of a graph G = (V, E), with V vertices and E edges, such that it is possible to partition G into smaller components with specific properties [82]. Graph partitioning arises as a pre-processing step to divide and conquer algorithms, where it is often viewed as an effective way to divide things into roughly equal-sized pieces [83]. Partitioning a graph into equal-sized components while minimizing the number of edges between the different components can contribute to a balanced task assigning strategy in parallel computing [84,85]. Thus, the computational load can be divided among CPU cores. Moreover, some typical spatial operations have been optimized using the graph-based divide and conquer method by representing the geometry-level computation as vertices and dependence as edges, and then partitioning and encapsulating the sub-graphs into sub-tasks [80]. Specifically, each processing task over a spatial partition is viewed as a sub-task that may have a dependence on the other tasks over the neighboring spatial partitions.

After transforming the task of the LUCC analysis into a graph, it can thus be partitioned into several parts that can be simultaneously processed on different CPU cores. Two requirements should be met to complete the task efficiently, i.e., inter-independence and balanced load allocation. The inter-independence requires that each partitioned part can be processed independently without referencing the neighboring partitions in the computation, while balanced load allocation requires different partitions be processed in approximately equal time [73,86]. As geospatial science problems often have constraints in spatial and temporal dimensions (e.g., topological relations and order relations), relevant principles should be systematically considered in exploiting the parallelism toward optimizing the problems [86–89]. Two general rules can be concluded: (1) the independence among different parts from the physical phenomena are more intensive for those that are closer (localization); (2) the data access, computation and communication all contribute to the spatial distribution of the computational loads (heterogeneity). The localization enables the geographical domains to be divided and conquered, while the heterogeneity in the distribution of the computational loads requires that a reasonable load balancing strategy is adopted to partition the problem.

Many algorithms have been developed to find a reasonably good partition. Typical methods include spectral portioning methods [90], geometric partition algorithms [91–93], and the multilevel graph partitioning algorithms [94–97]. The multilevel graph partitioning algorithms work by applying one or more stages. Each stage reduces the size of the graph by collapsing the vertices and edges, partitioning the smaller graph, then mapping back and refining this partition of the original graph [97]. In many cases, this approach can provide both fast execution speed and very high-quality results. One widely used example of such an approach is called METIS [84], which was developed in the Karypis lab (http://glaros.dtc.umn.edu/gkhome/). METIS provides a set of programs for partitioning graphs, partitioning finite element meshes, and producing fill-reducing orderings for sparse matrices. The algorithms implemented in METIS are based on the multilevel recursive bisection, multilevel k-way, and multi-constraint partitioning schemes. Herein, the multilevel k-way partitioning scheme is selected as the partitioner in the proposed method. METIS was chosen because METIS's multilevel k-way partitioning algorithm provides the advantageous capabilities of the minimized resulting sub-domain connectivity graph, enforced contiguous partitions and minimized alternative objectives.

2.3. *Architecture and Scheduling*

2.3.1. Master-Slave Architecture for Distributed Computers

We implemented a system employing a dynamic load balancing strategy, as shown in Figure 4. The system enables a workflow to support load measuring on the master side, spatial decomposition on the master side, data moving from the master side to multiple slave sides, data clipping on the slave sides, overlay operation on the slave sides, area calculation on the slave sides and the final matrix building on the master side.

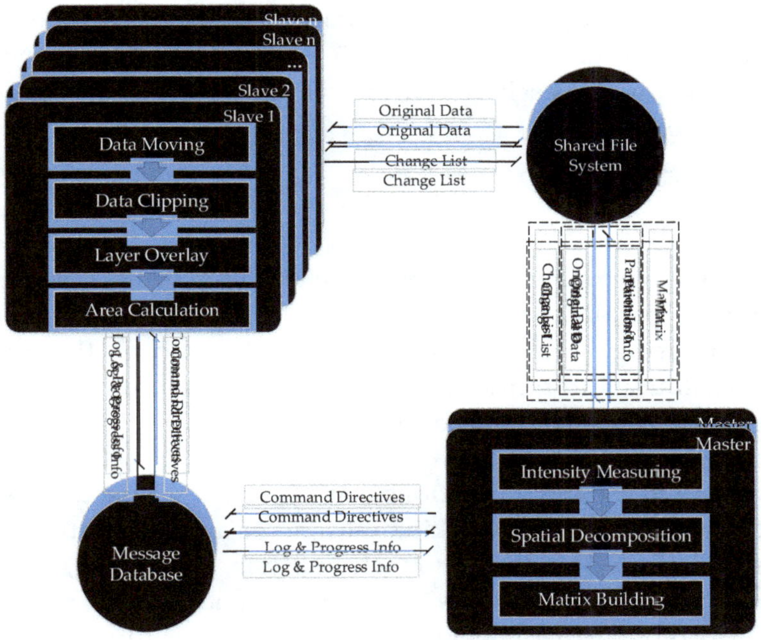

Figure 4. Master-slave architecture for distributed workstations.

Measuring the computational loads in the original chunks is performed on the server side. After this step, the LUCC analysis graph is constructed on the master side, and then partitioned to a certain number equal to that of the slaves. Next, the data partitions are moved from the master to the slave sides. Thus, data clipping, overlay operation and area calculation can be performed concurrently on each of the slave sides. Finally, the resultant data will be transferred to the master side.

2.3.2. Stream Scheduling of Tasks

With the above system, communications between any pair of workstations are supported. Furthermore, we adopted a stream scheduling method to exploit the parallelism among the slave workstations. Herein, each administrative division was viewed as an independent compute unit for scheduling, as illustrated in Figure 5.

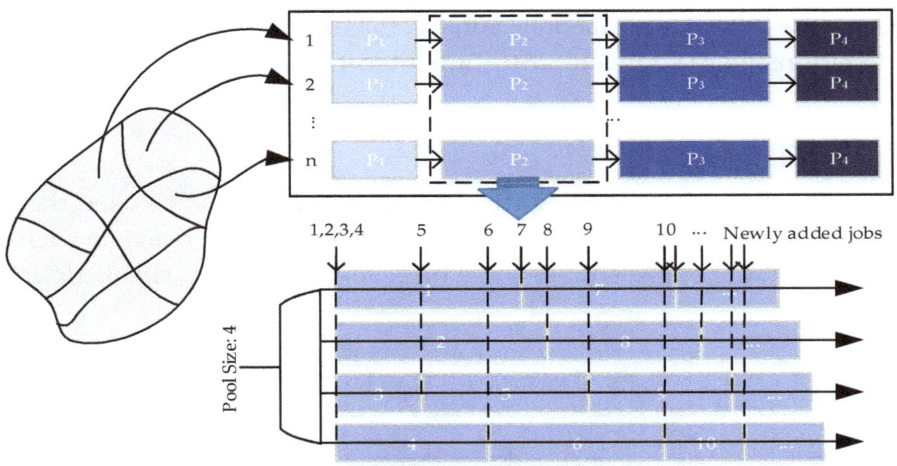

Figure 5. Stream scheduling of asynchronous compute units (P_1, P_2, P_3 and P_4 are four different procedures; P_1 is data moving, P_2 is data clipping, P_3 is overlay operation, and P_4 is area calculation).

Different administrative divisions often have different computational loads. The different computational loads mean the time usage for different compute units may vary greatly due to load differences. A process pool, whose size is set to the number of CPU cores on each of the slave workstations, is first initialized to support the computation. Moreover, in the pool, different units are scheduled asynchronously. As illustrated in Figure 5, four compute units were initiated at startup. When a unit finished, a new one was scheduled to push into the pool. When all the units finished, the whole task was terminated.

2.3.3. Implementation

On the side of each slave workstation, the stream scheduling was realized in the Python language and used to fully exploit the parallel performance in the process pool, which was supported in the package named Multiprocessing. A shared table was established on the server side to distribute the jobs to different workstations to work smoothly. At the same time, the working status of the workstations is continuously transferred to the server side.

Moreover, to support all the spatial operations from P_1 to P_3, we used the ArcGIS API for Python, which is a widely used Python library for working with maps and geospatial data. It provides simple and efficient tools for sophisticated vector and raster analysis, geocoding, map making, routing and directions. Furthermore, a series of error checks were developed to guarantee that all the procedures had correct inputs and outputs. Error checking enabled the workflow to work smoothly. For P_4, an area calculation was developed in the C++ language, and the File Geodatabase API provided by ESRI Company was used to access the spatial data. In comparison with the Python language, a very optimal performance could be obtained.

In addition, an MS-SQL database was used as the message database for exchanging the command directives, and the log and progress information between the master and the slave workstations.

3. Performance Evaluation

3.1. Testing Environment

A computer cluster was used to support the LUCC analysis, which included 15 workstations running Microsoft Windows 7 (×64) with a six-core Intel Xeon E5-1650 CPU (3.50 GHz) and 16 GB DDR4-2133. The total number of CPU cores was 90. In addition, Hyper-Threading Technology makes a single physical core appear as two logical cores, and the maximum number of threads for a workstation was 12. These workstations were connected through a ten-thousand-megabyte bandwidth switch.

3.2. Determination of the Factors in Estimation

To select the proper factors in measuring the computational loads, we used the PCC coefficient to measure the linear correlation between the possible factors and the time usage in each of the procedures. Values near 1 indicated that the curve shapes have high comparability. This is defined as follows.

$$\rho = \frac{\text{cov}(X+Y)}{\sigma_X \sigma_Y} \tag{5}$$

where cov is the covariance, σ_X is the standard deviation of X, σ_Y is the standard deviation of Y. Herein, we randomly selected 89 counties in China, and the testing result is provided in the Supplementary

Materials. Shown in Table 3 is the correlation analysis result obtained through calculating the PCC between the time usage in each of procedures and the potential influencing factors.

Table 3. Comparison of PCC values from different factors (The green items indicated higher correlations).

Factors\Time	T_{P1}	T_{P2}	T_{P3}	T_{P4}	T
F	0.969167	0.530038	0.568205	0.740030	0.582032
N	0.718055	0.973142	0.994402	0.964224	0.990629
A	0.513562	0.790036	0.825275	0.779288	0.809465
$N \times log(N)$	0.706571	0.976781	0.994454	0.960296	0.991878

Herein, F is the count of polygons in the layers, N is the number of vertices that constitute the polygons. According to Table 3, F was the most influencing factor that determined T_{P1} in data movement, and thus it was selected for measuring the loads in this procedure. As the land cover data were stored in vector objects, the movement of the data layers was realized through applying the query condition to each of the vector features. Even N should be a more effective parameter in theory for assessing the data volume, F showed a closer linear relationship with time T_{P1}. Thus, f_1 was set to F. Herein, a linear curve-fitting method was used to capture the trend of time with the changes in F. As illustrated in Figure 6, the time usage in this procedure was modeled using the following formula.

$$T_{P1} = 0.0001743 \times F + 14.395 \tag{6}$$

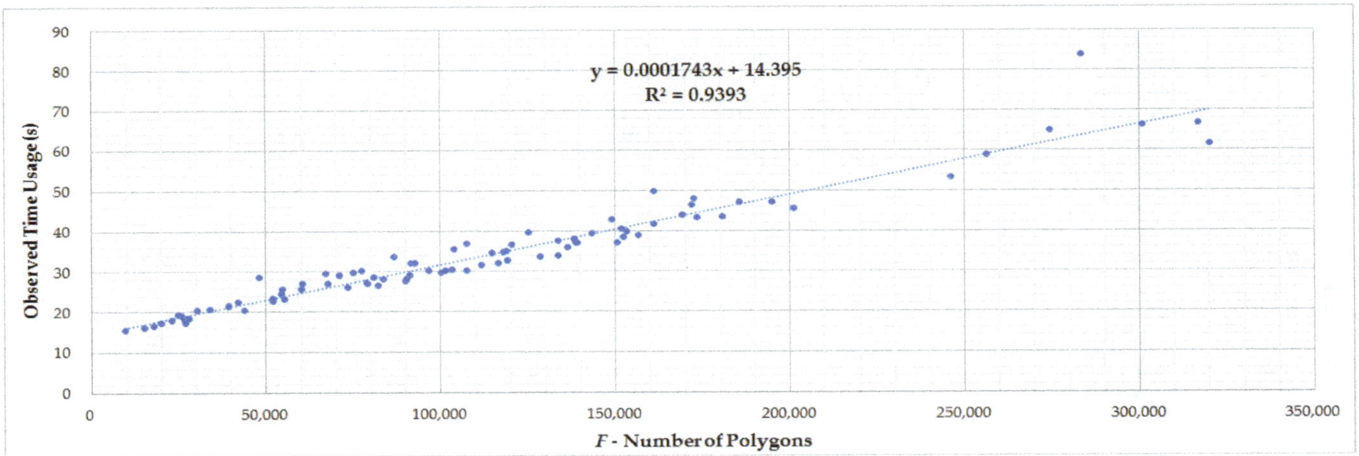

Figure 6. Linear regression of P_1 (This is the relation between the number of polygons and the time usage in data movement, where F is the number of polygons).

As the data clipping (P_2) and overlay (P_3) are essentially intersection operations, the time usage should be measured by the number of vertices. In P_2, the number of the vertices constituting the clipping boundaries was much less than that of the land cover polygons. Thus, factor N could be used in the time complexity format to measure the computational load, i.e., $f_2 = N \times log(N)$. In addition, clipping the two polygon layers involved more data I/O operations, and thus a linear fitting curve may not well match. In Figure 7, a second-order curve-fitting method was used to capture the trend of time with the change in the number of vertices.

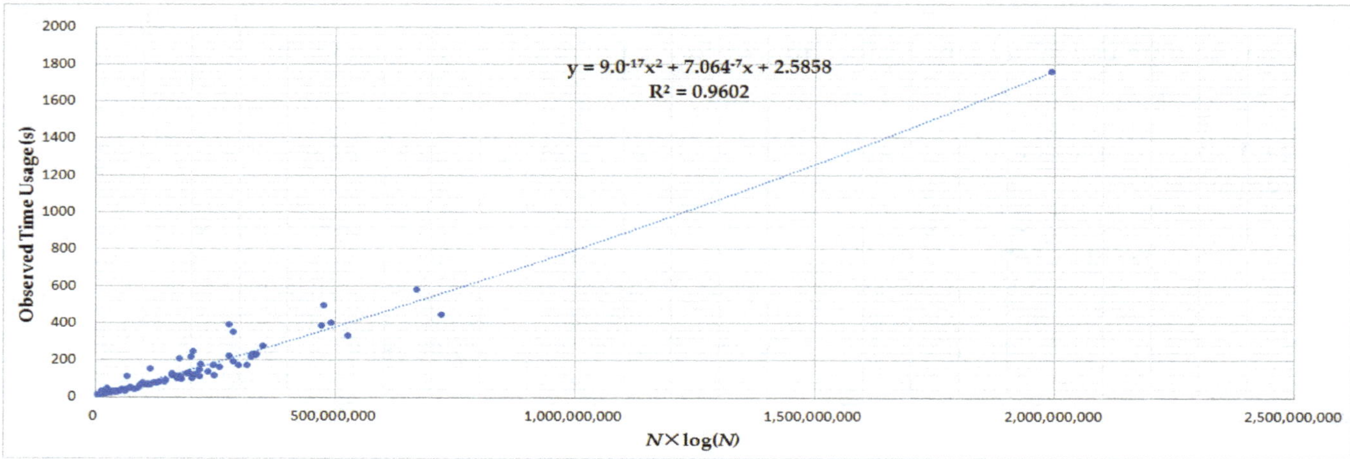

Figure 7. Second-order regression of P_2 (This is the relation between $N \times log(N)$ and the time usage in data clipping, where N is the number of vertices).

Then, we have

$$T_{P2} = 9.0^{-17} \times (N \times log(N))^2 + 0.0000007064 \times N \times log(N) + 2.5858 \quad (7)$$

Meanwhile, in P_3, the overlay of the different layers involved two polygon layers from different years. Therefore, the number of vertices from the two participating layers was used, i.e., $f_3 = N \times log(N)$. As the time usage was mainly determined by the computation, we found a linear fitting curve was more suitable than other curves, as illustrated in Figure 8.

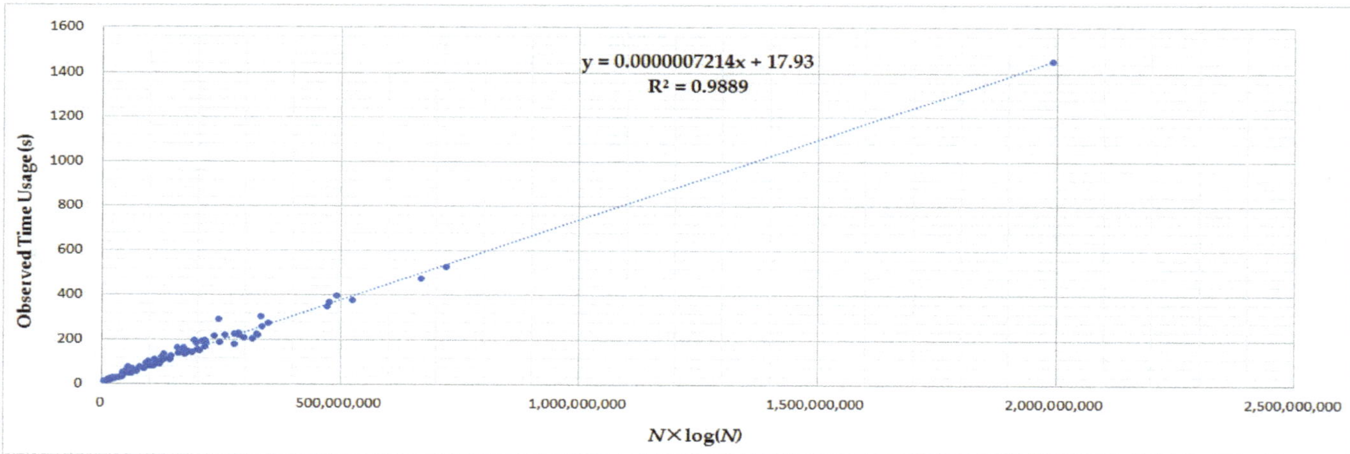

Figure 8. Linear order regression of P_3 (This is the relation between $N \times log(N)$ and the time usage in polygon overlay, where N is the number of vertices).

Then, we have

$$T_{P3} = 0.0000007214 \times N \times log(N) \quad (8)$$

In P_4, the computational load in the area calculation was positively proportional to the N, and a linear relation was most suitable, i.e., $of_4 = N$. The fitting curve was shown in Figure 9.

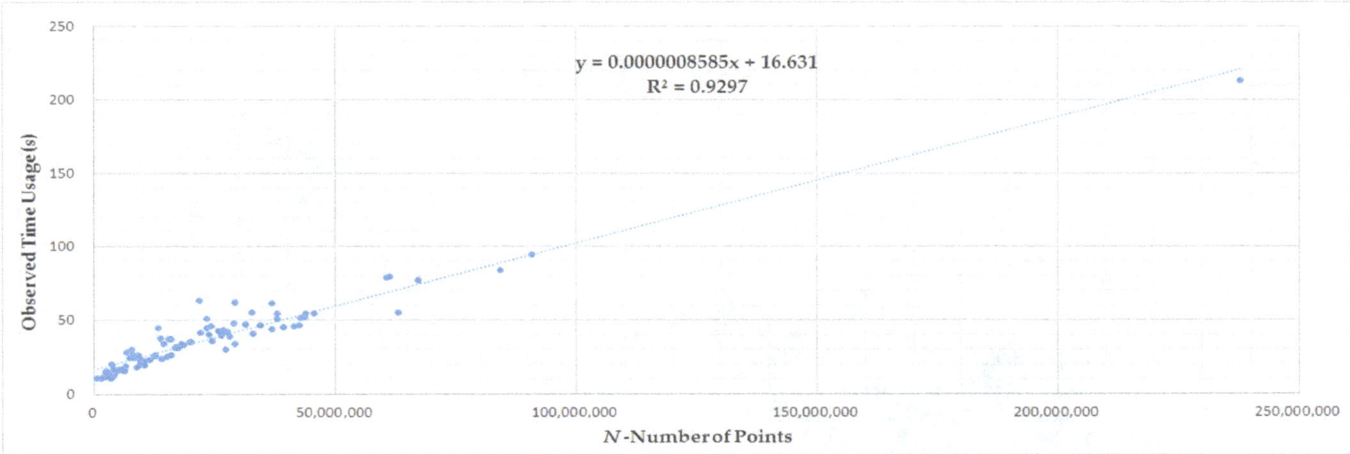

Figure 9. Linear regression of P_4 (This is the relation between N and the time usage in area calculation, where N is the number of vertices).

Then, we have

$$T_{P4} = 0.0000008585 \times N + 16.631 \quad (9)$$

Finally, the total time used in the four procedures can be measured as follows.

$$T = T_{P1} + T_{P2} + T_{P3} + T_{P4} \quad (10)$$

It is important to note that these time functions (T_{P1}, T_{P2}, T_{P3} and T_{P4}) were all obtained under a certain testing hardware environment, including the CPU, the physical memory, the disk, etc. Different hardware conditions, including the CPU cores, memory bandwidth, disk transfer rate, etc., may cause certain differences in these functions. After transforming the related factors into an estimable function of the time, the computational load can be measured according to the data size and the related procedures.

3.3. Experimental Results

3.3.1. Testing data

The testing data covered all 31 provinces and autonomous regions in mainland China (Hong Kong, Macao and Taiwan regions were not covered), and 2858 counties were involved. Specifically, the data were organized into 2858 chunks, each of which was stored in an individual File Geodatabase. Moreover, these data chunks were located in a shared-disk file system, and each computer could freely access any part of the data independently. According to the time functions proposed in Section 3.2, we estimated the computational loads for all these chunks. Shown in Figure 10 is the estimated time usage for all the counties. Clearly, there were obvious differences among these counties due to the differences in the number of geometries and number of vertices. It could be seen that the computation on some large counties cost more than two hours, while computation on some small counties cost less than one minute.

In general, larger counties often had higher loads, while for some areas with complicated land cover conditions, the time usage per square kilometer showed a large difference. In other words, the computational loads in different locations were different. As shown in Figure 11, it can be concluded that in eastern and southern China, the time density was relatively higher. In fact, these areas were usually fragmented, and the representative polygons were often narrow and twisted. Thus, more points were essential for constraining the boundaries of the land cover parcels.

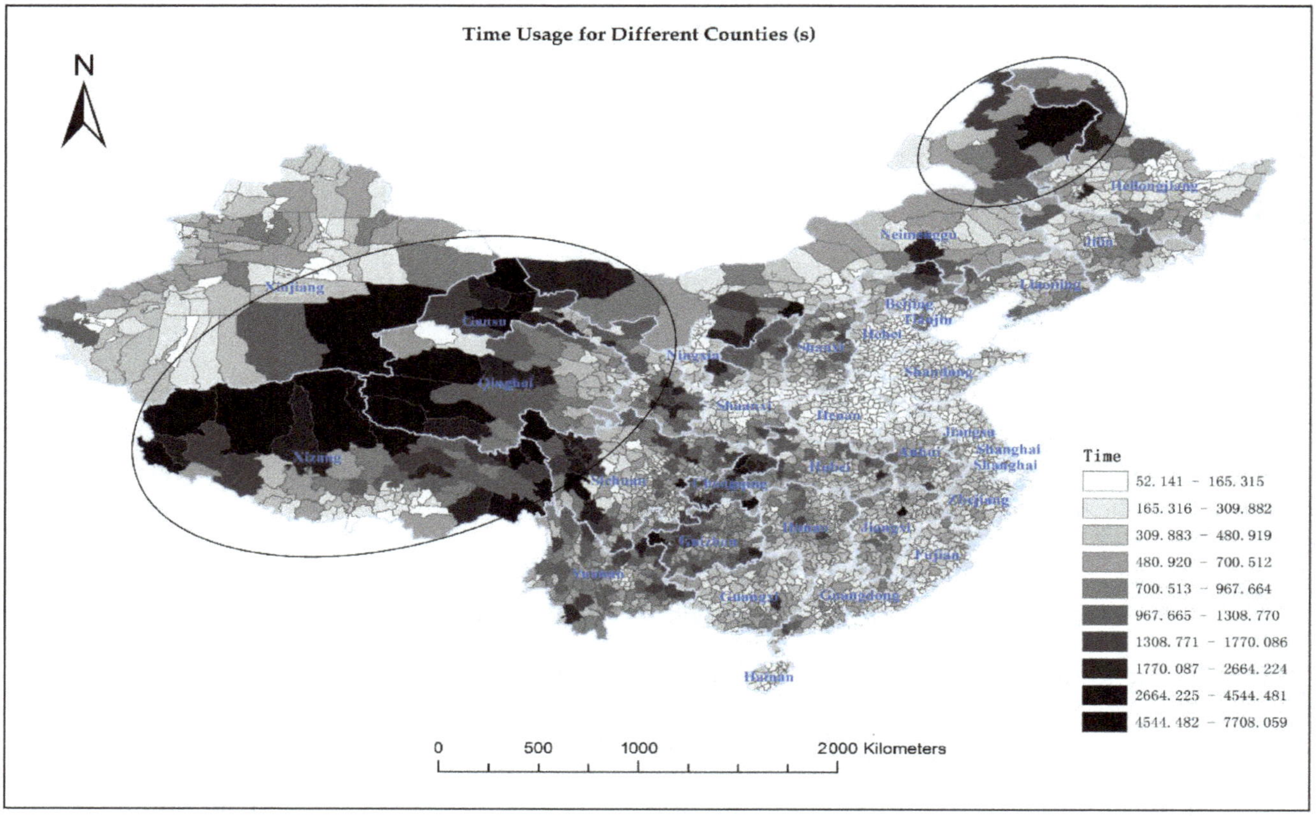

Figure 10. Estimated time usage in the 2858 counties in China.

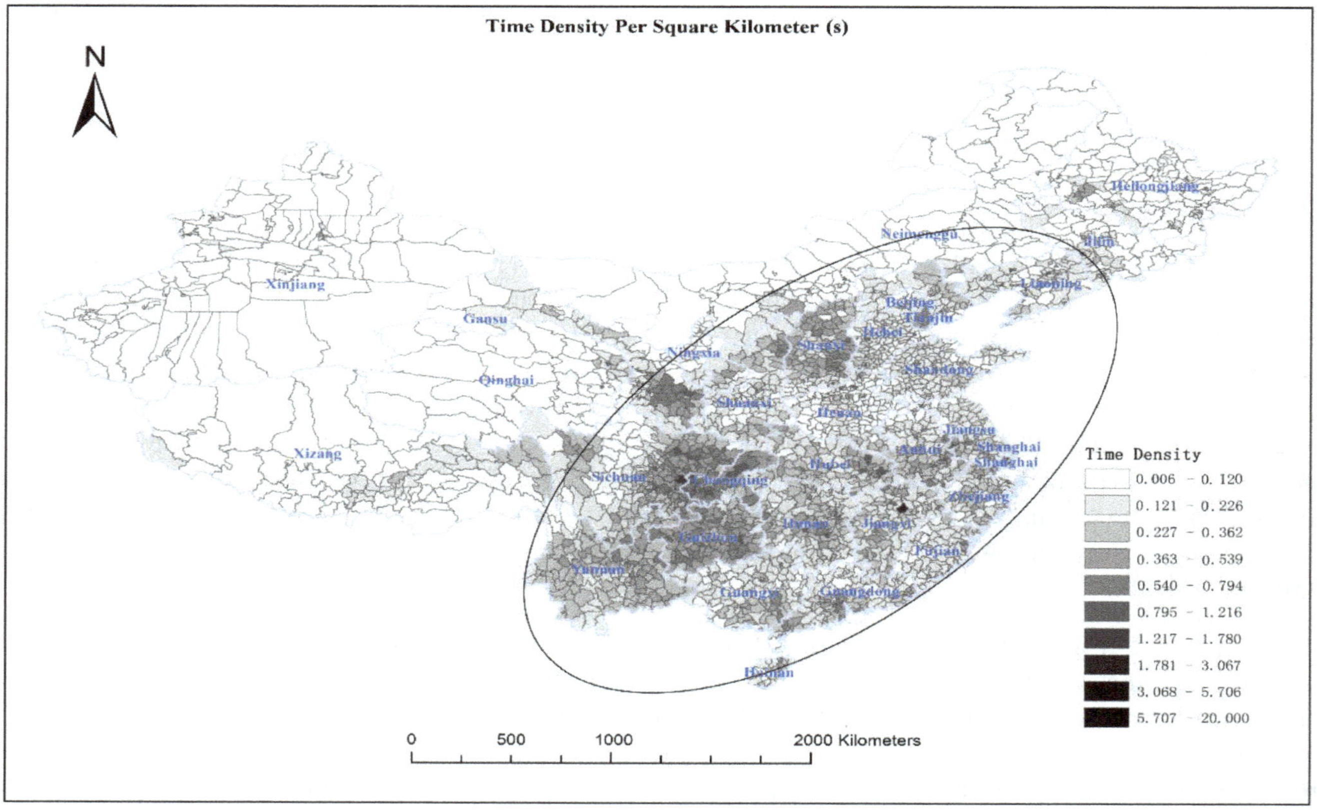

Figure 11. Estimated time density for the counties in China.

3.3.2. Results

Two individual experiments were conducted to verify the performance of the proposed method. The first one was to test the performance of the LUCC analysis over 7 provinces on a single workstation, which could validate the load measurement and the scalability of the program under a multi-core environment. The second one was performed in a cluster environment, which aimed at testing the validity of the spatial decomposition and the entire efficiency.

In the first experiment, Shanxi, Heilongjiang, Shanghai, Zhejiang, Shandong, Xizang and Qinghai were selected, and detailed testing results are provided in Table S3 in the Supplementary Materials. First, seven provinces were tested in a serial manner towards testing the time loads. Shown in Table 4 are the observed and estimated time usage.

Table 4. Comparison of the observed time and estimated time.

Order	Name	Number of Counties	Time Usage (s)		Difference	
			Observed	Estimated	Absolute	Ratio
1	Shanxi	117	54,158.173	51,664.532	2493.640	4.6%
2	Heilongjiang	128	39,550.255	37,914.468	1635.787	4.1%
3	Shanghai	16	2743.127	2529.132	213.995	7.8%
4	Zhejiang	90	28,432.039	21,489.041	6942.998	24.4%
5	Shandong	137	35,075.643	27,635.252	7440.391	21.2%
6	Xizang	74	74,659.131	77,058.161	2399.029	3.2%
7	Qinghai	46	32,129.256	34,644.451	2515.195	7.8%
-	Total/Avg	-	266,747.624	252,935.037	23,641.035	8.9%

The maximum difference ratio was approximately 24.4%, and the minimum difference ratio was 3.2%. On the whole, the average difference ratio was approximately 8.9%. Considering some unstable factors, such as the disk transfer rate and geometry complexity, a further improvement in accuracy for measuring the loads should be difficult to achieve. Actually, the absolute differences could be further reduced when performing the test in a multicore CPU. Furthermore, we tested the performance improvement when giving different computing processes (CPU cores), as illustrated in Table 5, and the performance curves are shown in Figure 12.

Table 5. Performance improvement in a multicore CPU environment (P is the number of processes).

Order	Name	Time (s)					Speedup
		P = 1	P = 2	P = 4	P = 6	P = 12	
1	Shanxi	54,158.173	29,733.984	18,042.815	12,955.973	8994.652	6.021
2	Heilongjiang	39,550.255	27,368.083	19,710.915	13,951.313	10,218.074	3.871
3	Shanghai	2743.127	1658.247	1083.171	877.641	681.561	4.025
4	Zhejiang	28,432.039	15,515.316	8185.116	6340.825	4595.972	6.186
5	Shandong	35,075.643	19,257.271	10,199.049	7279.027	5601.673	6.262
6	Xizang	74,659.131	43,597.761	26,888.392	18,922.136	15,557.339	4.799
7	Qinghai	32,129.256	18,202.262	12,600.972	9032.178	6538.687	4.914
-	Total/Avg	266,747.624	155,332.924	96,710.430	69,359.093	52,187.958	5.111

In the second experiment, the proposed method was tested on a cluster. As all the 2858 counties would be processed, spatial decomposition should be conducted first. The target of partitioning the task of the LUCC analysis was to distribute it to different slave workstations, and then coordinate them to complete the entire task. Therefore, a partitioning method that can both maintain the balancing loads and the smallest possible cost of cutting should be pursued. Before partitioning the data, the LUCC analysis graph was first constructed by estimating the loads in the focused counties and the neighboring counties. The loads in the counties focused upon contributed to the major computational loads from P_1 to P_4, which were represented as graph vertices, while the loads in neighboring counties mainly contributed to the data moving load in P_1, which were represented as graph edges.

Thus, we obtained the LUCC graph that completely described the task. Before partitioning the graph, we stored the vertices and edges as a graph file by referencing the METIS soft manual (http://glaros.dtc.umn.edu/gkhome/views/metis). Moreover, to minimize the edge cut costs and the total communication volume, multilevel k-way partitioning was used. The LUCC graph was partitioned into 15 parts, as shown in Figure 13, to be in accordance with the testing environment.

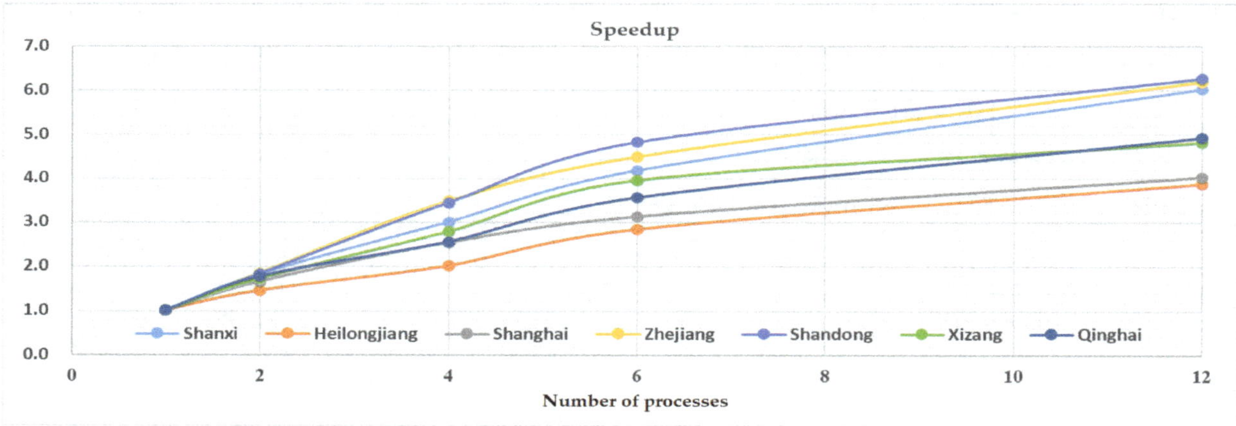

Figure 12. Performance improvement in different numbers of processes.

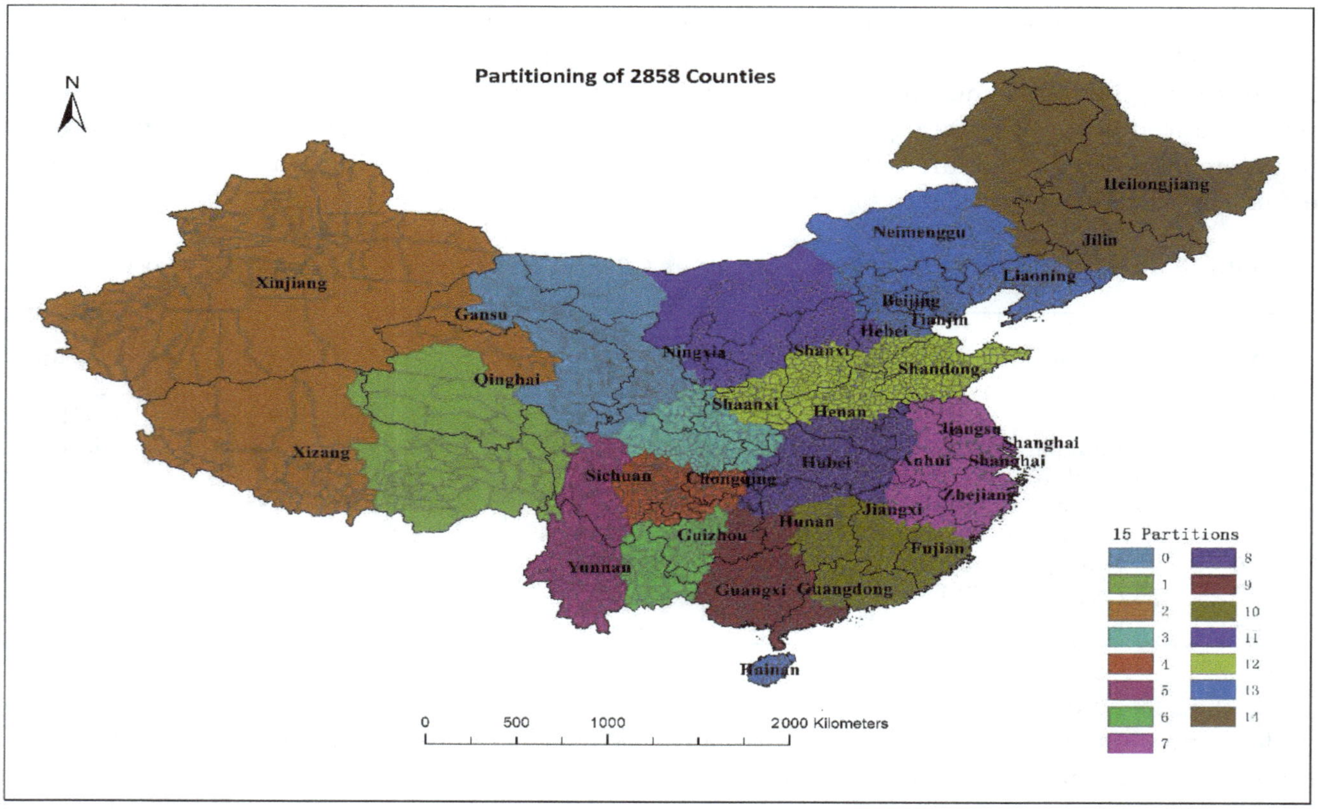

Figure 13. Partitioning of the LUCC analysis graph (the graph was partitioned into 15 parts, each of which was labeled with a single color).

After partitioning the graph, 15 sub-graphs were formed, and the related cutting edges were also recorded. As shown in Table 6, 15 sub-tasks were built. Then, we estimated the time loads in the sub-graphs and the cutting edges, respectively. Moreover, the total loads for each sub-task were obtained by summing the sub-graph loads and the cutting-edge loads.

Table 6. Computing time loads for the partitioning domains and total loads.

Computer ID	Graph Loads (s)	Cutting Loads (s)	Costs Ratio	Task Loads (s)	Balance
192.168.3.1	78,737.800	7834.684	9.95%	86,572.484	6.52%
192.168.3.2	78,945.698	2513.631	3.18%	81,459.329	6.13%
192.168.3.3	78,444.883	6984.443	8.90%	85,429.326	6.43%
192.168.3.4	83,135.226	13,525.892	16.27%	96,661.118	7.28%
192.168.3.5	82,848.954	8524.572	10.29%	91,373.526	6.88%
192.168.3.6	80,668.012	7148.928	8.86%	87,816.940	6.61%
192.168.3.7	83,146.467	6905.600	8.31%	90,052.067	6.78%
192.168.3.8	83,042.980	6721.402	8.09%	89,764.382	6.76%
192.168.3.9	82,401.041	11,897.392	14.44%	94,298.433	7.10%
192.168.3.10	80,064.289	11,565.392	14.45%	91,629.681	6.90%
192.168.3.11	79,672.611	12,012.495	15.08%	91,685.106	6.90%
192.168.3.12	78,525.364	2017.754	2.57%	80,543.118	6.06%
192.168.3.13	78,805.905	3550.061	4.50%	82,355.966	6.20%
192.168.3.14	80,616.409	5116.888	6.35%	85,733.297	6.45%
192.168.3.15	82,511.703	10,519.652	12.75%	93,031.355	7.00%

According to Table 6, the LUCC analysis task was decomposed into 15 sub-tasks. These tasks were distributed in a relatively even manner. In addition, extra costs caused by decomposition were approximately 10%. The proportions of the computational loads ranged from 6.06% to 7.28%.

Table 7 shows the results of the entire testing data. Obviously, a large proportion of the time was reduced on each of the slave workstations, and the average time load was approximately 5.049 h. The performance improvement ranged from 3.834 to 5.573, and the maximum gap in time was approximately 4790 s, with reference to the maximum task load on workstation 192.168.3.4 and the minimum task load on workstation 192.168.3.10.

Table 7. Performance improvement in a cluster (All the CPU cores were used).

Slave ID	Counties	Task Loads (s)	Observed Time (s)	Speedup
192.168.3.1	176	86,572.484	16,713.703	5.180
192.168.3.2	212	81,459.329	21,245.131	3.834
192.168.3.3	189	85,429.326	18,003.370	4.745
192.168.3.4	327	96,661.118	18,944.093	5.102
192.168.3.5	294	91,373.526	17,064.811	5.354
192.168.3.6	333	87,816.940	18,688.152	4.699
192.168.3.7	207	90,052.067	17,794.495	5.061
192.168.3.8	268	89,764.382	18,774.252	4.781
192.168.3.9	197	94,298.433	19,458.336	4.846
192.168.3.10	106	91,629.681	16,454.875	5.569
192.168.3.11	88	91,685.106	18,048.192	5.080
192.168.3.12	137	80,543.118	17,363.678	4.639
192.168.3.13	71	82,355.966	18,177.326	4.531
192.168.3.14	111	85,733.297	19,216.351	4.461
192.168.3.15	142	93,031.355	16,692.415	5.573
Avg	190.5	88,560.409	18,175.945	4.872

When the total task was performed on a single workstation without adopting any parallel optimization method, the estimated time was more than 14 days by summing all the estimated graph loads (i.e., the total task was performed on a single workstation, and no cutting loads were incurred).

Whereas, in a cluster with the proposed method, the total time usage was approximately 5.901 h considering the maximum task load. On the whole, the speedup was approximately 57.028 in the cluster in comparison with the estimated time on a single workstation.

4. Discussion

Generally, time complexity is simply a measure of the time it takes for a function or expression to complete its task, as well as the name of the process to measure that time. However, the hardware environment should also be considered. This is because different CPU frequencies, memory bandwidth and disk transfer rates will no doubt exert an influence on the time usage. Therefore, there exists a gap between time complexity and actual time usage. Through a polynomial regression analysis, we built the relevance between the actual time usage and the data volume factors, such as the number of geometries and vertices. Moreover, the estimated time usage from different procedures could be transformed into a unified function. According to the testing results in the Supplementary Materials, the PCC coefficient for the observed time and the estimated time was more than 0.99. Thus, the estimated time function can exactly reflect the actual time usage by referencing the number of geometry vertices or the number of polygons. According to the analysis of the testing data, spatial heterogeneity is a common phenomenon that is determined both by the data and the corresponding procedures. Herein, two individual experiments were provided. The first one was performed on a single workstation, while the second one was performed on a cluster.

The results from the first experiment indicated that an average speed improvement of 5.1 times could be achieved over the sequential version with 12 processes. When the number of allocating processes was fewer than 6, a stable improvement could be achieved. However, there was no continuous performance improvement when the allocating processes were more than 6. The main reason was due to the I/O bottlenecks. When multiple processes read the data from disk or dump data to the disk with conflicts, the overall performance can easily decline due to the I/O speed limit of the storage drive. In addition, multiple procedures that were working on different CPU cores may also have caused some bad effects on the computation speed when assessing the shared memory space. Therefore, to alleviate the competition from slow disks, a flexible method is to adopt a faster drive, such as a solid-state drive (SSD). If conditions allowed, loading all the data into the main memory could definitely provide an ideal way of conquering large-scale data processing. According to our early tests [79], using a streaming framework that can coordinate the I/O events with the computation may also provide an effective way of alleviating this bottleneck. Thus, an individual procedure should be allocated to continuously reading the data and producing a continuous data block stream; then, other procedures can concurrently process these available data blocks acquired from the stream.

When performing the task on a single workstation, we focused on how to fully exploit the parallelism from multicore CPUs. In a cluster environment, the difficulty lies in how to allocate the computational loads to different workstations. Therefore, a quantitative representation of the task should first be accomplished. According to the results from the second experiment, the average difference ratio of the partitioned loads maintained a relatively low level, which demonstrated that the estimation and the partitioning method could be used to decompose the computational loads. Furthermore, two temporal land cover layers covering mainland China were decomposed into multiple sub-datasets with even loads. Thus, this extremely time-consuming task was solved in a very short period of time.

In addition, two improvements should also be made in the near future. One regards the lack of assessment on how the geometry complexity influences the time consumption. Though time functions have given an approximate estimation of the observed time, some deep rules might have been covered up. Another improvement is in terms of applicability. All the tests were performed in a homogeneous environment, and all the workstations had an identical configuration. Actually, heterogeneous environments are very common, and some adjustments should be made.

As a matter of fact, the earth's surface is changing at an amazing speed, and thus, more and more large-scale, short-period and high-resolution data are becoming available with the development of sensor technology. The proposed method was only used to achieve a two-temporal-layer LUCC analysis, and no geomorphic layers were involved for a deep-level analysis. When dealing with more temporal layers and thematic layers, the time costs will definitely increase sharply. Therefore, methods for how to perform the change analysis for a large-scale region, such as a country, a continent or even the globe, especially using high-performance computing facilities, is still a challenge that should be addressed.

5. Conclusions

This work presented a novel graph-based spatial decomposition to conquer the very heavy computational load when overlaying multi-temporal land cover data. This approach quantitatively represented the computational loads as graph vertices and edges, and balanced subtasks were formed through graph partitioning. Basically, PCC analysis was first conducted towards selecting the determining factors that influence the time usage. Then, stream scheduling was used to achieve an efficient parallel LUCC method. Together with a master-slave architecture connecting the distributed workstations and a stream scheduling method, we performed two experiments for testing the performance improvements on a single workstation and a cluster.

Clearly, selecting the proper factors and regression methods can contribute to a unified representation that quantitatively transforms each of the procedures in a complex workflow to a time function with the data size as measurable variables. On this basis, graph-based spatial decomposition gives a balanced load allocation method, which not only considers the volume of the data to be processed, but also the spatial operations performed on the data. As a main research object, the LUCC analysis is thus optimized efficiently. It is worth mentioning that the consumed time was reduced from a two-week period to less than 6 h, for detecting the two-temporal changes over China.

In the near future, research should also be done to validate the feasibility of approach to raster maps and explore the potential influence of the geometric complexity that may also governs the parallel efficiency to some degree. In addition, a heterogeneous environment should also be considered.

Author Contributions: X.K. conceived the original idea and designed the experiments; X.K., J.L. and C.D. performed and analyzed the experiments; X.K. and S.X. wrote the paper. All authors read and approved the final manuscript.

Acknowledgments: We thank the editors and the anonymous reviewers for their insightful comments which have helped to improve the quality of the paper.

References

1. Reid, R.S.; Kruska, R.L.; Muthui, N.; Taye, A.; Wotton, S.; Wilson, C.J.; Mulatu, W. Land-use and land-cover dynamics in response to changes in climatic, biological and socio-political forces: The case of Southwestern Ethiopia. *Landsc. Ecol.* **2000**, *15*, 339–355. [CrossRef]
2. Lambin, E.F.; Turner, B.L.; Geist, H.J.; Agbola, S.B.; Angelsen, A.; Bruce, J.W.; Coomes, O.T.; Dirzo, R.; Fischer, G.; Folke, C. The causes of land-use and land-cover change: Moving beyond the myths. *Glob. Environ. Chang.* **2001**, *11*, 261–269. [CrossRef]
3. Bürgi, M.; Hersperger, A.M.; Schneeberger, N. Driving forces of landscape change-current and new directions. *Landsc. Ecol.* **2005**, *19*, 857–868. [CrossRef]
4. Zhang, J.; Zhang, Y. Remote sensing research issues of the national land use change program of China. *ISPRS J. Photogramm. Remote Sens.* **2007** *62*, 461–472. [CrossRef]
5. Yin, H.; Pflugmacher, D.; Li, A.; Li, Z.; Hostert, P. Land use and land cover change in Inner Mongolia-understanding the effects of China's re-vegetation programs. *Remote Sens. Environ.* **2018**, *204*, 918–930. [CrossRef]

6. Liu, J.; Hull, V.; Batistella, M.; DeFries, R.; Dietz, T.; Fu, F.; Hertel, T.; Izaurralde, R.C.; Lambin, E.; Li, S. Framing sustainability in a telecoupled world. *Ecol. Soc.* **2013**, *18*. [CrossRef]
7. Eakin, H.; Defries, R.; Kerr, S.; Lambin, E.F.; Liu, J.; Marcotullio, P.J.; Messerli, P.; Reenberg, A.; Rueda, X.; Swaffield, S.R. Significance of telecoupling for exploration of land-use change. In *Rethinking Global Land Use in an Urban Era*; Seto, K.C., Reenberg, A., Eds.; MIT Press: Cambridge, MA, USA, 2014; pp. 141–162.
8. Liu, J. Forest sustainability in China and implications for a telecoupled world. *Asia Pac. Policy Stud.* **2014**, *1*, 230–250. [CrossRef]
9. Friis, C.; Nielsen, J.O. Exploring the potential of the telecoupling framework for understanding land change. *IRI THESys Discuss. Pap.* **2014**, *1*, 1–32.
10. Liu, J.; Hull, V.; Luo, J.; Yang, W.; Liu, W.; Viña, A.; Vogt, C.; Xu, Z.; Yang, H.; Zhang, J. Multiple telecouplings and their complex interrelationships. *Ecol. Soc.* **2015**, *20*. [CrossRef]
11. Liu, J.; Deng, X. Progress of the research methodologies on the temporal and spatial process of LUCC. *Chin. Sci. Bull.* **2010**, *55*, 1354–1362. [CrossRef]
12. Liu, J.; Mooney, H.; Hull, V.; Davis, S.J.; Gaskell, J.; Hertel, T.; Lubchenco, J.; Seto, K.C.; Gleick, P.; Kremen, C. Systems integration for global sustainability. *Science* **2015**, *347*, 1258832. [CrossRef] [PubMed]
13. Liu, J.; Zhang, Z.; Xu, X.; Kuang, W.; Zhou, W.; Zhang, S.; Li, R.; Yan, C.; Yu, D.; Wu, S. Spatial patterns and driving forces of land use change in China during the early 21st century. *J. Geogr. Sci.* **2010**, *20*, 483–494. [CrossRef]
14. Plieninger, T.; Draux, H.; Fagerholm, N.; Bieling, C.; Bürgi, M.; Kizos, T.; Kuemmerle, T.; Primdahl, J.; Verburg, P.H. The driving forces of landscape change in Europe: A systematic review of the evidence. *Land Use Policy* **2016**, *57*, 204–214. [CrossRef]
15. Lunetta, R.S.; Knight, J.F.; Ediriwickrema, J.; Lyon, J.G.; Worthy, L.D. Land-cover change detection using multi-temporal modis ndvi data. *Remote Sens. Environ.* **2006**, *105*, 142–154. [CrossRef]
16. Wu, W.; Yu, Q.; You, L.; Chen, K.; Tang, H.; Liu, J. Global cropping intensity gaps: Increasing food production without cropland expansion. *Land Use Policy* **2018**, *76*, 515–525. [CrossRef]
17. Yang, C.; Yu, M.; Hu, F.; Jiang, Y.; Li, Y. Utilizing cloud computing to address big geospatial data challenges. *Comput. Environ. Urban Syst.* **2016**, *61*, 120–128. [CrossRef]
18. Liu, J.; Dou, Y.; Batistella, M.; Challies, E.; Connor, T.; Friis, C.; Millington, J.D.; Parish, E.; Romulo, C.L.; Silva, R.F.B. Spillover systems in a telecoupled anthropocene: Typology, methods, and governance for global sustainability. *Curr. Opin. Environ. Sustain.* **2018**, *33*, 58–69. [CrossRef]
19. Li, X. A review of the international researches on land use/land cover change. *Acta Geogr. Sin.* **1996**, *51*, 553–558.
20. Chen, J.; Gong, P.; He, C.; Pu, R.; Shi, P. Land-use/land-cover change detection using improved change-vector analysis. *Photogramm. Eng. Remote Sens.* **2003**, *69*, 369–379. [CrossRef]
21. Xiao, J.; Shen, Y.; Ge, J.; Tateishi, R.; Tang, C.; Liang, Y.; Huang, Z. Evaluating urban expansion and land use change in Shijiazhuang, China, by using gis and remote sensing. *Landsc. Urban Plan.* **2006**, *75*, 69–80. [CrossRef]
22. Shalaby, A.; Tateishi, R. Remote sensing and gis for mapping and monitoring land cover and land-use changes in the Northwestern coastal zone of Egypt. *Appl. Geogr.* **2007**, *27*, 28–41. [CrossRef]
23. Parker, D.C.; Manson, S.M.; Janssen, M.A.; Hoffmann, M.J.; Deadman, P. Multi-agent systems for the simulation of land-use and land-cover change: A review. *Ann. Assoc. Am. Geogr.* **2003**, *93*, 314–337. [CrossRef]
24. Gibon, A.; Sheeren, D.; Monteil, C.; Ladet, S.; Balent, G. Modelling and simulating change in reforesting mountain landscapes using a social-ecological framework. *Landsc. Ecol.* **2010**, *25*, 267–285. [CrossRef]
25. Ralha, C.G.; Abreu, C.G.; Coelho, C.G.; Zaghetto, A.; Macchiavello, B.; Machado, R.B. A multi-agent model system for land-use change simulation. *Environ. Model. Softw.* **2013**, *42*, 30–46. [CrossRef]
26. Liu, X.; Liang, X.; Li, X.; Xu, X.; Ou, J.; Chen, Y.; Li, S.; Wang, S.; Pei, F. A future land use simulation model (flus) for simulating multiple land use scenarios by coupling human and natural effects. *Landsc. Urban Plan.* **2017**, *168*, 94–116. [CrossRef]
27. Walker, R.; Solecki, W. Theorizing land-cover and land-use change: The case of the Florida everglades and its degradation. *Ann. Assoc. Am. Geogr.* **2004**, *94*, 311–328. [CrossRef]
28. Rindfuss, R.R.; Entwisle, B.; Walsh, S.J.; Mena, C.F.; Erlien, C.M.; Gray, C.L. Frontier land use change:

Synthesis, challenges, and next steps. *Ann. Assoc. Am. Geogr.* **2007**, *97*, 739–754. [CrossRef]

29. Dong, L.; Wang, W.; Ma, M.; Kong, J.; Veroustraete, F. The change of land cover and land use and its impact factors in upriver key regions of the yellow river. *Int. J. Remote Sens.* **2009**, *30*, 1251–1265. [CrossRef]
30. Wang, K.; Zhang, Q.; Chen, Y.D.; Singh, V.P. Effects of land-use/cover change on hydrological processes using a GIS/RS-based integrated hydrological model: Case study of the east river, China. *Hydrol. Sci. J.* **2015**, *60*, 1724–1738. [CrossRef]
31. Meneses, B.; Reis, R.; Vale, M.; Saraiva, R. Land use and land cover changes in zêzere watershed (portugal)—Water quality implications. *Sci. Total Environ.* **2015**, *527*, 439–447. [CrossRef] [PubMed]
32. Yu, W.; Zang, S.; Wu, C.; Liu, W.; Na, X. Analyzing and modeling land use land cover change (LUCC) in the daqing city, China. *Appl. Geogr.* **2011**, *31*, 600–608. [CrossRef]
33. Turner, B.; Meyer, W.B.; Skole, D.L. Global land-use/land-cover change: Towards an integrated study. *Ambio Stockholm.* **1994**, *23*, 91–95.
34. Bartholomé, E.; Belward, A.S. Glc2000: A new approach to global land cover mapping from earth observation data. *Int. J. Remote Sens.* **2005**, *26*, 1959–1977. [CrossRef]
35. Verburg, P.H.; Crossman, N.; Ellis, E.C.; Heinimann, A.; Hostert, P.; Mertz, O.; Nagendra, H.; Sikor, T.; Erb, K.-H.; Golubiewski, N. Land system science and sustainable development of the earth system: A global land project perspective. *Anthropocene* **2015**, *12*, 29–41. [CrossRef]
36. Ringeval, B.; Augusto, L.; Monod, H.; Apeldoorn, D.; Bouwman, L.; Yang, X.; Achat, D.L.; Chini, L.P.; Van Oost, K.; Guenet, B. Phosphorus in agricultural soils: Drivers of its distribution at the global scale. *Glob. Chang. Biol.* **2017**, *23*, 3418–3432. [CrossRef] [PubMed]
37. Li, X.; Chen, G.; Liu, X.; Liang, X.; Wang, S.; Chen, Y.; Pei, F.; Xu, X. A new global land-use and land-cover change product at a 1-km resolution for 2010 to 2100 based on human–environment interactions. *Ann. Assoc. Am. Geogr.* **2017**, *107*, 1040–1059. [CrossRef]
38. Munroe, D.K.; Müller, D. Issues in spatially explicit statistical land-use/cover change (LUCC) models: Examples from western honduras and the central highlands of Vietnam. *Land Use Policy* **2007**, *24*, 521–530. [CrossRef]
39. Defries, R.S.; Chan, J.C.W. Multiple criteria for evaluating machine learning algorithms for land cover classification from satellite data. *Remote Sens. Environ.* **2000**, *74*, 503–515. [CrossRef]
40. Cracknell, M.J.; Reading, A.M. Geological mapping using remote sensing data: A comparison of five machine learning algorithms, their response to variations in the spatial distribution of training data and the use of explicit spatial information. *Comput. Geosci.* **2014**, *63*, 22–33. [CrossRef]
41. Jensen, J.R. *Introductory Digital Image Processing: A Remote Sensing Perspective*, 3rd ed.; Prentice Hall: Upper Saddle River, NJ, USA, 2005.
42. Lu, D.; Batistella, M.; Moran, E. Multitemporal spectral mixture analysis for amazonian land-cover change detection. *Can. J. Remote Sens.* **2004**, *30*, 87–100. [CrossRef]
43. Coppin, P.; Jonckheere, I.; Nackaerts, K.; Muys, B.; Lambin, E. Review articledigital change detection methods in ecosystem monitoring: A review. *Int. J. Remote Sens.* **2004**, *25*, 1565–1596. [CrossRef]
44. Weng, Q. Land use change analysis in the Zhujiang delta of China using satellite remote sensing, GIS and stochastic modelling. *J. Environ. Manag.* **2002**, *64*, 273–284. [CrossRef]
45. Mas, J.-F. Change estimates by map comparison: A method to reduce erroneous changes due to positional error. *Trans. GIS* **2005**, *9*, 619–629. [CrossRef]
46. Zhang, H.; Shu, N. Land use/cover change detection using feature database based on vector-image data conflation. In Proceedings of the 8th International Symposium on Spatial Accuracy Assessment in Natural Resources and Environmental Sciences, Shanghai, China, 25–27 June 2008; pp. 255–262.
47. Lu, Y.; Cao, M.; Zhang, L. A vector-based cellular automata model for simulating urban land use change. *Chin. Geogr. Sci.* **2015**, *25*, 74–84. [CrossRef]
48. Rozenstein, O.; Karnieli, A. Comparison of methods for land-use classification incorporating remote sensing and GIS inputs. *Appl. Geogr.* **2011**, *31*, 533–544. [CrossRef]
49. Khatami, R.; Mountrakis, G.; Stehman, S.V. A meta-analysis of remote sensing research on supervised pixel-based land-cover image classification processes: General guidelines for practitioners and future research. *Remote Sens. Environ.* **2016**, *177*, 89–100. [CrossRef]
50. Cheng, G.; Han, J.; Guo, L.; Liu, Z.; Bu, S.; Ren, J. Effective and efficient midlevel visual elements-oriented land-use classification using VHR remote sensing images. *IEEE Trans. Geosci. Remote Sens.* **2015**, *53*, 4238–4249. [CrossRef]

51. Huang, X.; Lu, Q.; Zhang, L. A multi-index learning approach for classification of high-resolution remotely sensed images over urban areas. *ISPRS J. Photogramm. Remote Sens.* **2014**, *90*, 36–48. [CrossRef]
52. Hu, F.; Xia, G.-S.; Hu, J.; Zhang, L. Transferring deep convolutional neural networks for the scene classification of high-resolution remote sensing imagery. *Remote Sens.* **2015**, *7*, 14680–14707. [CrossRef]
53. Tewkesbury, A.P.; Comber, A.J.; Tate, N.J.; Lamb, A.; Fisher, P.F. A critical synthesis of remotely sensed optical image change detection techniques. *Remote Sens. Environ.* **2015**, *160*, 1–14. [CrossRef]
54. Ahlqvist, O. Extending post-classification change detection using semantic similarity metrics to overcome class heterogeneity: A study of 1992 and 2001 us national land cover database changes. *Remote Sens. Environ.* **2008**, *112*, 1226–1241. [CrossRef]
55. Tiede, D. A new geospatial overlay method for the analysis and visualization of spatial change patterns using object-oriented data modeling concepts. *Cartogr. Geogr. Inf. Sci.* **2014**, *41*, 227–234. [CrossRef] [PubMed]
56. Yang, X. Satellite monitoring of urban spatial growth in the atlanta metropolitan area. *Photogramm. Eng. Remote Sens.* **2002**, *68*, 725–734.
57. Yuan, F.; Sawaya, K.E.; Loeffelholz, B.C.; Bauer, M.E. Land cover classification and change analysis of the twin cities (minnesota) metropolitan area by multitemporal landsat remote sensing. *Remote Sens. Environ.* **2005**, *98*, 317–328. [CrossRef]
58. Im, J.; Jensen, J.; Tullis, J. Object-based change detection using correlation image analysis and image segmentation. *Int. J. Remote Sens.* **2008**, *29*, 399–423. [CrossRef]
59. Zhou, W.; Huang, G.; Troy, A.; Cadenasso, M. Object-based land cover classification of shaded areas in high spatial resolution imagery of urban areas: A comparison study. *Remote Sens. Environ.* **2009**, *113*, 1769–1777. [CrossRef]
60. Dingle Robertson, L.; King, D.J. Comparison of pixel- and object-based classification in land cover change mapping. *Int. J. Remote Sens.* **2011**, *32*, 1505–1529. [CrossRef]
61. Myint, S.W.; Gober, P.; Brazel, A.; Grossman-Clarke, S.; Weng, Q. Per-pixel vs. Object-based classification of urban land cover extraction using high spatial resolution imagery. *Remote Sens. Environ.* **2011**, *115*, 1145–1161. [CrossRef]
62. Chrisman, N.R. *Exploring Geographic Information Systems*; Wiley: Hoboken, NJ, USA, 1997.
63. Deren, D.L.; Shao, Z. Reflections on issues in national geographical conditions monitoring. *Geomat. Inf. Sci. Wuhan Univ.* **2016**, *41*, 143–147.
64. Openshaw, S.; Alvanides, S. Applying geocomputation to the analysis of spatial distributions. In *Geographic Information Systems: Principles and Technical Issues*; Longley, P.A., Michael, M., Maguire, D.J., Rhind, D.W., Eds.; John Wiley and Sons Inc.: New York, NY, USA, 1999; Volume 1, pp. 267–282.
65. Wang, S.; Liu, Y. Teragrid giscience gateway: Bridging cyberinfrastructure and giscience. *Int. J. Geogr. Inf. Sci.* **2009**, *23*, 631–656. [CrossRef]
66. Yang, C.; Raskin, R.; Goodchild, M.; Gahegan, M. Geospatial cyberinfrastructure: Past, present and future. *Comput. Environ. Urban Syst.* **2010**, *34*, 264–277. [CrossRef]
67. Zhang, J. Towards personal high-performance geospatial computing (HPC-G): Perspectives and a case study. In Proceedings of the ACM SIGSPATIAL International Workshop on High Performance and Distributed Geographic Information Systems, San Jose, CA, USA, 2 November 2010; ACM: New York, NY, USA, 2010; pp. 3–10.
68. Lee, C.A.; Gasster, S.D.; Plaza, A.; Chang, C.I.; Huang, B. Recent developments in high performance computing for remote sensing: A review. *IEEE J. STARS* **2011**, *4*, 508–527. [CrossRef]
69. Plaza, A.; Benediktsson, J.A.; Boardman, J.W.; Brazile, J.; Bruzzone, L.; Camps-Valls, G.; Chanussot, J.; Fauvel, M.; Gamba, P.; Gualtieri, A. Recent advances in techniques for hyperspectral image processing. *Remote Sens. Environ.* **2009**, *113*, S110–S122. [CrossRef]
70. Jin, X.; Sieber, R.E. A land use/land cover change geospatial cyberinfrastructure to integrate big data and temporal topology. *Int. J. Geogr. Inf. Sci.* **2016**, *30*, 573–593.
71. Pijanowski, B.C.; Tayyebi, A.; Doucette, J.; Pekin, B.K.; Braun, D.; Plourde, J. A big data urban growth simulation at a national scale: Configuring the gis and neural network based land transformation model to run in a high performance computing (HPC) environment. *Environ. Model. Softw.* **2014**, *51*, 250–268. [CrossRef]
72. Wang, S.; Armstrong, M.P. A Theory of the Spatial Computational Domain. In Proceedings of the 8th International Conference on GeoComputation University of Michigan, Ann Arbor, MI, USA, 31 July–3 August 2005.

73. Wang, S.; Armstrong, M.P. A theoretical approach to the use of cyberinfrastructure in geographical analysis. *Int. J. Geogr. Inf. Sci.* **2009**, *23*, 169–193. [CrossRef]
74. Becker, L.; Giesen, A.; Hinrichs, K.H.; Vahrenhold, J. Algorithms for performing polygonal map overlay and spatial join on massive data sets. In Proceedings of the International Symposium on Spatial Databases, Hong Kong, China, 20–23 July 1999; Springer: Berlin/Heidelberg, Germany, 1999; pp. 270–285.
75. Agarwal, D.; Puri, S.; He, X.; Prasad, S.K. A system for GIS polygonal overlay computation on linux cluster —An experience and performance report. In Proceedings of the 2012 IEEE 26th International Parallel and Distributed Processing Symposium Workshops & PhD Forum, Shanghai, China, 21–25 May 2012; pp. 1433–1439.
76. Zhou, C.; Chen, Z.; Liu, Y.; Li, F.; Cheng, L.; Zhu, A.-X.; Li, M. Data decomposition method for parallel polygon rasterization considering load balancing. *Comput. Geosci.* **2015**, *85*, 196–209. [CrossRef]
77. Liu, J.; Kang, X.; Dong, C.; Xu, S. A stream tilling approach to surface area estimation for large scale spatial data in a shared memory system. *Open Geosci.* **2017**, *9*, 622–634. [CrossRef]
78. Dorier, M.; Antoniu, G.; Cappello, F.; Snir, M.; Orf, L. Damaris: How to efficiently leverage multicore parallelism to achieve scalable, jitter-free I/O. In Proceedings of the 2012 IEEE International Conference on Cluster Computing (CLUSTER), Beijing, China, 24–28 September 2012; IEEE: New York, NY, USA, 2012; pp. 155–163.
79. Kang, X.; Liu, J.; Lin, X. Streaming progressive tin densification filter for airborne lidar point clouds using multi-core architectures. *Remote Sens.* **2014**, *6*, 7212–7232. [CrossRef]
80. Kang, X.; Lin, X. Graph-based divide and conquer method for parallelizing spatial operations on vector data. *Remote Sens.* **2014**, *6*, 10107–10130. [CrossRef]
81. Hendrickson, B.; Kolda, T.G. Graph partitioning models for parallel computing. *Parallel Comput.* **2000**, *26*, 1519–1534. [CrossRef]
82. Lipton, R.J.; Tarjan, R.E. A separator theorem for planar graphs. *SIAM J. Appl. Math.* **1979**, *36*, 177–189. [CrossRef]
83. Skiena, S. *The Algorithm Design Manual*, 2nd ed.; Springer Science & Business Media: New York, NY, USA, 2008.
84. Karypis, G.; Kumar, V. A fast and high quality multilevel scheme for partitioning irregular graphs. *SIAM J. Sci. Comput.* **1998**, *20*, 359–392. [CrossRef]
85. Andreev, K. Balanced graph partitioning. *Theor. Comput. Syst* **2006**, *39*, 929–939. [CrossRef]
86. Yang, C.; Goodchild, M.; Huang, Q.; Nebert, D.; Raskin, R.; Xu, Y.; Bambacus, M.; Fay, D. Spatial cloud computing: How can the geospatial sciences use and help shape cloud computing? *Int. J. Digit. Earth* **2011**, *4*, 305–329. [CrossRef]
87. Goodchild, M.F.; Yuan, M.; Cova, T.J. Towards a general theory of geographic representation in GIS. *Int. J. Geogr. Inf. Sci.* **2007**, *21*, 239–260. [CrossRef]
88. De Smith, M.J.; Goodchild, M.F.; Longley, P. *Geospatial Analysis: A Comprehensive Guide to Principles, Techniques and Software tools*, 2nd ed.; Troubador: Leicester, UK, 2008.
89. Yang, C.; Wu, H.; Huang, Q.; Li, Z.; Li, J. Using spatial principles to optimize distributed computing for enabling the physical science discoveries. *Proc. Natl. Acad. Sci. USA* **2011**, *108*, 5498–5503. [CrossRef] [PubMed]
90. Hendrickson, B.; Leland, R. An improved spectral graph partitioning algorithm for mapping parallel computations. *SIAM J. Sci. Comput.* **1995**, *16*, 452–469. [CrossRef]
91. Nour-Omid, B.; Raefsky, A.; Lyzenga, G. Solving finite element equations on concurrent computers. In Proceedings of the 1986 Symposium on Parallel Computations and Their Impact on Mechanics, Boston, MA, USA, 13–18 December 1987; pp. 209–227.
92. Miller, G.L.; Teng, S.H.; Vavasis, S.A. A unified geometric approach to graph separators. In Proceedings of the 1991 Symposium on Foundations of Computer Science, San Juan, PR, USA, 1–4 October 1991; pp. 538–547.
93. Miller, G.L.; Teng, S.; Thurston, W.; Vavasis, S.A. *Automatic Mesh Partitioning*; Springer: New York, NY, USA, 1993; pp. 57–84.
94. Garbers, J.; Promel, H.J.; Steger, A. Finding clusters in vlsi circuits. In Proceedings of the 1990 IEEE International Conference on Digest of Technical Papers, Santa Clara, CA, USA, 11–15 November 1990; pp. 520–523.

95. Cheng, C.K.; Wei, Y.C.A. An improved two-way partitioning algorithm with stable performance [VLSI]. *IEEE Trans. Comput. Aided Des. Integr. Circuits Syst.* **1991**, *10*, 1502–1511. [CrossRef]
96. Bui, T.N.; Jones, C. A heuristic for reducing fill-in in sparse matrix factorization. *Proc. Parallel Process. Sci. Comput.* **1993**, *1*, 445–452.
97. Hendrickson, B.; Leland, R. A multi-level algorithm for partitioning graphs. In Proceedings of the 1995 ACM/IEEE Conference on Supercomputing, New York, NY, USA, 4–8 December 1995; pp. 28–41.

2

Geospatial Data Management Research: Progress and Future Directions

Martin Breunig [1], Patrick Erik Bradley [1,*], Markus Jahn [1], Paul Kuper [1,*], Nima Mazroob [1], Norbert Rösch [1], Mulhim Al-Doori [2], Emmanuel Stefanakis [3] and Mojgan Jadidi [4]

[1] Geodetic Institute, Karlsruhe Institute of Technology, 76128 Karlsruhe, Germany; martin.breunig@kit.edu (M.B.); markus.jahn@kit.edu (M.J.); nima.mazroob@kit.edu (N.M.); norbert.roesch@kit.edu (N.R.)

[2] College of Information Technology, University of Fujairah, P.O. Box 1207 Fujairah, UAE; maldoori@uof.ac.ae

[3] Department of Geomatics Engineering, Schulich School of Engineering, University of Calgary, Calgary, AB T2N 1N4, Canada; emmanuel.stefanakis@ucalgary.ca

[4] Department of Earth & Space Science & Engineering, Lassonde School of Engineering, York University, Toronto, ON M3J 1P3, Canada; mjadidi@yorku.ca

* Correspondence: patrick.bradley@kit.edu (P.E.B.); paul.kuper@kit.edu (P.K.)

Abstract: Without geospatial data management, today's challenges in big data applications such as earth observation, geographic information system/building information modeling (GIS/BIM) integration, and 3D/4D city planning cannot be solved. Furthermore, geospatial data management plays a connecting role between data acquisition, data modelling, data visualization, and data analysis. It enables the continuous availability of geospatial data and the replicability of geospatial data analysis. In the first part of this article, five milestones of geospatial data management research are presented that were achieved during the last decade. The first one reflects advancements in BIM/GIS integration at data, process, and application levels. The second milestone presents theoretical progress by introducing topology as a key concept of geospatial data management. In the third milestone, 3D/4D geospatial data management is described as a key concept for city modelling, including subsurface models. Progress in modelling and visualization of massive geospatial features on web platforms is the fourth milestone which includes discrete global grid systems as an alternative geospatial reference framework. The intensive use of geosensor data sources is the fifth milestone which opens the way to parallel data storage platforms supporting data analysis on geosensors. In the second part of this article, five future directions of geospatial data management research are presented that have the potential to become key research fields of geospatial data management in the next decade. Geo-data science will have the task to extract knowledge from unstructured and structured geospatial data and to bridge the gap between modern information technology concepts and the geo-related sciences. Topology is presented as a powerful and general concept to analyze GIS and BIM data structures and spatial relations that will be of great importance in emerging applications such as smart cities and digital twins. Data-streaming libraries and "in-situ" geo-computing on objects executed directly on the sensors will revolutionize geo-information science and bridge geo-computing with geospatial data management. Advanced geospatial data visualization on web platforms will enable the representation of dynamically changing geospatial features or moving objects' trajectories. Finally, geospatial data management will support big geospatial data analysis, and graph databases are expected to experience a revival on top of parallel and distributed data stores supporting big geospatial data analysis.

Keywords: geo-data management; big geospatial data; *nD* geo-database; topology; graph database; artificial intelligence

1. Introduction

The massive use of geo-referenced data sets in many fields of science and economy including earth observation [1], environmental sciences [2], city planning [3], BIM [3,4], real-time processing [5,6], and analytics for geospatial data [5] makes geospatial data management increasingly a central task in the workflow of geospatial data processing [1,5,7]. Recent approaches consider open data platforms and containers to handle big geospatial raster and vector data efficiently. However, little benchmarking has been done to compare the efficiency of the different big data stores especially in respect of their spatial functionalities [7], and big data support is not the whole story. For example, geospatial data management is also responsible for the cumbersome task of data integration from heterogeneous data sources [3,4,8].

In this article, we use the term "geospatial data management" for the insertion, deletion, update, and retrieval of geo-referenced geometric, i.e., vector or raster data sets, as well as topological data sets. However, the close connection of geospatial data management with other research fields within geo-information science such as data modelling, data visualization, and data analysis is the focus of our interest. The main contributions of this study are first to give an overview of past research during the last decade, coming up with five milestones, and secondly to highlight future directions of geospatial data management research that have the potential to become key research fields in the next decade.

Geospatial data management is also heavily influenced by the fusion of application fields such as BIM and GIS. As reported by the Web of Science and presented in Figure 1, the number of publications dealing with BIM and GIS integration has increased substantially since 2015, which reflects the significance of BIM and 3D GIS integration by both AEC and OGC communities. Alongside, a new term for integrated BIM and GIS data is coined as GeoBIM, and a benchmark (https://3d.bk.tudelft.nl/projects/geobim-benchmark/) was initiated by ISPRS Commission IV communities to foster the emergence of BIM and GIS integration in research and development engaging academia, stakeholders, government, and industries.

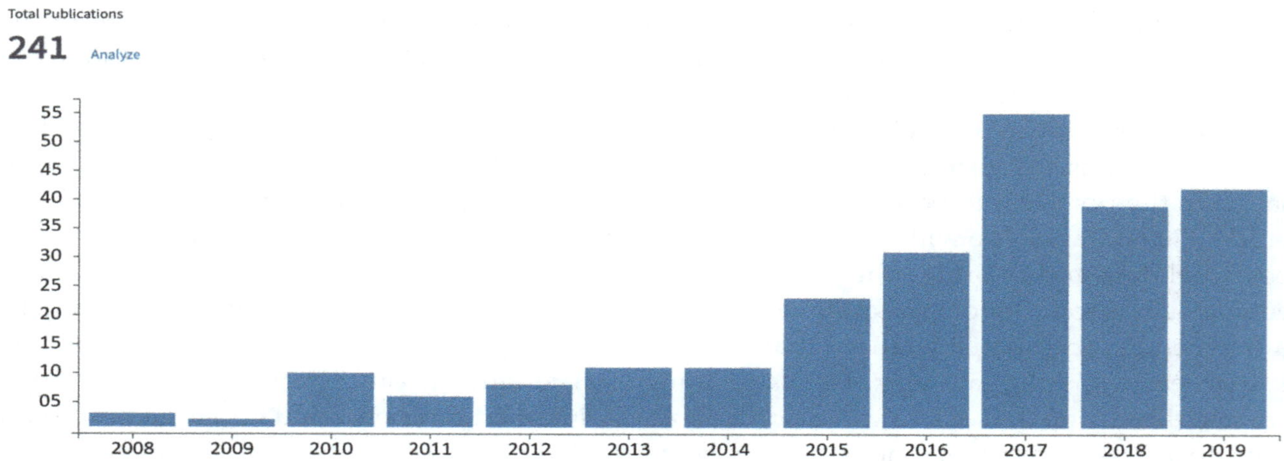

Figure 1. Publication report by the Web of Science for BIM and GIS integration.

This paper is structured as follows: In Section 2, we refer to research in the field of geospatial data management closely referring to ISPRS progression during the last decade, followed by future directions of geospatial database research in Section 3. Finally, Section 4 presents the conclusions of the paper.

2. Progress during the Last Decade

During the last decade, significant progress has been achieved in geospatial data management research. This includes application-near research as well as theoretical results. In the following we present five milestones that reflect the wide spectrum of this research.

2.1. Milestone 1: Advancing GIS/BIM Integration at Data, Process, and Application Levels

For geospatial data management, the fusion of GIS and BIM [9–12] means that two different "philosophies" of spatial data representations are growing together: composition models such as triangle networks used for surfaces and tetrahedron networks used for solids on the one (GIS) side, and constructive solid geometry (CSG) and sweep models on the other (BIM) side. Thus, both kinds of spatial data models have to be considered as geometric and topological data types in geospatial databases. Additionally, point clouds from terrestrial laser scanning [12] have to be considered as important data sources for BIM. They additionally have to be handled in geospatial databases.

Recently, 3D virtual city space has been noticed as a key characteristic of smart sustainable cities [4]. To foster dynamic city applications such as business growth and solutions for monitoring infrastructure, buildings, and the built environment life cycle, many cities around the world have launched 3D open data to the public [3]. Geospatial communities have been highly active in the collection of 3D data using laser point cloud and imaging and in providing 3D GIS models of building envelopes and city spaces [13]. On the other hand, construction communities have offered BIM and rich 3D geometric and semantic information of internal building structure. Also, BIM is being considered as a rich intelligent digital repository that uses an object-oriented (OO) approach to describe the characteristics, i.e., semantics, geometry, and relationships in the architecture, engineering and construction (AEC) domain [4]. Although there are many open BIM standards such as BIMXML and COINS, industry foundation classes (IFC) have become the most popular open standard BIM format for interoperability purposes in the AEC industry developed by the BuildingSMART community [8].

In practice, building elements in BIM are represented and stored as CSG for simple objects, sweep volumes for paths to define a solid, and boundary representation (B-Rep) for its bounding surfaces with implicit topology either in EXPRESS-based or XML-based formats [8,14]. In the industrial foundation classes, BIM models are divided into five levels of development (LOD) marking the course of the construction within the life cycle of each building information model: LOD 100 stands for "conceptual", 200 for "approximate geometry", 300 for "precise geometry", 400 for "fabrication", and 500 for "as-built".

Thus, LOD 100 is a simpler model than LOD 200, etc. This means that the resolution of the model increases during its construction process. Figure 2a shows some of the different LODs for a special building part. In contrast to this, 3D geospatial information systems (GIS) focus on collecting, storing, and analyzing geospatial data at a small scale. More than two decades ago, the Open Geospatial Consortium (OGC) developed the City Geography Markup Language (CityGML) standard data model, an Extensible Markup Language (XML-based) standard for geospatial data interoperability and exchange using the BRep geometry model. This geometry model can also be used for the integration of GIS and BIM geometries [15].

Similar to IFC for BIM, CityGML divides 3D GIS data into four levels of detail (LoDs), from LoD 1 to LoD4 as presented in Figure 2b. Here, the building or city model usually is provided at the highest resolution (e.g., LoD4), and abstractions from this high-resolution model are made by changing the level of detail from LoD4 to LoD3, etc. Furthermore, LoD1-3 describes the model from the outside, whereas LoD4 presents the interior of the model (indoor model).

collection of layers with enabled feature selection and controllable visibility. Obviously, the consistent mapping of geospatial features, such as roads, rivers, lakes, forests, seashores or vessel trajectories on

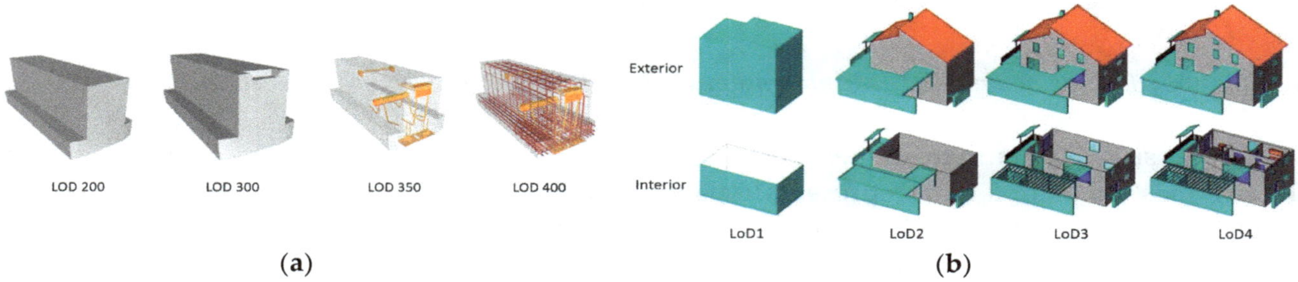

Figure 2. Different LODs in IFC for a precast structural inverted T beam (extracted from https://bimforum.org/lod/) (**a**) and LoDs for CityGML of a residential building (courtesy of Karlsruhe Institute of Technology) (**b**).

It is obvious that both BIM and 3D GIS are developed with fundamentally different purposes and sometimes contradictory geometries, semantics, and relationship data models. Although these differences between BIM and GIS became the target of many studies, they do share common characteristics such as data error checking (clash detection and topology analysis), *nD* simulation, data contents, and extensibility [8]. Therefore, over the last decade, the requirements for linking cross-domain specific information have increased [4]. Obviously, multiple applications in the AEC domain require adjoining information for pre- and post-construction phases, involving GIS data. On the contrary, GIS is enhancing from world and city level information to specific entity or detailed building level information which was originally in the AEC domain [15]. However, a building in the GIS domain even in LoD4 is less complete than in BIM. Thus, these vital conditions have accelerated the focus of bringing the BIM and GIS domains together, and during the last decade, a large number of studies have been conducted to achieve an integration of these heterogeneous systems [4,8,15–18]. In addition, BIM and GIS integration can provide essential capabilities of quantitative and semantics analysis as well as visualization opportunities for knowledge discovery and informed decision making in smart sustainable cities and the built environment.

Based on the recent review by [8], the integration of BIM and GIS has been conducted at the fundamental level to the application level. The fundamental level focuses on data exchange standards and interoperability at the data level, while the application level deals with developing new methods with full potential to exploit the BIM and GIS systems together [16]. Also, this can be divided into five groups such as schema-based, service-based, ontology-based, process-based, and system-based methods [18]. Furthermore, a comprehensive three-level data, process, and application level framework was developed by [17] as presented schematically in Figure 3. At the data level, models and structures are converted, modified or extended to meet the requirements by using extract, transform and load (ETL) techniques. Process-level integration encompasses data standards from BIM and GIS to be simultaneously adopted in a cooperative way stipulating the flexibility of data interoperability among systems. Application-level integration deals with the development of new applications with integrated BIM-GIS or extending existing applications such as plugins. The application approach is stated to be costly and time consuming due to its inflexibility in different domains where no software so far can read and process both BIM and GIS data at once without having required plugins [18].

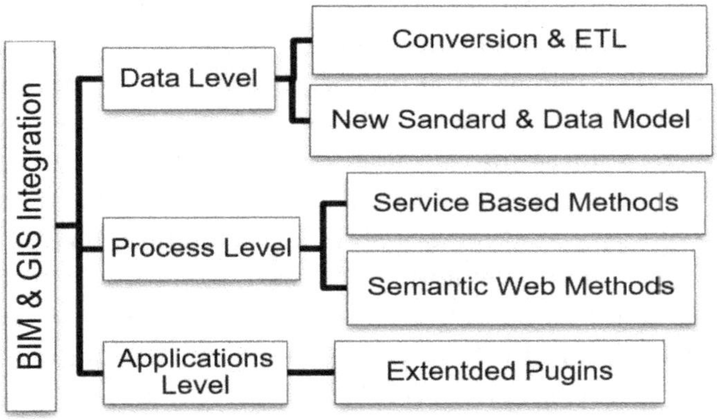

Figure 3. BIM and 3D GIS integration approaches at three levels: data, process, and application.

However, a wide range of studies has been conducted at the data level to achieve interoperability among BIM-GIS data formats with promising results [19]. These include linking, translation/conversion, extension, and meta-models (mediation) and can be further divided into geometric level and semantic level integration [18]. The geometric-based integration addressed the problem of reference systems (local vs. geodetic coordinate systems), 3D geometry (B-Rep, CSG, sweep volumes vs. B-Rep or ESRI® multi patch), and level of details (LOD vs. LoD). Although there are huge achievements for such conversion or integration even with commercial software packages such as Feature Manipulation Engine (FME®) from Safe Software®, BIMServer®, and IfcExplorer®, all these studies focused only on a single building structure rather than the context of the standard GeoBIM having multiple built structures at the city scale [20]. Also, most studies focused on IFC to CityGML exchange where the opposite direction is important to be considered [20]. Indeed, the exchange from CityGML to IFC seems more complex since there are more classes of objects in IFC than in CityGML. Thus, the transformation from CityGML leads to many empty classes in the IFC domain. The semantic-based integration methods include schema extension by ADEs, simplification, or new intermediate schema creation. Among the modification and introducing new models at semantic-level integrations, Semantic Web Technology has proven to be promising with its flexibility of integrating heterogeneous data formats.

Recent studies by [15] suggested that integrating BIM and GIS requires an understanding of the ontology of each data model first, then studying the similarity of the geometry and semantics for each data item. The core concept of the semantic web methods relies on the ontologies of two domains via a hierarchical graph structure. This usually consists of three steps: Ontology generation for Open Building Information Management (OBIM) and Open Geospatial Open Geodata Interoperability Specification (OGIS), ontology mapping to link the similar relationships or concepts, OBIM-GIS, and querying OBIM-GIS in the application domain using the SPARQL protocol and RDF query language (SPARQL) in order to retrieve the information needed from the model [4,15]. The main challenge in semantic web methods is ontology generation where there is no unified and accepted method to achieve this with a solid quality measurement matrix. Also, a lack of automatic generation of ontology is another burden in this method. Therefore, this leads to potential opportunities for future research in BIM and GIS integration using the Web Ontology Language (OWL).

Finally, topology has been proven to be an important concept for unifying GIS and BIM concepts as well as to describe GIS and BIM data structures and methods.

2.2. Milestone 2: Advancing Topology as a Key Concept for Geospatial Data Management

Topology, as it is used in the GIS/BIM world, models the relationships between spatial entities, e.g. it can express the fact that a certain object is part of the border of another object. This example relationship can be labeled bounded by. Certain types of spatial analysis, e.g., such that rely on connectivity queries or path finding, require topology to be modeled. Certain relationships bear a "topological" notion, such as contains, covers, bounded by, and touch. On the other hand, as

Alexandrov already observed in 1937, any binary relationship defines a topology in the mathematical sense, and any topology of a finite set of entities comes from a binary relation [21]. Consequently, the topology of the GIS/BIM world is a part of the mathematical domain named topology. Furthermore, this kind of topology can model the relationships between any kind sof entities, whether they are spatial or not. The geo-information community has not seemed to be aware of this fact yet.

A famous example of the aforementioned kinds are Egenhofer's 9-intersection matrices which are nothing but a means of classifying spatial relationships between regions [22] and which were extended to raster data in [23]. However, due to Alexandrov's result, topology is not restricted to relationships which are viewed as "spatial". A special case is given by binary relations which form a partial order, i.e., which do not allow cyclic iterated relationships between distinct objects. Obviously, GIS/BIM topologies exclusively deal with topologies given by partial orders.

An important example for topology in GIS or BIM is given by the adjacency graph whose nodes model GIS/BIM objects such as regions or rooms, and an edge is defined whenever two objects are adjacent. The first time that this concept from GIS was used in BIM is probably the article [24], cf. also [25] for the three-dimensional case. [26] contains an application to emergency response in buildings. Also, indoor networks from IFC models rely on correctly capturing the topology in buildings [27].

Undirected graphs are well-known examples of one-dimensional Alexandrov topological spaces. The introduction of simplicial complexes has been an important step towards generalizing graph topology to higher dimensions for GIS/BIM. These are used in [28] to model spatial or space-time scenarios. Here, there is the important relation face of, as simplices have faces which are also simplices. This defines the so-called incidence graph. Its reflexive and transitive closure is a partial order of all simplexes. A natural generalization of a simplicial complex is given by cell complexes or cw-complexes. They have a boundary operator between the vector spaces spanned by the cells (where each cell produces a new dimension). This boundary operator has a natural representation in relational databases, called relational chain complex, and the face topology (i.e., incidence graph) can be easily derived from this representation [29].

Manifold-like spaces are often modeled as so-called generalized maps (G-maps) [30] whose entities are the darts which are nothing but maximal relationship chains in the underlying partially ordered set. These darts are also called cell tuples, as their elements are cells at the boundary of other cells. A certain structure is imposed on the set of darts with the help of a family of involutions. In [31], it was observed that G-maps for higher dimension tend to be verbose, as the worst-case number of maximal chains in a partially ordered set is exponential to its size. This is in contrast with the worst-case storage complexity of the incidence graph itself, which is quadratic in the number of objects, and this is also a worst-case lower bound for storing topology [32]. In other words, the incidence graph is storage efficient. Nevertheless, combinatorial maps which are similar to G-maps have been recently used for the topological reconstruction of CityGML data [33].

Topology, geometry, and semantics come together in 3D plus time, i.e., 3D/4D applications that play a major role in planning scenarios and geo-scientific applications.

2.3. Milestone 3: Advancing 3D/4D Geospatial Data Management

In the last decade there have been significant advancements in the field of 3D/4D geospatial databases. The parallelization of queries using n-dimensional space-filling curves [34] has been examined [35] and applied to massive point clouds [35–38]. Furthermore, following the concepts of Langran and Stuart [39], spatio-temporal data models including moving surface and solid geometries have been implemented in object-relational [40–43] and object-oriented [44,45] geospatial database management systems. The latter have been implemented by time-dependent net components containing simplicial complexes [46,47]. As applications for these spatio-temporal data models, geological strata and faults of open mines have been used extensively [44,45,47].

One approach to classify the various lines of development for spatio-temporal data models is to distinguish them by their data models [47]. For 3D/4D geospatial databases, four different prominent data models can be identified depending on if they are focusing on geometry, topology or semantics: point clouds, simplicial complexes, other topological approaches, and CityGML.

For an efficient management of massive point cloud data, space-filling curves [34] are frequently used on top of object-relational databases [35,36]. Massive data sets were tested within different environments and point cloud benchmarks were developed [37]. This research includes solutions to handle the level of detail (LoD) of various point cloud properties [38] by discrete global grid systems (DGGSs) [48].

Usually, simplicial complexes [46] are used as a data model for the geometric modelling of natural structures [49,50] in the geosciences because they allow a better approximation of the geometry in 3D space than grids and ease geo-computation, as the algorithms can be broken down to basic geometric operations on triangles and tetrahedra. Whenever the data are modelled over a period of time, multiple time steps can be combined and form a spatio-temporal object. The so-called snapshot model is a well-known concept for the handling of such data [51]. Nearly three decades ago, Worboys introduced an object-oriented approach for handling simplicial complexes in a geospatial database [52]. Based on the data models of GIS pioneers such as Worboys, Le et al. developed an approach for a spatio-temporal DBMS that is based on the object-relational geospatial database PostGIS® [40–43]. This approach is capable of handling spatio-temporal data based on the snapshot model and is linked to the 3D modelling software GOCAD® [53] that is rampant in the oil industry. The geospatial database architecture DB4GeO has been realized as an object-oriented approach [44,45,54]. Both approaches focus mainly on the management of 3D models, but also manage the composition and storage of simple 4D (3D space plus 1D time) models. Recently, an approach based on integrating the concept of point tubes with net components has been developed [47]. This approach can be used for data-saving and high-performance management of spatio-temporal data based on simplicial complexes.

Finally, the management of digital twins based on CityGML within a DBMS has been examined [55,56]. With 3DCityDB® [57] a robust DBMS solution based either on Oracle or PostgreSQL is available as free and open source software [45]. Latest developments focused on the support of CityGML application domain extensions (ADEs) and the 3D web viewer 3DCityDB-Web-Map-Client [57]. For both, subsurface and city applications, 3D/4D geospatial data management is indispensable, as for planning purposes it is no more sufficient to store only the current city model in the geospatial database, but the history of the 3D city model as well [55,56]. This includes the integration of subsurface and city 3D/4D objects [47].

2.4. Milestone 4: Modelling and Visualization of Massive Geospatial Features on Web Platforms

With the increasing popularity of web-mapping applications and the rapid growth of map data availability, the pre-computation and caching of map image tiles (a.k.a. tile maps) has become a common practice in online map services as these processes use far fewer server resources than maps rendered on demand [58,59]. Online map service providers, such as Google Maps, Bing Maps, ArcGIS® Online, and OpenStreetMap, organize and deliver their content in tile maps that correspond to different areas (multi-location mapping), are displayed at various zoom levels (multi-resolution representation), and include various themes (multi-theme mapping) in different annotation languages (multi-language mapping) [59]. Obviously, rendering these products on the fly is not an option due to the cartographic limitations, such as the lack of a reliable mechanism for automated map generalization [60,61].

Google was one of the first major mapping providers to adopt map tiles [59]. Others, such as Bing and OpenStreetMap, followed the same practice. Geographic information systems (GIS) software vendors, such as Esri® and Oracle®, provide functionality for map tiling and caching of both vector layers and raster images. They also support single fused and multi-layer tiles. In the former, a group of layers is combined into a single image per tile, while in the latter, these layers appear at the client as a all these tiles and zoom levels involves a huge investment of resources, provided that the simplification of linear features is a semi-automated process [61].

Discrete global grid systems (DGGS) offer an alternative geospatial reference framework with increasing popularity over the last few years [62]. The growth in the number of ground, airborne, and satellite sensors has led to a significant increase in the amount, variety, and rate of collection of geospatial data. However, combining huge volumes of heterogeneous geospatial data on the geographic grid is computationally expensive and time consuming [60,61]. A DGGS enables the rapid assembly of geospatial data without the difficulties of working with projected coordinate reference systems. It offers a base information multi-resolution grid capable of efficient management, storage, integration, exploration, mining, and visualization of geospatial big data.

In October 2017, OGC officially created the DGGS Abstract Specification [63], which provides a concise definition of the DGGS conceptual model and essential characteristics of a conformant DGGS. With the recent explosion in volume and variety of geospatial data, DGGSs consisting of equal area uniform grids provide a way to store and manage such data while enabling efficient integration, analysis, and visualization [64].

Another emerging method for delivering web maps is vector tiles. Vector tiles are packets of geographic data in vector format that have been clipped to the boundaries of predefined tiles, instead of pre-rendered map images. In summer 2018, OGC launched the Vector Tiles Pilot [65] in an effort to standardize vector tiles and promote interoperability. The combination of vector tiles with map (raster) tiles and DGGS data has also become an emerging need. Obviously, rendering these products on the fly is not an option due to the cartographic limitations, such as the lack of a reliable mechanism for automated map generalization [60,61]. The visualization of cartographic features at various scales (also known as multiple representation) has been an open issue in the digital era. Over the last decade, many advantages have been made in effective and meaningful visualization of mapping content in various environments and over different devices [66].

2.5. Milestone 5: Extensive Use of Geosensor Data Sources

Over the last decade, sensor technology developed rapidly, which has many advantages concerning the availability of geo-referenced data, but also includes privacy issues to be researched [67,68]. Devices have gotten smaller, cheaper, and need less energy. A sensor with a high interest for the geospatial community is the smartphone [67,69]. Being equipped with global navigation satellite system (GNSS) sensors, the location of the device is fairly well known, depending on the quality of the sensor and the signal [68,69]. Location in this context implicitly means geo-reference, but even if the phone has no GNSS sensor or the sensor is switched off, at least the phone cell delivers raw knowledge about the position of the device [69].

A GNSS track is an excellent example to explain the possibilities given by these data [70]. At first, a track represents nothing but a collection of points, each initially defined by a 3D coordinate and a time stamp; hence, a 4D data set. This set would be needless if we ignore the track as a whole because based on the consecutive points, further information can be derived, such as velocity, acceleration or azimuth, to name but a few. All these are valuable input to get information about the travel mode of citizens, e.g., walk, bicycle, etc., which is a highly interesting issue in urban planning [71]. Based on the properties of the trip (velocity, acceleration, length, etc.) in combination with sophisticated algorithms, the pattern can be evaluated to get an idea how the user moved [71]. Furthermore, singular point positioning ("navigation solution") can be used for applications where the GNSS signal is not available for tracking [69].

Almost all smart devices communicate with the net. The term Internet of Things (IoT) was coined for this [72–74], and data mining in the IoT has become a field of research on its own [75,76]. The purposes of the sensors are various. Some of them have been dedicated to security purposes such as fire, gas or humidity [77]. Others have been installed to get additional data such as private meteorological stations [78], or so-called wildlife traps have been placed in remote areas to get information about shy or rare animals, etc. [79]. All these examples have in common that they produce a huge amount of data which have to be managed in geospatial databases, and, most of all, these data can no longer be analyzed in the traditional way. This has led immediately to approaches such as machine learning or artificial intelligence and approaches to combine them with parallel data stores such as SpatialHadoop® [80], ST-Hadoop® [81] or Hadoop-GIS® [82]. These approaches, as mentioned before, have developed to the key science for big data [83].

As a consequence of the available geosensor data, "LinkedGeoData" has been established as a concept to add the spatial reference, i.e., coordinates, to the Web of Data/Semantic Web [73]. As an example, LinkedGeoData has used data of the OpenStreetMap project [84,85] and thus has made the OpenStreetMap data available as a resource description framework (RDF) knowledge base following the linked data principles [73]. Furthermore, OpenStreetMap started to interlink this data with knowledge bases of the Linking Open Data initiative [86].

It is obvious that the collection of geosensor data has led to thousands of gigabytes in a short time period. Without geospatial data management, these large data sets would be no more manageable, especially if the replicability of the examinations is required. For big geosensor data applications, parallel data storage platforms such as SpatialHadoop® [80], ST-Hadoop® [81] or Hadoop-GIS® [57] have been used to support geospatial data analysis. This facilitates the use of artificial intelligence (AI) algorithms [81].

Following Michael Goodchild´s concept of "citizens as sensors" [87], which opened the world of "volunteered geography" and "volunteered geographic information" (VGI) [88,89], researchers analyzed large crowd source data sets such as the OpenStreetMap project [85,90]. Furthermore, online data sources of citizens generated by social media have been included in the examinations [88,89,91]. As an example, the collection of VGI from social media and other VGI sensors [88,89,91] has been identified as having enormous potential to get a detailed real-time picture of a scene after hazards such as earthquakes or tsunamis [92].

These improvements in geospatial data-collecting technology have helped non-experts to share and aggregate geospatial data in a progressively accessible way [88,89]. However, this "revolution" of data collection has also posed new questions for geospatial data management: Opposite to traditional spatial data production processes, data can be easily generated by each citizen, i.e., non-experts. This has led to badly structured data and has caused serious quality issues as observed in the OpenStreetMap project [85,90]. Hitherto, traditional geospatial database management systems are not well prepared to manage data with heterogeneous schemes for the same data type or data with dynamically changing schemes. The same is true for the management of unstructured data, i.e., data that neither have a pre-defined data model nor are organized in another pre-defined manner. Unstructured data usually are found in texts or text messages and have been analyzed e.g., in the context of hazard monitoring [91,92]. Using today´s parallel data stores such as Spatial Hadoop® [80], ST-Hadoop® [81] or Hadoop GIS® [57], the storage of these data has not been the problem. Opposite to the traditional retrieval of structured data in geospatial data management systems, the retrieval of unstructured data often has focused on the retrieval of spatial or spatio-temporal patterns or clusters. Thus, the retrieval of the data has not been realized directly in geospatial database management systems, but by statistical methods and methods of artificial intelligence such as unsupervised neural networks [91].

3. Future Directions

Based on the five milestones of geospatial data management achieved during the last decade presented in Section 2, in this section, fields of research are outlined that have the potential to become key research fields of geospatial data management in the next decade.

3.1. Geospatial Data Science

Data science has been developed as an interdisciplinary field of science that enables scientifically sound methods, processes, algorithms, and systems to extract knowledge, patterns, and conclusions from both unstructured and structured data [93]. Thus, in future geospatial data science will have to deal with the integration of unstructured and structured data extracting geospatial knowledge, i.e., data and patterns, from unstructured data such as text messages or tweets. On the one hand, methods for the extraction of knowledge from unstructured data have to be improved concerning data quality and the comparability of different data sources. On the other hand, for structured data, the heart of data science will retain the model or the structure of the data [94]. With knowledge about the structure of the data and how it should be available, a data scientist will be able to plan a suitable system to handle the needs. Without some sort of, e.g., relational schema, the data themselves are only a string of bits and bytes without any meaning. One of the bottlenecks for the moment, the amount of data transfer between different software components, has to be eliminated in future. Cloud-based solutions are reaching their limits shown by the problems raised by the Internet of Things [73]. Edge- and fog-based solutions mentioned by Cisco® tend to solve those problems, but they are still not implemented widely [95].

It is the duty of geospatial data management to solve those problems by bridging the gap between modern information technology concepts and the geo-related sciences such as geography, geo-sciences or civil engineering to provide tools and models to ease geo-related work across all disciplines [56]. Geospatial data themselves are by nature large in quantity and extremely heterogeneous in quality. Although there are optimized and specialized solutions for sub-disciplines of geo-related sciences such as CityGML for city models [55], there is a lack of general solutions. To process geospatial data quickly and to generate added value (information) in foreseeable time, the models have to adapt to parallel and distributed network systems on the one hand, and on the other hand, they have to enrich a common communication medium. The integration of unstructured with structured geospatial data using parallel and distributed network systems, realized by ontologies, linking, and a common communication medium, is likely to become the greatest challenge geospatial data management is facing. All together, there is a need for concepts and systems consisting of related and linked ontologies to describe the semantic contents of all the intervening components such as geospatial, topological, thematic, and temporal. Finally, a communication standard has to be developed for n-dimensional data exchange between systems processing the different components.

3.2. Topology

Topology is a powerful and general concept to analyze GIS and BIM data structures and spatial relations that will be of great importance in emerging applications such as smart cities [96] and digital twins [97]. Besides semantics and geometry, topology is likely to become a key concept to support geospatial data modelling and management efficiently [98].

For example, the use of several levels of details leads to the storage of aggregation maps which are nothing but a binary relation between objects at different levels of detail [60]. Including urban infrastructure, such as water pipes, leads to relationships between buildings or land parts and infrastructure parts in a combined topological model [99]. DB4GeO allows the management of topological space-time data. All objects are stored as triangulated or polytope meshes in different dimensions at different levels of detail [45].

A spatial model consisting of points, (open) segments, (open) polygons, (open) polytopes, etc. (called objects) is topologically consistent if all objects are pairwise disjoint. This definition appears first in [100] and differs from previous definitions of topological consistency in the literature. The advantage is that it relates geometry and topology in a meaningful way.

In [100], it was observed that this notion of consistency is equivalent with specifications in the ISO 19107 standard, and that CityGML (Version 2.*) does not comply with this standard, as there are correctly modeled data sets which are not topologically consistent. This is remarkable, as CityGML claims to be based on OGC standards, cf. Figure 4.

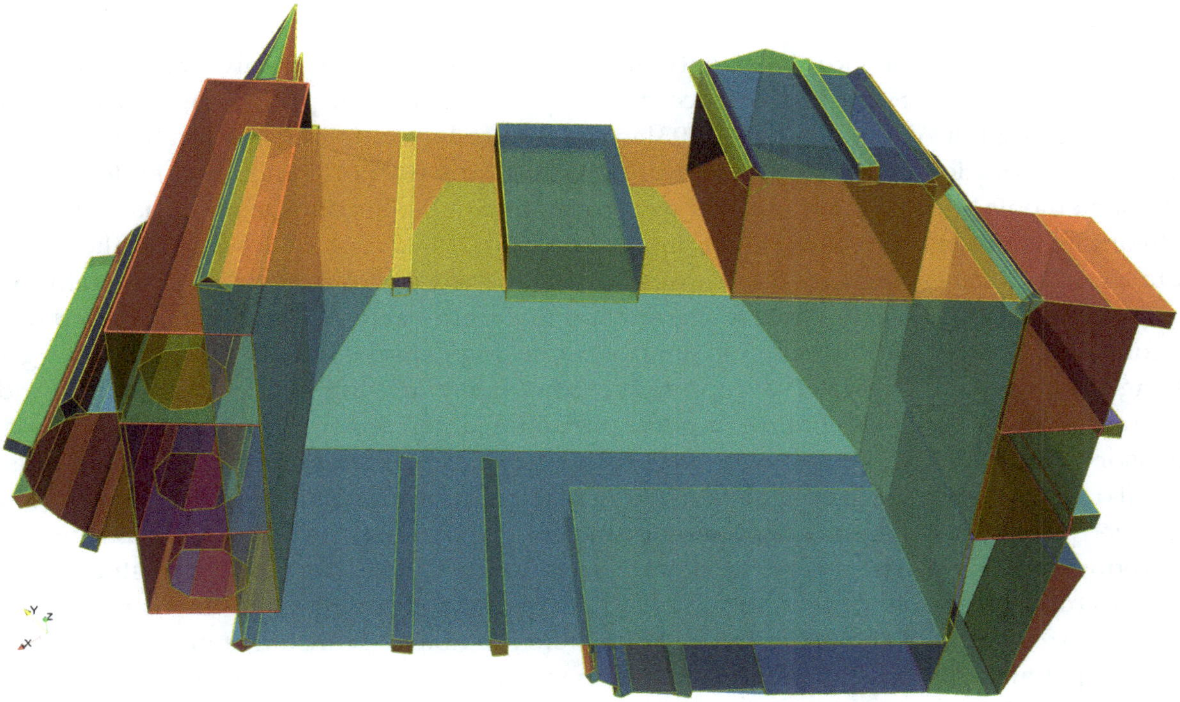

Figure 4. Example of non-consistent CityGML data.

Wireframe/Linestring-based boundary representations used in CityGML are not trivial to be triangulated in order to retrieve higher dimensional geo-objects (surface boundary representations or true volumes, e.g., tetrahedron nets). If the focus to create the model is analysis or even simulation rather than only visualization, all n-dimensional geometrical solids and their topological relations must be declared or calculated in an efficient and unequivocal way. Figure 4 shows some of the topological inconsistencies. The multi-surface of the bottom plane shows intersections which also lead to penetrations within the defined multi-volume. Recalculating all topological inconsistencies by finding all n-dimensional intersections in runtime is a complex task which should be avoided. The model is also not unequivocal due to different ways of triangulating Wireframe/Linestring-based boundary representations to create surfaces for a surface boundary representation and different ways of triangulating surface-based boundary representations to create the final volumes as a tetrahedron net. The problem gets more complex by adding further dimensions such as time coordinates or any other thematic/semantic coordinates.

Topological consistency is a requirement for analysis and simulation of geospatial data models, as cumbersome geometric operations such as intersections can be avoided, and the simulation can be run on the topological model itself. In order to obtain a topologically consistent model, an overlay method can be applied, which breaks the overlapping objects into disjoint atomic objects and retains the boundary relationships. A precise mathematical definition of this is deferred to a future publication.

The topology model has a minimal representation in the form of the so-called Hasse diagram, a graph in which only direct boundary relationships are explicitly modeled. This can be used as the core model for analyzing flows or running simulations. This means that in future work, partial differential equations need to be solved numerically on graphs and the solutions compared with corresponding solutions for the underlying 3D model. The mathematical theory for this approach is still quite recent, cf. e.g., [101] and the references therein, but lacks a thorough numerical treatment. Its potential lies in the applicability of the topology of geospatial data stored in NoSQL databases and possible parallelization of numerical algorithms.

The tasks of parallelisation and streaming of big topological data sets are the consequent next steps for the future, as by now the means for this are at hand.

3.3. Bridging Geospatial Data Management with Data Streaming Libraries and "In-Situ" Geo-Computing

In the future, different kinds of network architectures which involve edge- and fog-based solutions [95] will have to be tested to connect online transaction processing (OLTP) and online analysis processing (OLAP) systems [102,103] in order to reduce data transfer and to establish efficient geo-computing solutions. Recent programming paradigms support functional-style operations on streams of elements such as map-reduced transformations of collections, e.g., the java streaming api. Data streaming is not only a technology to move data from A to B, but it allows us to filter and analyze the data as well. While some functional-style operations work within one OLAP-system to retrieve certain added value, it is also possible to use streaming libraries across a largely distributed network to set up some real-time streaming pipeline or application, e.g., apache kafka®. Examples for geospatial streaming data are point streams that represent continuously moving points, but also line and region streams representing complex spatio-temporal phenomena derived from geo-referenced data [104].

Sensor networks, geospatial data warehouses or high-performance computing clusters for geospatial data may differ in geospatial data modelling and network architectures. Each network/cluster is optimized to handle certain collection, simulation, processing, analysis, and presentation tasks and depends strongly on its geospatial data model. In the future, the modelling steps from sensor networks to high-performance computing could be managed through a real-time streaming pipeline or application. If some of those networks or clusters are able to translate a geospatial dataset into another geospatial data model in parallel or even shares the same geospatial data model, then "in-situ"-approaches could become feasible and tested. "In-situ" geo-computing of objects executed directly on the sensors will revolutionize geo-information science and bridge geo-computing closer to geospatial data management.

3.4. Geospatial Data Visualization on Web Platforms

The representation of geospatial features (e.g., roads, rivers), outlines of areas (e.g., municipal boundaries, lake banks) or moving objects trajectories (e.g., of humans, vehicles) on paper or digital maps comprises a massive number of vertices. To facilitate the processing, analysis or mapping of these geometries at a small scale, beyond semi-automated cartographic methods, new developments in Geospatial Web and Digital Earth have to be introduced to model and map voluminous geospatial data. To deliver the content of map services in a standardized tile format [64] for high-speed dissemination of voluminous data over the Web and the adaptation of discrete global grid systems (DGGS) to understand the planet model by offering an analysis-ready-information-grid [63,65] are only the first steps in this way of development. The rapidly growing use of these reference frameworks (i.e., tile maps and DGGS) has urged the development of new approaches to an efficient, consistent, and compliant mapping of massive polyline geometries representing geographic features or trajectories.

Future research will aim to introduce new approaches to the fundamental problem of multiple representation in geospatial data handling. These approaches will offer sophisticated modelling and visualization of massive linear geospatial features in modern geospatial reference frameworks. Industry and government map service providers will be able to generate massive map products in a faster and more accurate mode.

Finally, geospatial data management and data visualization have to grow together to enable more efficient solutions for geospatial data analysis. This means that the visualization of large amounts of data should be executed by visualization clusters. A good presentation often needs some analysis to filter information which will be visualized. Those filters in extremely large datasets need to work on distributed memory computing resources. The visualization of simulation results or field data in real-time analysis of extremely large datasets is still challenging. The amount of data to be transferred to the visualization cluster from data-warehouses and/or a sensor networks is as large as the extremely large datasets which should be visualized. It is in question if a visualization cluster can be merged with data warehouses or even sensor networks. One approach of merging simulation clusters and visualization clusters at a large scale is implemented by Kitware within ParaView® and its "in-situ"-technology called "catalyst®" [105]. The visualization cluster is able to calculate simulations if the source code of the simulation has been provided. After a simulation step has been calculated, the presentation calculations can be done in memory without moving the data to a different cluster. With this approach, it is possible to visualize each simulation result at simulation/presentation-calculation time. Certainly, new research will be based upon this research and will adopt it for new applications.

3.5. Database Support for Big Geospatial Data Analysis

Obviously, big geospatial data analysis will be increasingly supported by artificial intelligence (AI) for geospatial data [106], also called geoAI [107], and the retrieval of the data will be facilitated by parallel geospatial database architectures. GeoAI will closely combine geo-information science, AI methods in machine learning, data mining, and parallel computing to extract knowledge from big geospatial data [107]. To give an example, in future, geoAI methods will be used for geospatial data cleansing, i.e., to detect and learn errors in geospatial data sets. These methods will complement the existing data consistency checking mechanisms of geospatial database management systems. Parallel geospatial database management systems will serve as a "preliminary stage" to pre-select geospatial and spatio-temporal data before AI-supported data analysis will take place. Thus, geospatial analysis tools will be supported by geospatial data management systems that split the data into geospatial or spatio-temporal partitions to organize parallel database accesses.

Many approaches for geospatial data management have been implemented on top of object-relational [55,108] or on object-oriented databases [45,109]. However, graph databases promise to be well suited for this task, as has been first proven by [110]. For future research, we expect that graph databases will experience a revival in the context of distributed and parallel geospatial data analysis. One of the major topics within this research will be how to integrate known geospatial-, spatio-temporal- or *nD*-access methods into the property graph system.

A further research topic will be how to use graph algorithms to quickly solve standard geo-tasks known from standard GISs on the feature graph.

Increasingly, software and methodologies will be imported primarily from intelligent structures [107] for the aim of developing shrewd GIServices. It is the ability to understand, plan, analyze, and behave in an innovative way that is crucial to intelligent structures. Intelligent geospatial services will serve as a kind of intelligent structure from the AI point of view [111–113]. Intelligent geospatial services will be intelligently aware, net-connected, and regulated by IoT software use [114]. The service-oriented architecture (SOA) strategy with relation to the IoT [73] will have to be efficiently implemented, which allows support sharing and recycling through services-oriented characteristics [115]. The creation of smooth ontology will help smart geospatial services to resolve confusion [116]. One of the key cutting-edge gaps in geospatial data transformation is an "intelligent" mechanism that facilitates the discovery of facts and integration in the Internet [117]. Ontologies and geospatial metadata will play a central role in this process by linking data from different data sets. Thus, the dissemination of such mechanisms will lead to a better understanding and integration of heterogeneous geospatial data sources.

4. Conclusions

This overview article has made two contributions: to highlight research in geospatial data management during the last decade and to formulate five future directions of research that have the potential to become key research fields of geospatial data management research in the next decade.

Five milestones of research in the last decade have been identified that range from application to theory. On the application side, especially the progress in interdisciplinary research has been emphasized, such as the growing together of the BIM and the GIS geospatial data management. This has opened hitherto closed doors on the data, process, and application side on the way to integrated BIM/GIS models. This three-level approach also considers for the first time the lifecycle of buildings and cities. Looking at the other (theoretical) spectrum of the milestones, the theory of geospatial data modelling and management has been improved by considering topology as a sound mathematical framework which is not limited to topological relationships as formulated in the 4- or 9-intersection model. Thus, topology is seen as a key concept of geospatial data management to describe and analyze data structures and algorithms presuming that geometry and semantics of objects can be added to topology. As basis for BIM and GIS applications, 3D/4D geospatial data management has been identified as a key concept to support the geometric, topological, thematic, and temporal modeling of buildings and cities and subsurface models. 3D/4D geospatial data management also has started to serve as an integration platform for different geospatial and spatio-temporal data representations. As geospatial data management is closely connected to and dependent on data modelling and data visualization, it is important that in the last decade, progress in modelling and visualization of massive geospatial features on web platforms has been made. In particular, the application of discrete global grid systems as an alternative geospatial reference framework and the development of visualization clusters on top of parallel data stores have provided significant research progress for the geospatial community. Finally, the extensive use of geo-sensor data sources has been identified as a milestone in research, not least because Michael Goodchild´s paradigm of "citizens as sensors" has revolutionized the handling of geospatial data, leading to new requirements including geospatial data management for unstructured and structured data to be solved in the next decade.

Geospatial data science is expected to become a key research field for the next decade to solve the integration of unstructured and structured data as well as the extraction of geospatial knowledge, i.e., data and patterns, in interdisciplinary applications. Furthermore, geospatial data science will bridge the gap between modern information technology concepts and the geo-related sciences. In this context, topology will continue to be a powerful and general concept to analyze GIS and BIM data structures and spatial relations that will be of great importance in emerging applications such as smart cities and digital twins. Streaming of big geospatial data and "in-situ" geo-computing will play a central role in the next decade. Thus, the bridging of geospatial data management with data streaming libraries and "in-situ" geo-computing on objects executed directly on the sensors has been identified as a future direction of research. This will revolutionize geo-information science and bridge geo-computing with geospatial data management. In future, visualization will be more integrated into web applications. Thus, the development of advanced geospatial data visualization on web platforms has been presented as a future direction of research. This will enable the representation of dynamically changing geospatial features or moving objects trajectories directly loaded from geospatial databases. Finally, geospatial data management will increasingly support big geospatial data analysis. This future direction of research will include the management of unstructured and structured geospatial data as well as the extraction of knowledge and patterns in geospatial data. Graph databases are expected to experience a revival on top of parallel and distributed data stores supporting big geospatial data analysis. Finally, raster data gained from satellites will be managed efficiently in geospatial data management systems. This includes the realization of seamless geospatial and temporal access to satellite data.

Author Contributions: Conceptualization & Methodology for the overall paper. Writing parts of Section 2.3, Section 2.5 (unstructured data, citizens as sensors, Linked GeoData, Spatial Hadoop), Section 3.1 (part with structured and unstructured data). Revision of Section 3.5. Conceptualization and Writing Abstract, Introduction and Conclusions, writing Original Draft Preparation; Supervision, Revisions of whole paper: M.B.; Conceptualization & Methodology of Sections 2.2 and 3.2; English editing: P.E.B.; Conceptualization, Methodology and implementation of sections Section 2.2 (together with P.E.B.), Section 3.1, Section 3.2 (together with P.E.B.), Section 3.3, Section 3.5 incl. Figure 4: M.J.; Conceptualization and writing Section 2.3, editing references: P.K.; Conceptualization and writing earlier versions of Sections 2.5 and 3.5: N.M.; Conceptualization and writing Sections 2.5 and 3.5, first parts, respectively: N.R.; Conceptualization and writing Sections 2.5 and 3.5, second parts, respectively: M.A.-D.; Conceptualization and writing sections Sections 2.4 and 3.4: E.S.; Conceptualization and writing Section 2.1: M.J., incl. Figures 1–3. All authors have read and agreed to the published version of the manuscript.

Acknowledgments: The authors appreciate the ISPRS research exchange, especially within the working groups of commission IV on Spatial Information Science, excellently led by Sisi Zlatanova.

References

1. Tan, Z.; Yue, P.; Gong, J. An Array Database Approach for Earth Observation Data Management and Processing. *ISPRS Int. J. Geo Inf.* **2017**, *6*, 220. [CrossRef]
2. Kim, H.-S.; Sun, C.-G.; Cho, H.-I. Geospatial Big Data-Based Geostatistical Zonation of Seismic Site Effects in Seoul Metropolitan Area. *ISPRS Int. J. Geo Inf.* **2017**, *6*, 174. [CrossRef]
3. Arroyo Ohori, K.; Diakité, A.; Krijnen, T.; Ledoux, H.; Stoter, J. Processing BIM and GIS Models in Practice: Experiences and Recommendations from a GeoBIM Project in The Netherlands. *ISPRS Int. J. Geo Inf.* **2018**, *7*, 311. [CrossRef]
4. Hor, A.-H.; Jadidi, A.; Sohn, G. Bim-Gis Integrated Geospatial Information Model Using Semantic Web and RDF Graphs. *ISPRS Ann. Photogramm. Remote Sens. Spat. Inf. Sci.* **2016**, *3*, 73–79. [CrossRef]
5. Wu, C.; Zhu, Q.; Zhang, Y.; Du, Z.; Ye, X.; Qin, H.; Zhou, Y. A NoSQL–SQL Hybrid Organization and Management Approach for Real-Time Geospatial Data: A Case Study of Public Security Video Surveillance. *ISPRS Int. J. Geo Inf.* **2017**, *6*, 21. [CrossRef]
6. Ge, M.; Dohnal, V. Quality Management in Big Data. *Informatics* **2018**, *5*, 19. [CrossRef]
7. Hu, F.; Xu, M.; Yang, J.; Liang, Y.; Cui, K.; Little, M.M.; Lynnes, C.S.; Duffy, D.Q.; Yang, C. Evaluating the Open Source Data Containers for Handling Big Geospatial Raster Data. *ISPRS Int. J. Geo Inf.* **2018**, *7*, 144. [CrossRef]
8. Zhu, J.; Wright, G.; Wang, J.; Wang, X. A Critical Review of the Integration of Geographic Information System and Building Information Modelling at the Data Level. *IJGI* **2018**, *7*, 66. [CrossRef]
9. Sacks, R.; Ma, L.; Yosef, R.; Borrmann, A.; Daum, S.; Kattel, U. Semantic Enrichment for Building Information Modeling: Procedure for Compiling Inference Rules and Operators for Complex Geometry. *J. Comput. Civ. Eng.* **2017**, *31*, 04017062. [CrossRef]
10. Borrmann, A.; König, M.; Koch, C.; Beetz, J. (Eds.) *Building Information Modeling*; Springer International Publishing: Berlin, Germany, 2015.
11. Borrmann, A. From GIS to BIM and back again—A spatial query language for 3D building models and 3D city models. In Proceedings of the 5th International 3D Geoinfo Conference, Berlin, Germany, 3–4 November 2010.
12. Reshetyuk, Y. *Terrestrial Laser Scanning: Error Sources, Self Calibration and Direct Georeferencing*; VDM: Stockholm, Sweden, 2009.
13. El-Mekawy, M.; Östman, A.; Hijazi, I. A Unified Building Model for 3D Urban GIS. *IJGI* **2012**, *1*, 120–145. [CrossRef]
14. Dave, B.; Buda, A.; Nurminen, A.; Främling, K. A framework for integrating BIM and IoT through open standards. *Autom. Constr.* **2018**, *95*, 35–45. [CrossRef]
15. Hor, A.-H.; Sohn, G.; Claudio, P.; Jadidi, M.; Afnan, A. A Semantic Graph Database For Bim-Gis Integrated Information Model for an Intelligent Urban Mobility Web Application. *ISPRS Ann. Photogramm. Remote Sens. Spat. Inf. Sci.* **2018**, *4*, 89–96. [CrossRef]
16. Irizarry, J.; Karan, E.P.; Jalaei, F. Integrating BIM and GIS to improve the visual monitoring of construction supply chain management. *Autom. Constr.* **2013**, *31*, 241–254. [CrossRef]
17. Amirebrahimi, S.; Rajabifard, A.; Mendis, P.; Ngo, T. A BIM-GIS integration method in support of the assessment and 3D visualisation of flood damage to a building. *J. Spat. Sci.* **2016**, *61*, 317–350. [CrossRef]
18. Kang, T.W.; Hong, C.H. A study on software architecture for effective BIM/GIS-based facility management data integration. *Autom. Constr.* **2015**, *54*, 25–38. [CrossRef]
19. Fosu, R.; Suprabhas, K.; Rathore, Z.; Cory, C. Integration of Building Information Modeling (BIM) and Geographic Information Systems (GIS)—A Literature Review and Future Needs. In Proceedings of the 32nd CIB W78 Conference 2015, Eindhoven, The Netherlands, 27–29 October 2015; p. 9.
20. Geiger, A.; Benner, J.; Haefele, K.-H. Generalization of 3D IFC Building Models. In *3D Geoinformation Science*; Springer: Cham, Switzerland, 2015; pp. 19–35.
21. Alexandrov, P.S. Diskrete Räume. *Matematicheskii Sbornik* **1937**, 501–518.
22. Egenhofer, M.J. Reasoning about binary topological relations. In *Advances in Spatial Databases*; Günther, O., Schek, H.-J., Eds.; Springer: Berlin/Heidelberg, Germany, 1991; Volume 525, pp. 141–160, ISBN 978-3-540-54414-2.
25. Kohler, N.; Lockemann, P. *Arbeits- und Entwicklungsbericht: Planungsplattform für Dynamische Gebäude*; Institut für industrielle Bauproduktion (IFIB): Karlsruhe, Germany, 2003.

26. Boguslawski, P.; Mahdjoubi, L.; Zverovich, V.; Fadli, F.; Barki, H. BIM-GIS Modelling in Support of Emergency Response Applications. *WIT Trans. Built Environ.* **2015**, *149*, 381–391.
27. Liu, L.; Li, B.; Zlatanova, S.; Liu, H. The Path from Bim to a 3d Indoor Framework—A Requirement Analysis. *Int. Arch. Photogramm. Remote Sens. Spat. Inf. Sci.* **2018**, *XLII–4*, 373–378.
28. Breunig, M. *On the Way to Component-Based 3D/4D Geoinformation Systems*; Lecture Notes in Earth Sciences; Springer International Publishing: Berlin, Germany, 2001; Volume 94.
29. Bradley, P.E.; Paul, N. Using the Relational Model to Capture Topological Information of Spaces. *Comput. J.* **2010**, *53*, 69–89. [CrossRef]
30. Lienhardt, P. N-Dimensional Generalized Combinatorial Maps and Cellular Quasi-Manifolds. *Int. J. Comput. Geom. Appl.* **1994**, *4*, 275–324. [CrossRef]
31. Bradley, P.E.; Paul, N. Comparing G-maps with other topological data structures. *Geoinformatica* **2014**, *18*, 595–620. [CrossRef]
32. Paul, N. Topologische Datenbanken für Architektonische Räume. Ph.D. Thesis, Universität Karlsruhe, Karlsruhe, Germany, 2008.
33. Vitalis, S.; Ohori, K.; Stoter, J. Incorporating Topological Representation in 3D City Models. *IJGI* **2019**, *8*, 347. [CrossRef]
34. Sagan, H. *Space Filling Curves*; Springer: New York, NY, USA, 1994. [CrossRef]
35. Guan, X.; van Oosterom, P.; Cheng, B. A Parallel N-Dimensional Space-Filling Curve Library and Its Application in Massive Point Cloud Management. *IJGI* **2018**, *7*, 327. [CrossRef]
36. Wang, J.; Shan, J. Space filling curve based point clouds index. In Proceedings of the 8th International Conference on GeoComputation, Ann Arbor, MI, USA, 31 July–3 August 2005; pp. 551–562.
37. Van Oosterom, P.; Martinez-Rubi, O.; Ivanova, M.; Horhammer, M.; Geringer, D.; Ravada, S.; Tijssen, T.; Kodde, M.; Gonçalves, R. Massive point cloud data management: Design, implementation and execution of a point cloud benchmark. *Comput. Graph.* **2015**, *49*, 92–125. [CrossRef]
38. Sirdeshmukh, N.; Verbree, E.; Oosterom, P.V.; Psomadaki, S.; Kodde, M. Utilizing a Discrete Global Grid System for Handling Point Clouds with Varying Locations, Times, and Levels of Detail. *Cartographica: Int. J. Geogr. Inf. Geovis.* **2019**, *54*, 4–15. [CrossRef]
39. Langran, G.; Stuart, N. Time in Geographic Information Systems. *Int. J. Geograph. Inf. Syst.* **1992**, *6*, 547. [CrossRef]
40. Le, H.H.; Gabriel, P.; Gietzel, J.; Schaeben, H. An object-relational spatio-temporal geoscience data model. *Comput. Geosci.* **2013**, *57*, 104–115. [CrossRef]
41. Le, H.H.; Schaeben, H.; Jasper, H.; Görz, I. Database Versioning and its Implementation in Geoscience Information Systems. *Comput. Geosci.* **2014**, *70*, 44–54. [CrossRef]
42. Le, H.H. Spatio-temporal Information System for the Geosciences: Concepts, Data models, Software, and Applications. Ph.D. Thesis, Technische Universität Bergakademie Freiberg, Freiberg, Germany, 2014.
43. Gabriel, P.; Gietzel, J.; Le, H.H.; Schaeben, H. GST: A Network Based Datastore for Geoscience Data and Geomodels and its Implementation—ProMine's Contribution Towards Interoperability. In *3D, 4D and Predictive Modelling of Major Mineral Belts in Europe*; Weihed, P., Ed.; Springer International Publishing: Cham, Switzerland, 2015; pp. 59–71, ISBN 978-3-319-17427-3.
44. Breunig, M.; Schilberg, B.; Thomsen, A.; Kuper, P.V.; Jahn, M.; Butwilowski, E. DB4GeO, a 3D/4D Geodatabase and Its Application for the Analysis of Landslides. In *Geographic Information and Cartography for Risk and Crisis Management*; Konecny, M., Zlatanova, S., Bandrova, T.L., Eds.; Springer: Berlin/Heidelberg, Germany, 2010; pp. 83–101, ISBN 978-3-642-03441-1.
45. Breunig, M.; Kuper, P.V.; Butwilowski, E.; Thomsen, A.; Jahn, M.; Dittrich, A.; Al-Doori, M.; Golovko, D.; Menninghaus, M. The story of DB4GeO—A service-based geo-database architecture to support multi-dimensional data analysis and visualization. *ISPRS J. Photogramm. Remote Sens.* **2016**, *117*, 187–205. [CrossRef]
46. Hopf, H.; Alexandroff, P. *Topologie*; Springer International Publishing: Berlin, Germany, 1935; p. 158.
47. Kuper, P.V. Efficient and Practical Handling of Spatio-Temporal Data Based on Time-Dependent Net Components. *Int. Arch. Photogramm. Remote Sens. Spat. Inf. Sci.* **2018**, *XLII–4*, 321–327. [CrossRef]
48. Dutton, G.H. *A Hierarchical Coordinate System for Geoprocessing and Cartography, Lecture Notes in Earth Sciences*; Springer: New York, NY, USA, 1999.
49. Peuquet, D.J. A Conceptual Framework and Comparison of Spatial Data Models. *Cartogr. Int. J. Geogr. Inf. Geovis.* **1984** *21*, 66–113. [CrossRef]

50. Mallet, J.L. *Geomodeling*; Oxford University Press: New York, NY, USA, 2002.
51. Langran, G.; Chrisman, N.R. A Framework for Temporal Geographic Information. *Cartographica* **1988**, *25*, 1–14. [CrossRef]
52. Worboys, M. A generic model for planar geographical objects. *Int. J. Geogr. Inf. Syst.* **1992**, *6*, 353–372. [CrossRef]
53. Mallet, J.L. GOCAD: A computer aided design program for geological applications. In *Three-Dimensional Modeling with Geoscientific Information Systems*; Turner, A.K., Ed.; NATO ASI; Kluwer Academic Publishers: Dordrecht, The Netherlands, 1992; Volume 354, pp. 123–142.
54. Thomsen, A.; Breunig, M.; Butwilowski, E. Towards the Unified Modelling and Management of Topology in Multiple Representation Databases. *Schweizerbart'sche Verlagsbuchhandlung* **2008**, *3*, 176–185.
55. Yao, Z.; Nagel, C.; Kunde, F.; Hudra, G.; Willkomm, P.; Donaubauer, A.; Adolphi, T.; Kolbe, T.H. 3DCityDB—A 3D geodatabase solution for the management, analysis, and visualization of semantic 3D city models based on CityGML. *Open Geospat. Data Softw. Stand.* **2018**, *3*, 5. [CrossRef]
56. Yao, Z.; Kolbe, T.H. A new Approach to Model Transformation using Graph Transformation System. In *PFGK18-Photogrammetrie-Fernerkundung-Geoinformatik-Kartographie, 37. Jahrestagung in München 2018*; Deutsche Gesellschaft für Photogrammetrie, Fernerkundung und Geoinformation e.V.: Munich, Germany, 2018; pp. 831–834.
57. 3dcitydb. Available online: https://github.com/3dcitydb/3dcitydb (accessed on 12 November 2019).
58. Peterson, M.P. (Ed.) *Online Maps with APIs and WebServices*; Lecture Notes in Geoinformation and Cartography; Springer: Berlin/Heidelberg, Germany, 2012; ISBN 978-3-642-27484-8.
59. Stefanakis, E. Web mercator and raster tile maps: Two cornerstones of online map service providers. *Geomatica* **2017**, *71*, 100–109. [CrossRef]
60. Keates, J.S. *Cartographic Design and Production*, 2nd ed.; Longman Scientific & Technical: Harlow, Essex, UK; Wiley: New York, NY, USA, 1989; ISBN 978-0-582-30133-7.
61. Robinson, A.H.; Morrison, J.L.; Muehrcke, P.C.; Kimerling, A.J.; Guptill, S.C. *Elements of cartography*, 6th ed.; John Wiley & Sons: New York, NY, USA, 1995; ISBN 978-0-471-55579-7.
62. Goodchild, M.F. Reimagining the history of GIS. *Ann. GIS* **2018**, *24*, 1–8. [CrossRef]
63. Purss, M. Topic 21: Discrete Global Grid Systems Abstract Specification. Available online: http://docs.opengeospatial.org/as/15-104r5/15-104r5.html (accessed on 12 November 2019).
64. OGC. OGC Announces a New Standard That Improves the Way Information Is Referenced to the Earth. Available online: https://www.opengeospatial.org/pressroom/pressreleases/2656 (accessed on 12 November 2019).
65. OGC. Vector Tiles Pilot. Available online: https://www.opengeospatial.org/projects/initiatives/vt-pilot-2018 (accessed on 13 November 2019).
66. ICA Commission on Generalisation and Multiple Representation. Available online: https://generalisation.icaci.org/ (accessed on 30 December 2019).
67. Annapurna, S.; Teja, P.K.V.S.; Murty, S. A Comparative Study on Mobile Platforms (Android vs. IOS). *Int. J. Adv. Res. Comput. Eng. Technol. IJARCET* **2016**, *5*, 547–553.
68. Nittel, S. A Survey of Geosensor Networks: Advances in Dynamic Environmental Monitoring. *Sensors* **2009**, *9*, 5664–5678. [CrossRef]
69. Huber, D. *Background Positioning for Mobile Devices—Android vs. iPhone*; Technical Report; School of Electrical and Computer Engineering: Berlin, Germany, 2011; p. 7.
70. Söderholm, S.; Bhuiyan, M.Z.H.; Thombre, S.; Ruotsalainen, L.; Kuusniemi, H. A Multi-GNSS Software-Defined Receiver: Design, Implementation, and Performance Benefits. *Ann. Telecommun.* **2016**, *71*, 399–410. [CrossRef]
71. Sauerländer-Biebl, A.; Brockfeld, E.; Suske, D.; Melde, E. Evaluation of a transport mode detection using fuzzy rules. *Transp. Res. Procedia* **2017**, *25*, 591–602. [CrossRef]
72. Li, S.; Xu, L.D.; Zhao, S. The internet of things: A survey. *Inf. Syst. Front.* **2015**, *17*, 243–259. [CrossRef]
73. Atzori, L.; Iera, A.; Morabito, G. The Internet of Things: A survey. *Comput. Netw.* **2010**, *54*, 2787–2805. [CrossRef]
74. Jara, A.J.; Olivieri, A.C.; Bocchi, Y.; Jung, M.; Kastner, W.; Skarmeta, A.F. Semantic Web of Things: An analysis of the application semantics for the IoT moving towar the IoT convergence. *IJWGS* **2014**, *10*, 244–272. [CrossRef]
75. Chen, F.; Deng, P.; Wan, J. Data Mining for the Internet of Things: Literature Review and Challenges. *Int. J. Distrib. Sens. Netw.* **2015**. [CrossRef]

76. Tao, F.; Cheng, Y.; Da Xu, L.; Zhang, L.; Li, B.H. CCIoT-CMfg: Cloud Computing and Internet of Things-Based Cloud Manufacturing Service System. *IEEE Trans. Ind. Inf.* **2014**, *10*, 1435–1442.
77. Bahrepour, M.; Meratnia, N.; Havinga, P. *Automatic Fire Detection: A Survey from Wireless Network Perspective*; Technical Report TR-CTIT-08-73; Centre for Telematics and Information Technology, University of Twente: Enschede, The Netherlands, 2008; ISSN 1381-362513p.
78. Muller, C.; Chapman, L.; Grimmond, C.S.B.; Young, D.; Cai, X. Sensors and the City: A Review of Urban Meteorological Networks. *Int. J. Climatol.* **2013**, *33*, 1585–1600. [CrossRef]
79. Moeller, A.K.; Lukacs, P.M.; Horne, J.S. Three Novel Methods to Estimate Abundance of Unmarked Animals Using Remote Cameras. *Ecosphere* **2018**, *9*, e02331. [CrossRef]
80. Eldawy, A.; Mokbel, M.F. SpatialHadoop: A mapreduce framework for spatial data. In Proceedings of the 2015 IEEE 31st International Conference on Data Engineering, Seoul, Korea, 13–17 April 2015.
81. Alarabi, L.; Mokbel, M.F.; Musleh, M. ST-Hadoop: A mapreduce framework for spatio-temporal data. *Geoinformatica* **2018**, *22*, 785–813. [CrossRef]
82. Aji, A.; Wang, F.; Vo, H.; Lee, R.; Liu, Q.; Zhang, X.; Saltz, J. Hadoop GIS: A high performance spatial data warehousing system over mapreduce. *Proc. VLDB Endow.* **2013**, *6*, 1009–1020. [CrossRef]
83. Günther, W.A.; Mehrizi, M.H.R.; Huysman, M.; Feldberg, F. Debating Big Data: A Literature Review on Realizing Value From Big Data. *J. Strateg. Inf. Syst.* **2017**, *26*, 191–209. [CrossRef]
84. Raifer, M.; Troilo, R.; Kowatsch, F.; Auer, M.; Loos, L.; Marx, S.; Przybill, K.; Fendrich, S.; Mocnik, F.B.; Zipf, A. OSHDB: A Framework for Spatio-Temporal Analysis of OpenstreetMap History Data. *Open Geospat. Data Softw. Stand.* **2019**, *4*, 3. [CrossRef]
85. Arsanjani, J.J.; Zipf, A.; Mooney, P.; Helbich, M. (Eds.) *OpenStreetMap in GIScience—Experiences, Research, and Applications, Lecture Notes in Geoinformation and Cartography*; Springer: Berlin/Heidelberg, Germany, 2015; p. 325.
86. Berners-Lee, T. Linked Open Data. 2006. Available online: https://www.w3.org/DesignIssues/LinkedData.html (accessed on 3 February 2020).
87. Goodchild, M.F. Citizens as sensors: The world of volunteered geography. *GeoJournal* **2007**, *69*, 211–221. [CrossRef]
88. Elwood, S. Volunteered geographic information: Key questions, concepts and methods to guide emerging research and practice. *GeoJournal* **2008**, *72*, 133–135. [CrossRef]
89. Elwood, S. Volunteered geographic information: Future research directions motivated by critical, participatory, and feminist GIS. *GeoJournal* **2008**, *72*, 173–183. [CrossRef]
90. Chilton, S. CrowdSourcing is radically changing the geodata landscape: Case study of OpenStreetMap. In Proceedings of the 24th International Cartographic Conference, Santiago, Chile, 15–21 November 2009.
91. Steiger, E.; Resch, B.; Zipf, A. Exploration of spatiotemporal and semantic clusters of Twitter data using unsupervised neural networks. *Int. J. Geogr. Inf. Sci.* **2016**, *30*, 1694–1716. [CrossRef]
92. Goodchild, M.F.; Glennon, J.A. Crowdsourcing Geographic Information for Disaster Response: A Research Frontier. *Int. J. Digit. Earth* **2010**, *3*, 231–241. [CrossRef]
93. Dhar, V. Data science and prediction. *Commun. ACM* **2013**, *56*, 64–73. [CrossRef]
94. Agarwal, R.; Dhar, V. Big data, data science, and analytics: The opportunity and challenge for IS research. *IS Res. J.* **2014**, *25*, 443–448. [CrossRef]
95. Omoniwa, B.; Hussain, R.; Javed, M.A.; Bouk, S.H. Fog/Edge Computing-based IoT (FECiOT): Architecture, Applications, and Research Issues. *IEEE Internet Things J.* **2018**. [CrossRef]
96. Meijer, A.; Bolivar, M.P.R. Governing the Smart City: A Review of the Literature on Smart Urban Governance. *Int. Rev. Adm. Sci.* **2016**, *82*, 392–408. [CrossRef]
97. Enders, M.R.; Hoßbach, N. Dimensions of Digital Twin Applications—A Literature Review. In Proceedings of the 25th Americas Conference on Information Systems, Cancún, Mexico, 15–17 August 2019; p. 10.
98. Bradley, P.E.; Paul, N. Topologically consistent space, time, version, and scale using Alexandrov topologies. In *Contemporary Strategies and Approaches in 3-D Information Modeling*; IGI Global: Pershey (PA), USA, 2018; pp. 52–82.
99. Ji, Q.; Barr, S.; James, P.; Fairbairn, D. A Geospatial Analysis Framework for Fine Scale Urban Infrastructure Networks. *Int. Arch. Photogramm. Remote Sens. Spat. Inf. Sci.* **2018**, *XLII–4*, 291–296. [CrossRef]
100. Giovanella, A.; Bradley, P.E.; Wursthorn, S. Evaluation of Topological Consistency in CityGML. *IJGI* **2019**, *8*, 278. [CrossRef]
101. Arioli, M.; Benzi, M. A finite element method for quantum graphs. *IMA J. Numer. Anal.* **2018**, *38*, 1119–1163. [CrossRef]

102. Gelinas, U.J.; Sutton, S.G.; Fedorowicz, J. Online Transaction Processing. In *Open Textbooks for Hong Kong*; Open University of Hong Kong: Hong Kong, China, 2019.
103. Conn, S.S. OLTP and OLAP data integration: A Review of Feasible Implementation Methods and Architectures for Real Time Data Analysis. In Proceedings of the IEEE Southeast Conference, Ft. Lauderdale, FL, USA, 8–10 April 2005; p. 6. [CrossRef]
104. Zhang, C.; Huang, Y.; Griffin, T. Querying Geospatial Data Streams in SECONDO. In Proceedings of the 17th ACM SIGSPATIAL International Symposium on Advances in Geographic Information Systems (ACM-GIS 2009), Seattle, WA, USA, 4–6 November 2009. [CrossRef]
105. Bauer, A.; Wissink, A.; Potsdam, M.; Jayaraman, B. ParaView Catalyst Computes Particle Paths In Situ. *Kitware Blog* **2016**, *39*.
106. Karimi, H.A. (Ed.) *BIG DATA: Techniques and Technologies in Geoinformatics*; CRC Press: Boca Raton, FL, USA, 2014; 312p, ISBN 978-1-138-07319-7.
107. VoPham, T.; Hart, J.E.; Laden, F.; Chiang, Y.-Y. Emerging Trends in Geospatial Artificial Intelligence (geoAI): Potential Applications for Environmental Epidemiology. *Environ. Health* **2018**, *17*, 40. [CrossRef]
108. Breunig, M.; Zlatanova, S. 3D geo-database research: Retrospective and future directions. *Comput. Geosci.* **2011**, *37*, 791–803. [CrossRef]
109. Balovnev, O.; Breunig, M.; Cremers, A.B. Geotoolkit: Opening Access to Object-Oriented Geodata Stores. In *Interoperating Geographic Information Systems*; Goodchild, M., Egenhofer, M., Fegeas, R., Kottman, C., Eds.; Springer: Boston, MA, USA, 1999; pp. 235–247, ISBN 978-1-4613-7363-6.
110. Güting, R.H. GraphDB: Modeling and Querying Graphs in Databases. In Proceedings of the 20th International Conference on Very Large Data Bases, Santiago de Chile, Chile, 12–25 September 1994; Morgan Kaufmann Publishers Inc.: San Francisco, CA, USA, 1994; pp. 297–308.
111. Russell, S.J.; Norvig, P. *Artificial Intelligence: A Modern Approach*; Prentice Hall Series in Artificial Intelligence; Prentice Hall: Englewood Cliffs, NJ, USA, 1995; ISBN 978-0-13-103805-9.
112. Baader, F.; Nutt, W. Basic description logics. In *The Description Logic Handbook. Theory, Implementation and Applications*; Baader, F., Calvanese, D., McGuinness, D., Nardi, D., Patel-Schneider, P., Eds.; Cambridge University Press: Cambridge, UK, 2003; pp. 43–95.
113. Rudas, I.; Fodor, J. Intelligent systems. *Int. J. Comput. Commun. Control* **2008**, *3*, 132–138.
114. Li, L.; Li, S.; Zhao, S. QoS-Aware Scheduling of Services-Oriented Internet of Things. *IEEE Trans. Ind. Inf.* **2014**, *10*, 1497–1505.
115. Viriyasitavat, W.; Xu, L.D. Compliance Checking for Requirement-Oriented Service Workflow Interoperations. *IEEE Trans. Ind. Inf.* **2014**, *10*, 1469–1477. [CrossRef]
116. He, L.; Yue, P.; Jiang, L.; Zhang, M. Fuzzy spatial relation ontology driven detection of complex geospatial features in a web service environment. *Earth Sci. Inf.* **2015**, *8*, 63–76. [CrossRef]
117. Yuan, J.; Yue, P.; Gong, J.; Zhang, M. A Linked Data Approach for Geospatial Data Provenance. *IEEE Trans. Geosci. Remote Sens.* **2013**, *51*, 5105–5112. [CrossRef]

3

Terrain Analysis in Google Earth Engine: A Method Adapted for High-Performance Global-Scale Analysis

José Lucas Safanelli [1], Raul Roberto Poppiel [1], Luis Fernando Chimelo Ruiz [1], Benito Roberto Bonfatti [2], Fellipe Alcantara de Oliveira Mello [1], Rodnei Rizzo [1] and José A. M. Demattê [1,*]

1 Department of Soil Science, Luiz de Queiroz College of Agriculture, University of São Paulo, Pádua Dias Av., 11, Piracicaba, Postal Box 09, São Paulo Postal Code 13416-900, Brazil; jose.lucas.safanelli@usp.br (J.L.S.); raulpoppiel@usp.br (R.R.P.); luisruiz@usp.br (L.F.C.R.); fellipeamello@usp.br (F.A.d.O.M.); rodnei.rizzo@usp.br (R.R.)

2 State University of Minas Gerais, 700 Colorado Street, Passos, Minas Gerais Code 37902-092, Brazil; benito.bonfatti@uemg.br

* Correspondence: jamdemat@usp.br

Abstract: Terrain analysis is an important tool for modeling environmental systems. Aiming to use the cloud-based computing capabilities of Google Earth Engine (GEE), we customized an algorithm for calculating terrain attributes, such as slope, aspect, and curvatures, for different resolution and geographical extents. The calculation method is based on geometry and elevation values estimated within a 3×3 spheroidal window, and it does not rely on projected elevation data. Thus, partial derivatives of terrain are calculated considering the great circle distances of reference nodes of the topographic surface. The algorithm was developed using the JavaScript programming interface of the online code editor of GEE and can be loaded as a custom package. The algorithm also provides an additional feature for making the visualization of terrain maps with a dynamic legend scale, which is useful for mapping different extents: from local to global. We compared the consistency of the proposed method with an available but limited terrain analysis tool of GEE, which resulted in a correlation of 0.89 and 0.96 for aspect and slope over a near-global scale, respectively. In addition to this, we compared the slope, aspect, horizontal, and vertical curvature of a reference site (Mount Ararat) to their equivalent attributes estimated on the System for Automated Geospatial Analysis (SAGA), which achieved a correlation between 0.96 and 0.98. The visual correspondence of TAGEE and SAGA confirms its potential for terrain analysis. The proposed algorithm can be useful for making terrain analysis scalable and adapted to customized needs, benefiting from the high-performance interface of GEE.

Keywords: topographic surface; terrain modeling; global terrain dataset

---

1. Introduction

Terrain analysis is essential for modeling environmental systems [1¯3]. The variability of landforms is frequently used to understand, map or model geomorphological, hydrological, and biological processes [4–7]. Elevation has a strong relationship with terrestrial temperature, vegetation type, and with the potential energy accumulated on a slope. The aspect and derived products, such as Northernness and Easternness attributes, can be linked to the potential solar irradiation on terrain. The Slope gradient, for example, controls the overland and subsurface flow velocity and runoff rate. Similarly, curvatures are associated with acceleration and dispersion of water and sediment flows, which impacts the erosion and soil water content [8].

The public availability of elevation data with global coverage, such as the digital elevation model (DEM) derived from NASA's Shuttle Radar Topography Mission (SRTM DEM, [9]) and the digital surface model from the Advanced Land Observing Satellite (AW3D30 DSM, [10]), has promoted the exploration of topographic features in different contexts using processing tools available in several geographic information systems (GIS) [4,11,12]. However, despite the popularization of many global elevation datasets, it is important to pay attention to their quality when used for modelling purposes, as the acquisition mean and other production aspects can significantly impact the outputs [13,14]. In addition, analyzing big geospatial datasets can pose some limitations to traditional GIS. This becomes more critical with the availability of new digital datasets, which are providing better temporal and spatial resolutions due to advances in sensor technologies [15].

The Global Multi-resolution Terrain Elevation Data 2010 [16] and the global suit of terrain attributes [2] are examples of datasets that were produced using large computational tasks for mapping the global extent and in different spatial resolutions, which demanded optimized processing architectures. In general, high performance architectures are based on splitting the data in smaller subsets (tiles) to take the advantage of distributed computing operations. Recently, with the advent and popularization of cloud-based interfaces for processing big geospatial data, e.g., Google Earth Engine [17], the Pangeo software packages [18], and Actinia REST service [19], computational tasks applied to terrain analysis could be scaled and customized directly by the user.

Earth Engine (GEE) is a cloud-based platform developed by Google that supports the global-scale analysis of big catalogs of Earth Observation data [17]. It has been used to map global forest change in the 21st century [20], Earth's surface water change [21], global urban areas [11], wildfire progression [22], global bare surface change [23], and others. In this sense, GEE becomes compelling not because the distributed processing tasks are executed on the server-side of Google, but also due to the increasing availability of many global geospatial datasets that could be explored in topographic mapping. There exist several available topographical data within GEE, such as the global SRTM DEM, AW3D30 DSM, Global 30 Arc-Second Elevation data (GTOPO30 DEM, [24]), and others. Thus, GEE characteristics could permit the customization of high-performance terrain analysis with minimal user input and any computational processing on the user side. In fact, GEE provides three algorithms for calculating slope, illumination, and aspect of terrain, but lacks in providing calculation methods of other terrain information, such as the curvatures and landscape characterization.

In addition, a common obstacle of global terrain analysis in common GIS is the need for projecting DEMs onto projected coordinate systems, which ensures the elevation data is equally spaced on a plane square grid [25]. This step is complicated because it is difficult to define a projected system that minimizes terrain distortions over a global extent [26]. Moreover, as many available global DEMs are referenced by geographical coordinate systems and some researchers continue to apply square-grid algorithms to them, the algorithms should consider the geometry and specificity of global spheroidal DEMs [25]. This aspect is important because the application of square-grid methods to spheroidal equal angular DEMs leads to substantial computational errors in models of morphometric variables [25].

In this paper, we aimed at describing and making available a user-friendly processing algorithm for performing terrain analysis in GEE. This algorithm takes advantage of GEE's high-performance architecture for making the computational analysis scalable, adapted to customized needs, and requiring minimal user input. For this, the proposed package takes advantage of a calculation method adapted for spheroidal elevation grids, which favors the global-scale analysis of different DEM resolutions without projecting elevation data.

2. Material and Methods

2.1. Algorithm Description

The Terrain Analysis in GEE (TAGEE) package use calculation methods adapted to spheroidal angular grids, i.e., the DEM can be referenced in a geographical coordinate system, e.g., the World

Geodetic System (WGS84). The following paragraphs briefly describe the calculation methods performed by TAGEE package. The readers are referred to [8] for the mathematical concepts of geomorphometry, a historical overview of the progress of digital terrain modelling, and the notion of the topographic surface and its limitations.

2.1.1. Topographic Surface

The land topography can be approximated by a topographic surface defined by a continuous, single-valued bivariate function (Equation (1)) [8]:

$$z = f(x, y) \tag{1}$$

where z is elevation (meters), and x and y are the coordinates in geographical coordinates (degrees).

The local morphometric variables are functions of the partial derivatives of elevation. Using the Evans–Young method, the function $z = f(x, y)$ is expressed as the second-order bivariate Taylor polynomial (Equation (2)):

$$z = \frac{rx^2}{2} + \frac{ty^2}{2} + sxy + px + qy + u \tag{2}$$

where r, t, s, p, and q are the partial derivatives, and u is the residual term.

Differently from a digital elevation model projected on a plane square grid, where the partial derivatives of terrain are estimated by finite differences, the processing and analysis of a spheroidal equal angular DEM must consider the spheroidal geometry. In such case, a grid spacing with approximately equal linear units along meridians and parallels exists only at the equator. To estimate the parameters of a spheroidal grid, a 3 × 3 moving window must retrieve both the geometry elements and the elevation values of the window nodes (Figure 1).

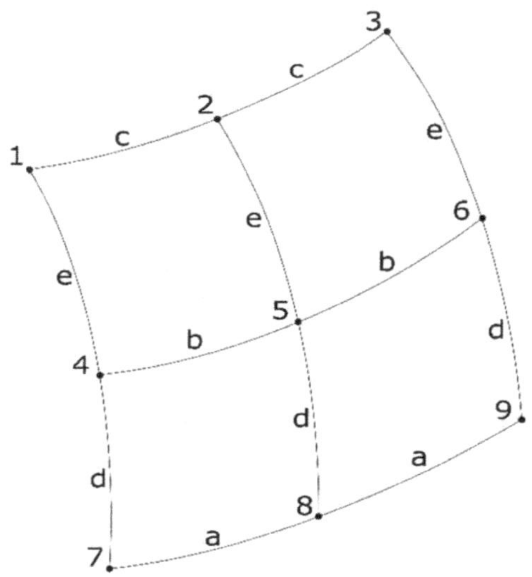

Figure 1. A 3 × 3 spheroidal equal angular grid with linear geometries a, b, c, d, and f, and nine elevation nodes—adapted from [8].

2.1.2. Terrain Parameters: Neighbor Elevations and Geometries

The elevation values of a 3 × 3 moving window are estimated by convolution kernels. For geometries, the Haversine formula is used to determine the great-circle distances between two neighbor nodes within the spheroidal window, given their latitude and longitude geographical positions (Equations (3)–(5)):

$$j = \sin^2\!\left(\frac{\Delta\phi}{2}\right) + \cos\phi_1 \cdot \cos\phi_2 \cdot \sin^2\!\left(\frac{\Delta\lambda}{2}\right) \tag{3}$$

$$k = 2 \cdot atan2\left(\sqrt{j}, \sqrt{(1-j)}\right) \tag{4}$$

$$l = R \cdot k \tag{5}$$

where ϕ_1 is latitude for the first given node in radians, ϕ_2 is the latitude for the second given node in radians, λ_1 is the longitude for the first given node in radians, λ_2 is the longitude for the second given node in radians, $\Delta\phi$ and $\Delta\lambda$ are the respective differences of latitude and longitude between the given nodes, and R is the mean radius of Earth equals to 6,371,000 meters. The linear distance l is given in meters.

Knowing the latitude and longitude of the window nodes (Figure 1), the Haversine formula allows the calculation of linear distances of a, b, c, d, and e, which are used with the neighbor elevation values (from z_1 to z_9) to calculate the partial derivatives of terrain.

2.1.3. Terrain Derivatives

To estimate the first and second-order partial derivatives r, t, s, p and q, the polynomial model is fitted by least squares and results in the following estimations (Equations (6)–(10)) [8]:

$$p = \frac{a^2cd(d+e)(z_3-z_1) + b\left(a^2d^2+c^2e^2\right)(z_6-z_4) + ac^2e(d+e)(z_9-z_7)}{2[a^2c^2(d+e)^2 + b^2(a^2d^2+c^2e^2)]} \tag{6}$$

$$\begin{aligned} q = \; & \tfrac{1}{3de(d+e)(a^4+b^4+c^4)} \\ & \cdot\left\{\left[d^2\left(a^4+b^4+b^2c^2\right) + c^2e^2\left(a^2-b^2\right)\right](z_1+z_3)\right. \\ & -\left[d^2\left(a^4+c^4+b^2c^2\right) - e^2\left(a^4+c^4+a^2b^2\right)\right](z_4+z_6) \\ & -\left[e^2\left(b^4+c^4+a^2b^2\right) - a^2d^2\left(b^2-c^2\right)\right](z_7+z_9) \\ & +d^2\left[b^4(z_2-3z_5) + c^4(3z_2-z_5) + \left(a^4-2b^2c^2\right)(z_2-z_5)\right] \\ & +e^2\left[a^4(z_5-3z_8) + b^4(3z_5-z_8) + \left(c^4-2a^2b^2\right)(z_5-z_8)\right] \\ & \left.-2\left[a^2d^2\left(b^2-c^2\right)z_8 + c^2e^2\left(a^2-b^2\right)z_2\right]\right\}. \end{aligned} \tag{7}$$

$$r = \frac{c^2(z_1+z_3-2z_2) + b^2(z_4+z_6-2z_5) + a^2(z_7+z_9-2z_8)}{a^4+b^4+c^4} \tag{8}$$

$$\begin{aligned} s = \; & \left\{c\left(a[^2(d+e) + b^2e\right](z_3-z_1)\right. \\ & -b\left(a^2d-c^2e\right)(z_4-z_6) \\ & \left.+a\left[c^2(d+e) + b^2d\right](z_7-z_9)\right\} \\ & \cdot\tfrac{1}{2[a^2c^2(d+e)^2+b^2(a^2d^2+c^2e^2)]} \end{aligned} \tag{9}$$

$$\begin{aligned} t = \; & \tfrac{2}{3de(d+e)(a^4+b^4+c^4)} \\ & \cdot\left\{\left[d\left(a^4+b^4+b^2c^2\right) - c^2e\left(a^2-b^2\right)\right](z_1+z_3)\right. \\ & -\left[d\left(a^4+c^4+b^2c^2\right) + e\left(a^4+c^4+a^2b^2\right)\right](z_4+z_6) \\ & +\left[e\left(b^4+c^4+a^2b^2\right) + a^2d\left(b^2-c^2\right)\right](z_7+z_9) \\ & +d\left[b^4(z_2-3z_5) + c^4(3z_2-z5) + \left(a^4-2b^2c^2\right)(z_2-z_5)\right] \\ & +e[a^4(3z_8-z_5) + b^4(z_8-3z_5) + c^4-2a^2b^2]\,(z_8-z_5)] \\ & \left.-2\left[a^2d\left(b^2-c^2\right)z_8 - c^2e\left(a^2-b^2\right)z_2\right]\right\} \end{aligned} \tag{10}$$

where the parameters a, b, c, d, and e are the linear distances calculated from the Haversine formula (Equations (3)–(5)), and the z values are elevation values from the neighbors of a moving window (Figure 1).

2.1.4. Terrain Attributes

Local attributes, such as slope, aspect, and curvatures, are calculated from the partial derivatives of terrain [8]. The slope gradient (G, Equation (11)) is a flow attribute that relates to the velocity of gravity-driven flows. For measuring the direction, the slope aspect is used (A, Equations (12) and (13)). Additionally, one can calculate the amount that a slope is faced to the North or East, resulting in the Northernness (A_N, Equation (14)) and Easternness (A_E, Equation (15)) derived from the aspect. The remaining flux attributes that can be calculated from the first and second-order partial derivatives are the horizontal (k_h, Equation (16)) and vertical curvatures (k_v, Equation (17)). While the horizontal curvature relates if a lateral flow converges ($k_h < 0$) or diverges ($k_h > 0$), the vertical curvature measures the relative acceleration ($k_v > 0$) and deceleration ($k_v < 0$) of a gravity-driven flow:

$$G = \arctan\sqrt{p^2 + q^2} \tag{11}$$

$$\begin{aligned} A = \; & -90[1 - \mathrm{sign}(q)](1 - |\mathrm{sign}(p)|) + 180[1 + \mathrm{sign}(p)] \\ & - \tfrac{180}{\pi}\mathrm{sign}(p)\arccos\left(\tfrac{-q}{\sqrt{p^2+q^2}}\right) \end{aligned} \tag{12}$$

$$\mathrm{sign}(x) = \begin{cases} 1 & \mathrm{for} x > 0 \\ 0 & \mathrm{for} x = 0 \\ -1 & \mathrm{for} x < 0 \end{cases} \tag{13}$$

$$A_N = \cos A \tag{14}$$

$$A_E = \sin A \tag{15}$$

$$k_h = -\frac{q^2 r - 2pqs + p^2 t}{(p^2 + q^2)\sqrt{1 + p^2 + q^2}} \tag{16}$$

$$k_v = -\frac{p^2 r + 2pqs + q^2 t}{(p^2 + q^2)\sqrt{(1 + p^2 + q^2)^3}} \tag{17}$$

Differently from flow attributes, which are gravity field-specific variables, form attributes are related to principal sections of terrain [8]. The mean curvature (H, Equation (18)) is a half-sum of any two orthogonal normal sections and represents two accumulation mechanisms of gravity-driven flows with equal weights: convergence and relative deceleration. Among the class of form attributes, the Gaussian curvature (K, Equation (19)) is a product of maximal (k_{max}) and minimal (k_{min}) curvatures. The two principal curvatures calculate the highest and lowest curvature for a given point of the topographic surface. The maximal curvature (k_{max}, Equation (20)) is useful for mapping rigdes ($k_{max} > 0$) and closed depressions ($k_{max} < 0$). Likewise, the minimal curvature (k_{min}, Equation (21)) is useful for identifying hills ($k_{min} > 0$) and valleys ($k_{min} < 0$) across the topographic surface. With the results of mean and Gaussian curvatures, a landform classification can be generated after [27] proposing the continuous form of the Gaussian classification [8,28]. Instead of providing categorical values, the shape index (SI, Equation (22)) ranges from –1 to 1 and map convex ($SI > 0$) and concave ($SI < 0$) landforms:

$$H = -\frac{(1 + q^2)r - 2pqs + (1 + p^2)t}{2\sqrt{(1 + p^2 + q^2)^3}} \tag{18}$$

$$K = \frac{rt - s^2}{(1 + p^2 + q^2)^2} \tag{19}$$

$$k_{max} = H + \sqrt{(H^2 - K)} \tag{20}$$

$$k_{min} = H - \sqrt{(H^2 - K)} \tag{21}$$

$$SI = \frac{2}{\pi}\arctan\frac{H}{\sqrt{H^2 - K}} \tag{22}$$

2.2. Package Description

Calculation methods presented in this paper were developed using the JavaScript programming interface available as the online code editor of GEE. TAGEE was developed by different modules of calculation, similarly to what was described in Methods. The first module, *calculateParameters*, uses convolution kernels and the Haversine formula to retrieve elevation values and the spheroidal geometries of a 3 × 3 moving window. In this module, a digital elevation model and a square polygon representing the bounding box (min. Longitude, min. Latitude, max. Longitude, and max. Latitude, in the WGS84 coordinate reference system) are required as input parameters to run. The bounding box is used both in this module and others for generating images with constant values and restrict the calculations to the study area. The first module returns an image with 14 bands, i.e., the neighbor elevation values (from Z_1 to Z_9) and the distances (a, b, c, d, and e) (Figure 1).

Once the basic parameters (elevation and distances) were established, the partial derivatives of terrain are calculated with the *calculateDerivatives* module. This second module requires the returned parameters from *calculateParameters* and also the bounding box of the study region. The second module adds the partial derivatives (r, t, s, p, and q) as new bands to the previous image. Then, terrain attributes are calculated by the module *calculateAttributes* (Figure 2).

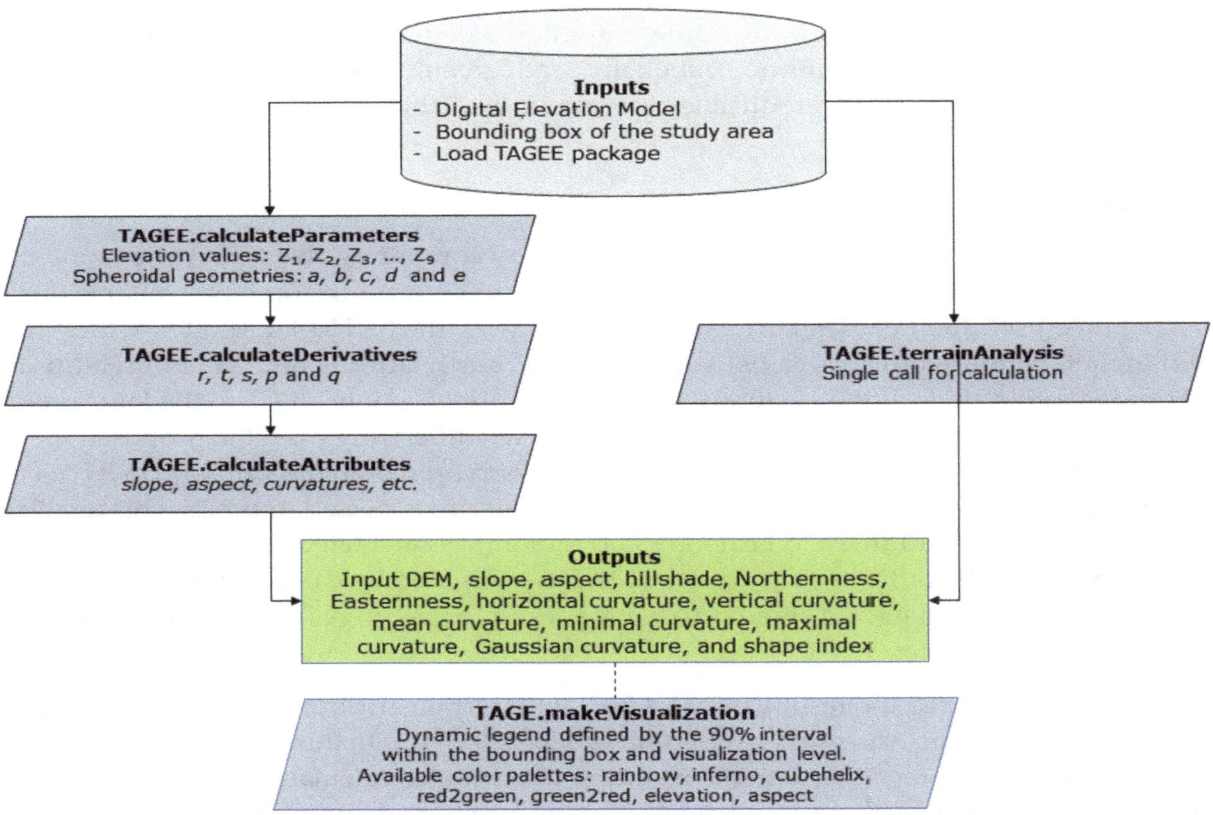

Figure 2. TAGEE modules for calculating terrain parameters, derivatives, and attributes.

Terrain attributes can also be calculated by a single function, without calling the intermediate modules. The final output, for both alternatives (Figure 2), is a multi band object containing the same data properties of the digital elevation model (resolution, data type, and coordinate reference system) with 13 bands (Table 1). The final attributes can be used for further modeling inside GEE or thematic mapping.

Table 1. Attributes of terrain, with their units and description, calculated by TAGEE package.

Attribute	Unit	Description
Elevation	meter	Height of terrain above sea level
Slope	degree	Slope gradient
Aspect	degree	Compass direction
Hillshade	dimensionless	Brightness of the illuminated terrain
Northernness	dimensionless	Degree of orientation to North
Easternness	dimensionless	Degree of orientation to East
Horizontal curvature	meter	Curvature tangent to the contour line
Vertical curvature	meter	Curvature tangent to the slope line
Mean curvature	meter	Half-sum of the two orthogonal curvatures
Minimal curvature	meter	Lowest value of curvature
Maximal curvature	meter	Highest value of curvature
Gaussian curvature	meter	Product of maximal and minimal curvatures
Shape Index	dimensionless	Continuous form of the Gaussian classification

The package has an additional feature that makes easier the visualization of terrain attributes. As the range of attribute values and the pixel resolution may vary according to the visualization level (zoom), which impacts the estimated geometries and elevation neighbor values, a module called *makeVisualization* automatically calculates the dynamic legend defined by the 0.05 and 0.95 percentiles within the bounding box. In addition, different color palettes for making the map legend are available in TAGEE: rainbow, inferno, cubehelix, red2green, green2red, elevation, and aspect. The package code and a minimal reproducible example are available in https://github.com/zecojls/tagee (Supplementary Materials).

2.3. Statistical Evaluation

We performed the evaluation of TAGEE attributes by comparing the aspect and slope derived from two available functions of GEE (ee.Terrain.aspect and ee.Terrain.slope) on a near-global scale. For this task, we used the Pearson correlation analysis with the SRTM DEM 30m, which contains elevation in meters limited to an area between about 60° north latitude and 56° south latitude. It is important to mention that for the currently available terrain functions of GEE, the local gradient is computed using the four-connected neighbors of each pixel, differently from the proposed method of TAGEE, which uses a 3×3 pixel window and also considers the spheroidal geometries in its calculation. Thus, minimal differences between the calculation methods are expected to occur. This analysis was performed in GEE and, in addition to Pearson's correlation, we calculated the relative mean absolute error (MAE) between the outputs. The relative MAE is estimated by calculating the mean absolute difference between two rasters and standardizing the result to the range (maximum minus minimum values) of the reference raster.

Similarly, we compared the results from TAGEE with terrain attributes calculated by the System for Automated Geoscientific Analyses (SAGA) GIS version 2.3.2 [12]. In this case, we downloaded from GEE the 30 m SRTM DEM together with the resulting 12 attributes calculated by TAGEE, all covering the Mount Ararat (located between 44.2° and 44.5° E, and 39.6° and 39.8° N). The Mount Ararat was selected due its high variability of landforms and the availability of published maps from previous works [8,29], allowing the visual comparison of spatial patterns. The Mount Ararat SRTM-DEM was processed in SAGA GIS using the "Slope, Aspect, Curvature" from the Morphometry module of

Terrain Analysis. The calculation method was the "Evan (1979)" based on six parameters and 2nd order polynomials, similarly to the TAGEE calculation method. The comparison was performed by calculating the Pearson's correlation coefficient (r) and the relative MAE, where the aspect, slope, horizontal curvature and vertical curvature from TAGEE were compared with aspect, slope, tangential, and profile curvature from SAGA GIS, respectively, following the equivalence described in [8].

3. Results and Discussion

The statistical analysis revealed a significant correlation ($p < 0.01$) of the TAGEE outputs with equivalent terrain attributes calculated from GEE and SAGA GIS (Table 2). The slope estimated over a near-global extent reached a correlation of 0.98 (error of 2%) between TAGEE and functions of GEE, while the aspect resulted in a Pearson's r of 0.89 (13% of error). The lower correlation of aspect can be associated to its dimension nature, i.e., a circular variable, as well as to the differences of calculation methods between TAGEE and GEE. Despite the small differences, TAGEE revealed the same spatial patterns and allowed the estimation of additional attributes at the global scale, such as the Northernness, horizontal and vertical curvatures (Figure 3A–C, respectively). The main mountain ranges of the Earth, such as the Rocky Mountains in North America, Andes in South America, Alps in Europe, Himalayas, and Tibetan plateau in Asia, etc., present the highest curvatures calculated by TAGEE. Conversely, the plains and flat surfaces had the lowest estimates for both curvatures. The degree of orientation to North (Figure 3A) also depict the main landforms of the Earth.

Table 2. Comparison of TAGEE attributes with outputs from GEE and SAGA GIS algorithms.

Attribute	Region	Reference	Pearson's r	rMAE [1]
Aspect	Near global SRTM DEM 30 m	GEE	0.89 *	13%
Slope	Near global SRTM DEM 30 m	GEE	0.98 *	2%
Aspect	Mount Ararat SRTM DEM 30 m	SAGA GIS	0.96 *	4%
Slope	Mount Ararat SRTM DEM 30 m	SAGA GIS	0.98 *	3%
Horizontal curvature	Mount Ararat SRTM DEM 30 m	SAGA GIS	0.98 *	4%
Vertical curvature	Mount Ararat SRTM DEM 30 m	SAGA GIS	0.98 *	4%

* Significant for $p < 0.01$; [1] relative mean absolute error.

TAGEE was developed in GEE to take advantage of the high-performance computing of the platform. As the cloud-based interfaces have created much enthusiasm and engagement in the remote sensing and geospatial fields, many processing algorithms have been adapted to make substantive progress on global challenges involving the processing of big geospatial data [30]. In this sense, GEE is providing petabytes of publicly available remote sensing imagery and other ready-to-use products. The high-speed parallel processing of GEE servers and the libraries of operators and machine learning algorithms available by Application Programming Interfaces (APIs) in popular coding languages, such as JavaScript and Python, are enabling users to discover, analyze, and visualize geospatial big data without needing access to supercomputers [30]. Within this framework, TAGEE supports the development of customized terrain analysis with different elevation data across large geographical extents.

When TAGEE outputs were compared to those from SAGA GIS (Table 2), the statistical evaluation resulted in a significant and high correlation for the slope, horizontal and vertical curvatures of terrain (Pearson's r of 0.98, with an error difference of 3 and 4%). Aspects from TAGEE and SAGA GIS had an inferior correlation coefficient, but the result was higher than the aspect from the algorithm of GEE. The region of Mount Ararat was also used to visually compare the slope, horizontal and vertical curvatures, calculated from both TAGEE and SAGA GIS (Figure 4). The 3D visualizations revealed a high similarity between both maps, but some small differences can be visualized by the color intensity. This is the case of the slope of the Mount Ararat calculated by TAGEE (Figure 4A), which had a higher intensity compared to the slope of SAGA GIS (Figure 4B). A slightly higher intensity for the vertical curvature calculated by SAGA GIS was also evident on an edge of the Mount Ararat (Figure 4F). Despite being small, these visual differences confirm the relative error of both methods (Table 2). In addition, the spatial patterns of aspect, slope, and curvatures from TAGEE presented a high correspondence with the terrain maps of Mount Ararat available in [8,29], reinforcing the confidence of the TAGEE calculation method.

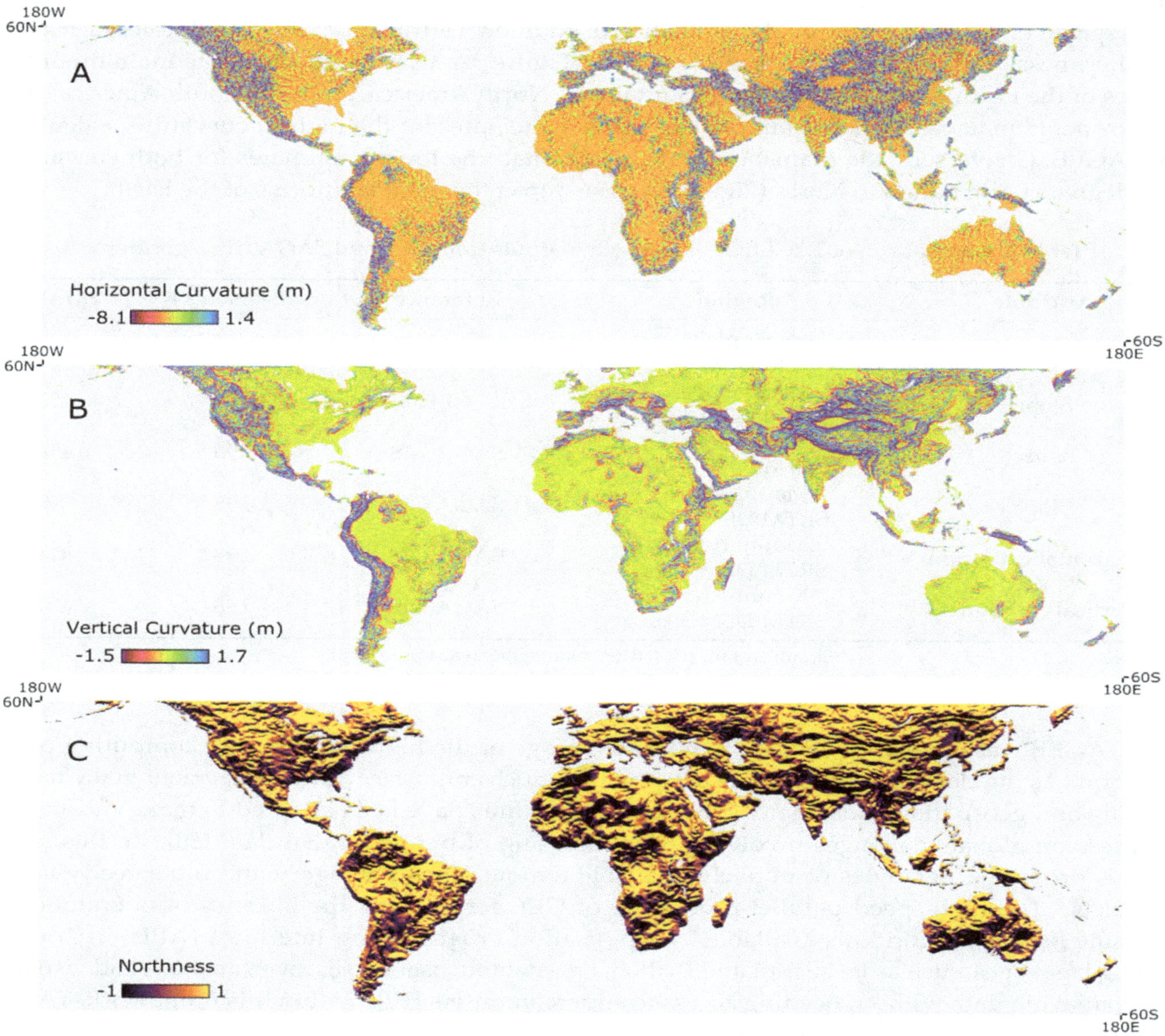

Figure 3. Example of terrain attributes calculated from TAGEE package and 1 arc-second SRTM DEM, displayed for the near-global extent at the visualization level 3 (~20 km pixel resolution): horizontal curvature (**A**), vertical curvature (**B**), and Northerness (**C**).

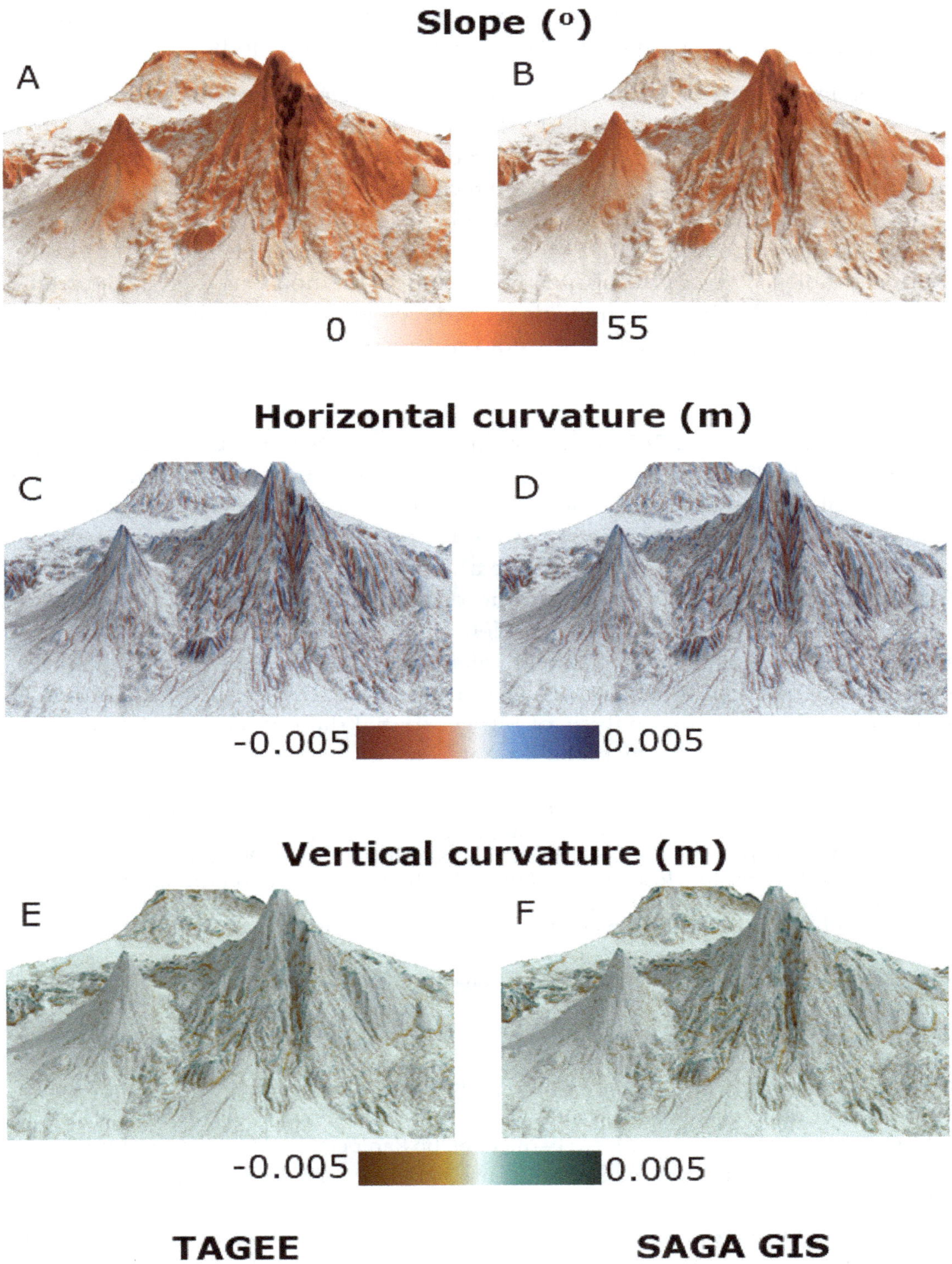

Figure 4. 3D visualizations of terrain attributes produced near Mount Ararat: slope, horizontal and vertical curvature from TAGEE (**A**,**C**,**E**, respectively) and SAGA GIS (**B**,**D**,**F**, respectively). 3D maps are displayed with a vertical exaggeration of 2.

In this work, the TAGEE algorithm was developed to consider spheroidal geometries in its calculation method. This approach diverges from the techniques available in traditional GIS, where TAGEE considers the great circle distances of the DEM defined by Latitude and Longitude positions. Common GIS software, such as SAGA GIS, requires the projection of the DEM to ensure the elevation data have the same pixel size. However, as identified by [25], some researchers continue to apply square-grid algorithms to spheroidal equal angular DEMs, which can lead to substantial computational errors in models of morphometric variables. The small relative errors between TAGEE and GEE or SAGA GIS could be linked to the differences in their calculation methods.

Finally, some limitations of TAGEE can also be noted. Only local morphometric variables can be calculated by the package, which includes flux and form attributes. Non-local attributes, such as specific catchment area, were not implemented due to the absence of a general analytical theory, which is still little developed [29], and due to the recursion processing that is still challenging within GEE [17]. Furthermore, a novel method became available to handle major problems of terrain analysis, which includes the approximation of DEM, generalization and denoising, and the computation of morphometric variables. The universal spectral analytical method based on high-order orthogonal expansions using the Chebyshev polynomials were developed by [31] to handle the aforementioned issues into an integrated framework, but was not implemented in this work.

4. Conclusions

The proposed package (TAGEE) can calculate terrain attributes using the high-performance platform of GEE with an accuracy equivalent to traditional GIS. The approach of using spheroidal geometries does not require the projection of input elevation data for terrain attributes calculation. The comparison between algorithms demonstrated that TAGEE estimates terrain slope and aspect similarly to the available functions of GEE. The advantage of TAGEE over the currently available functions is that additional outputs can be produced, such as curvatures and shape index, which can be useful for environmental mapping and modelling studies. In addition, a good agreement was also found when TAGEE was compared to equivalent outputs from SAGA GIS, reaching a Pearson's correlation coefficient between 0.96 and 0.98, and differences between 3–4 %. Thus, TAGEE becomes a feasible tool for making terrain analysis of big geospatial data, which can be customized to any spatial resolution and scaled up to the global extent.

Author Contributions: Conceptualization, José Lucas Safanelli; Methodology, José Lucas Safanelli, Raul Roberto Poppiel, and Luis Fernando Chimelo Ruiz; Software, José Lucas Safanelli; Validation, José Lucas Safanelli, Luis Fernando Chimelo Ruiz, Fellipe Alcantara de Oliveira Mello, and Rodnei Rizzo; Writing—Original Draft Preparation, José Lucas Safanelli; Writing—Review and Editing, all authors; Supervision, José A. M. Demattê; Funding Acquisition, José A. M. Demattê. Validation, Writing—Review & Editing, Benito Roberto Bonfatti. All authors have read and agreed to the published version of the manuscript.

Acknowledgments: The authors are grateful to the Geotechnologies in Soil Science (GEOCIS) group.

References

1. Moore, I.D.; Grayson, R.B.; Ladson, A.R. Digital terrain modelling: A review of hydrological, geomorphological, and biological applications. *Hydrol. Process.* **1991**, *5*, 3–30. [CrossRef]
2. Amatulli, G.; Domisch, S.; Tuanmu, M.-N.; Parmentier, B.; Ranipeta, A.; Malczyk, J.; Jetz, W. A suite of global, cross-scale topographic variables for environmental and biodiversity modeling. *Sci. Data* **2018**, *5*, 180040. [CrossRef]
3. Pike, R.J. Geomorphometry-diversity in quantitative surface analysis. *Prog. Phys. Geogr. Earth Env.* **2000**, *24*, 1–20. [CrossRef]
4. Bogaart, P.W.; Troch, P.A. Curvature distribution within hillslopes and catchments and its effect on the hydrological response. *Hydrol. Earth Syst. Sci.* **2006**, *10*, 925–936. [CrossRef]
5. Alexander, C.; Deák, B.; Heilmeier, H. Micro-topography driven vegetation patterns in open mosaic landscapes. *Ecol. Indic.* **2016**, *60*, 906–920. [CrossRef]
6. Oliveira, S.; Pereira, J.M.C.; San-Miguel-Ayanz, J.; Lourenço, L. Exploring the spatial patterns of fire density in Southern Europe using Geographically Weighted Regression. *Appl. Geogr.* **2014**, *51*, 143–157. [CrossRef]
7. McGuire, K.J.; McDonnell, J.J.; Weiler, M.; Kendall, C.; McGlynn, B.L.; Welker, J.M.; Seibert, J. The role of topography on catchment-scale water residence time. *Water Resour. Res.* **2005**, 41. [CrossRef]
8. Florinsky, I.V. *Digital Terrain Analysis in Soil Science and Geology*; Academic Press: Cambridge, MA, USA, 2016; ISBN 9780128046326.

9. USGS EROS. USGS EROS Archive-Digital Elevation-Shuttle Radar Topography Mission (SRTM) Void Filled. Available online: https://doi.org/10.5066/F7F76B1X (accessed on 4 April 2020).
10. JAXA EORC. ALOS Global Digital Surface Model "ALOS World 3D-30m (AW3D30)". Available online: https://www.eorc.jaxa.jp/ALOS/en/aw3d30/index.htm (accessed on 4 April 2020).
11. Liu, X.; Hu, G.; Chen, Y.; Li, X.; Xu, X.; Li, S.; Pei, F.; Wang, S. High-resolution multi-temporal mapping of global urban land using Landsat images based on the Google Earth Engine Platform. *Remote Sens. Env.* **2018**, *209*, 227–239. [CrossRef]
12. Olaya, V.; Conrad, O. Geomorphometry in SAGA. In *Developments in Soil Science*; Elsevier: Amsterdam, The Netherlands, 2009; Volume 33, pp. 293–308.
13. Miliaresis, G. The Landcover Impact on the Aspect/Slope Accuracy Dependence of the SRTM-1 Elevation Data for the Humboldt Range. *Sensors* **2008**, *8*, 3134–3149. [CrossRef]
14. Bindzárová Gergeľová, M.; Kuzevičová, Ž.; Labant, S.; Gašinec, J.; Kuzevič, Š.; Unucka, J.; Liptai, P. Evaluation of Selected Sub-Elements of Spatial Data Quality on 3D Flood Event Modeling: Case Study of Prešov City, Slovakia. *Appl. Sci.* **2020**, *10*, 820. [CrossRef]
15. Xia, J.; Yang, C.; Li, Q. Building a spatiotemporal index for Earth Observation Big Data. *Int. J. Appl. Earth Obs. Geoinf.* **2018**, *73*, 245–252. [CrossRef]
16. Danielson, J.J.; Gesch, D.B. *Global Multi-Resolution Terrain Elevation Data 2010 (GMTED2010)*; U.S. Geo-logical Survey Open-File Report 2011–1073; United States Geological Survey (USGS): Sioux Falls, SD, USA, 2011.
17. Gorelick, N.; Hancher, M.; Dixon, M.; Ilyushchenko, S.; Thau, D.; Moore, R. Google Earth Engine: Planetary-scale geospatial analysis for everyone. *Remote Sens. Env.* **2017**, *202*, 18–27. [CrossRef]
18. Abernathey, R.; Paul, K.; Hamman, J.; Rocklin, M.; Lepore, C.; Tippett, M.; Henderson, N.; Seager, R.; May, R.; Del Vento, D. Pangeo NSF Earthcube Proposal. Available online: https://figshare.com/articles/Pangeo_NSF_Earthcube_Proposal/5361094 (accessed on 10 March 2020).
19. mundialis GmbH & Co. KG. Actinia: Geoprocessing in the Cloud. Available online: https://actinia.mundialis.de/ (accessed on 10 March 2020).
20. Hansen, M.C.; Potapov, P.V.; Moore, R.; Hancher, M.; Turubanova, S.A.; Tyukavina, A.; Thau, D.; Stehman, S.V.; Goetz, S.J.; Loveland, T.R.; et al. High-Resolution Global Maps of 21st-Century Forest Cover Change. *Science* **2013**, *342*, 850–853. [CrossRef] [PubMed]
21. Donchyts, G.; Baart, F.; Winsemius, H.; Gorelick, N.; Kwadijk, J.; van de Giesen, N. Earth's surface water change over the past 30 years. *Nat. Clim. Chang.* **2016**, *6*, 810–813. [CrossRef]
22. Crowley, M.A.; Cardille, J.A.; White, J.C.; Wulder, M.A. Multi-sensor, multi-scale, Bayesian data synthesis for mapping within-year wildfire progression. *Remote Sens. Lett.* **2019**, *10*, 302–311. [CrossRef]
23. Demattê, J.A.M.; Safanelli, J.L.; Poppiel, R.R.; Rizzo, R.; Silvero, N.E.Q.; Mendes, W.S.; Bonfatti, B.R.; Dotto, A.C.; Salazar, D.F.U.; Mello, F.A.O.; et al. Bare Earth's Surface Spectra as a Proxy for Soil Resource Monitoring. *Sci. Rep.* **2020**, *10*, 4461. [CrossRef]
24. USGS EROS. GTOPO30-Global 1-km Digital Raster Data Derived from a Variety of Sources. Available online: https://doi.org/10.5066/F7DF6PQS (accessed on 10 March 2020).
25. Florinsky, I.V. Spheroidal equal angular DEMs: The specificity of morphometric treatment. *Trans. GIS* **2017**, *21*, 1115–1129. [CrossRef]
26. Brainerd, J.; Pang, A. Interactive map projections and distortion. *Comput. Geosci.* **2001**, *27*, 299–314. [CrossRef]
27. Koenderink, J.J.; van Doorn, A.J. Surface shape and curvature scales. *Image Vis. Comput.* **1992**, *10*, 557–564. [CrossRef]
28. Gauss, C.F. *General Investigations of Curved Surfaces of 1827 and 1825*; Princeton University Library: Princeton, NY, USA, 1902.
29. Florinsky, I.V. An illustrated introduction to general geomorphometry. *Prog. Phys. Geogr. Earth Env.* **2017**, *41*, 723–752. [CrossRef]
30. Tamiminia, H.; Salehi, B.; Mahdianpari, M.; Quackenbush, L.; Adeli, S.; Brisco, B. Google Earth Engine for geo-big data applications: A meta-analysis and systematic review. *ISPRS J. Photogramm. Remote Sens.* **2020**, *164*, 152–170. [CrossRef]
31. Florinsky, I.V.; Pankratov, A.N. A universal spectral analytical method for digital terrain modeling. *Int. J. Geogr. Inf. Sci.* **2016**, *30*, 2506–2528. [CrossRef]

4

A Novel Method of Missing Road Generation in City Blocks based on Big Mobile Navigation Trajectory Data

Hangbin Wu [1], Zeran Xu [1,*] and Guangjun Wu [2]

[1] College of Surveying and Geo-informatics, Tongji University, Shanghai 200092, China; hb@tongji.edu.cn
[2] 1RenData (ShangHai) Technology Co., Ltd., Shanghai 200092, China; sw4_sw4_119@126.com
* Correspondence: 1551171@tongji.edu.cn

Abstract: With the rapid development of cities, the geographic information of urban blocks is also changing rapidly. However, traditional methods of updating road data cannot keep up with this development because they require a high level of professional expertise for operation and are very time-consuming. In this paper, we develop a novel method for extracting missing roadways by reconstructing the topology of the roads from big mobile navigation trajectory data. The three main steps include filtering of original navigation trajectory data, extracting the road centerline from navigation points, and establishing the topology of existing roads. First, data from pedestrians and drivers on existing roads were deleted from the raw data. Second, the centerlines of city block roads were extracted using the RSC (ring-stepping clustering) method proposed herein. Finally, the topologies of missing roads and the connections between missing and existing roads were built. A complex urban block with an area of 5.76 square kilometers was selected as the case study area. The validity of the proposed method was verified using a dataset consisting of five days of mobile navigation trajectory data. The experimental results showed that the average absolute error of the length of the generated centerlines was 1.84 m. Comparative analysis with other existing road extraction methods showed that the F-score performance of the proposed method was much better than previous methods.

Keywords: missing road; city blocks; topology; big mobile navigation trajectory data

---

1. Introduction

With rapid road construction development in urban and rural areas, the consequent road changes result in lagging road-data updates that are a poor match to the current situation and have low integrity and accuracy. Traditional technologies used for detecting and updating missing roads, such as professional surveys, map downsizing, remote sensing image interpretation, etc., are more costly, require longer update cycles, are more complicated in data processing, and cannot easily adapt to the needs of rapid urban development [1].

Detecting new and missing roads on existing road networks has become a common concern in the fields of urban management, intelligent transportation, and driverless technology [2]. Imagery, GNSS (global navigation satellite system) trajectories, and multisource data fusion are some of the major data sources used for the renewal and repair of missing GIS (geographical information system) road network data [3]. With technological developments such as wireless communications, Big Data, and cloud computing, navigation trajectories are gradually becoming the main data source for urban road updates. Massive on-vehicle positioning trajectories have the characteristics of large data, multiple data sources, and heterogeneous structures. The use of VGI (volunteered geography information)-based trajectories, such as those collected by smart phones [4,5], and smart devices installed in vehicles [6] or held by pedestrians [7], to update road data has recently achieved substantial results.

In this paper, a new method of extracting missing roads in city blocks from big mobile navigation trajectory data is proposed, which mainly consists of the following steps: (1) Useless information from pedestrian users on existing urban roads was filtered out of the data; (2) After preprocessing, roadway centerlines were extracted using the proposed RSC (ring-stepping clustering) method; (3) The road topology of each block was then established, and connections between existing urban roads and missing roads in the blocks were built. Compared to traditional methods, the proposed method has the following advantages: First, compared to specialized vehicles equipped with mapping devices [8,9], the whole process can achieve a high level of automation and rarely requires manual operation, and the device surveying investment is also much smaller; second, compared to remote sensing image extraction [10], the influence on road extraction of the ubiquitous trees on city blocks will be appreciably less, and the satellite revisit period will no longer be a problem; finally, compared to existing road extraction methods using GNSS trajectory data [9,11–14], although the computational complexity of the proposed method ($\Theta(n^2)$) is higher than the method in [9] ($\Theta(n)$), the $F-$score of the generated road centerline is much higher. A case study with 9,944,710 trajectory GNSS points over an area of 5.76 square kilometers was selected to verify the feasibility of the proposed method. The results showed that the extracted road network was well matched with the real network.

In the next section, we briefly discuss the road network data updating method found in the literature. The rest of the paper is organized as follows: Section 3 describes the research methodology in detail; Section 4 presents a case study with the adopted data and evaluates the quality of the proposed method; and Section 5 derives and discusses the main conclusions and concerns related to the proposed method.

2. Literature Review

The current methods are mainly divided into two major aspects: updating of road geometry data and attribute data. The following subsections introduce the related works corresponding to these two aspects in detail.

2.1. Road Geometry Data Updating

Selecting GNSS trajectory points not located on existing roads is usually a necessary stage in road mapping in order to update road networks and refine the geometry of road segments or intersections [5,15–19]. Then, different algorithms, strategies, and methods for detecting new roads, extensions, and disappearances of existing roads are proposed based on the outlier trajectory points.

Considering that GNSS trajectories—or points—only appear on roads, clustering GNSS points is the most commonly used method for updating the geometric data of a road. For example, *K*-Means [13,20–23] has been used to cluster a large number of GNSS points at a certain position on the road to identify the center of the road. This kind of method proceeds by inputting GNSS points or trajectories and specifying an initial seed point to cluster the center of the road. Then, after iterations over a certain distance interval, the geometric information for the road network is obtained.

The trace merging method [12,24] is usually used for road extraction from massive GNSS trajectories. By using iterations of each GNSS trajectory, raw trajectory edges are added to the current road map according to the map-matching results. Edges of current road maps are given a weight to describe the repeat occurrences and the possibility of a road's presence. Those edges with lower weights are removed, and the remaining edges are regarded as newly found roads.

A large number of unordered GNSS points exhibit geometrical features distributed along the road. Based on this phenomenon, researchers proposed a kernel density estimation (KDE) method and extracted the most densely distributed areas of GNSS points, supplemented by a certain threshold, to achieve the purpose of road boundary extraction and road skeleton extraction [14,25,26]. The advantage of this method is that as the number of samples increases, the output is more reliable and robust. However, when the number of samples is insufficient, the results often have large deviations.

Recently, the total least squares [27] and turning-point detection methods [28] were proposed to segment and group GNSS trajectory data. After the segmentation and grouping, the intersection position and road segment were determined using the phenomena of intersecting and crossing of different groups. [28] A dynamic time-warping (DTW) algorithm was then used to align road segments from the connection matrix between intersections.

Moreover, [29] proposed the hidden Markov model (HMM) map matching method for topological reconstruction and intersection refinement. Reference [5] used a genetic algorithm to establish new road segments.

However, road-network extraction remains faced with difficulties in some places. For example, when a road is covered by a bridge or an underpass is present under a road, GNSS points are particularly chaotic because current mobile phone GNSS positioning is not good at heights; GNSS points can be particularly chaotic, thus making it difficult to obtain this part of the road using the existing method.

2.2. Road Attribute Data Updating

Another use for the movement trajectory is updating the attribute information of the road network. Changes in road attribute information mainly fall into eight categories: directionality, speed limit, number of lanes, access, average speed, congestion, importance, and geometric offset [30]. Winden et al. [30] proposed a decision-tree algorithm for deriving the above eight attributes for an open street map (OSM). The results show that whether a road is a one- or two-way road it is classified with an accuracy of 99% and the accuracy for the road speed limit is 69%.

Reference [31] modeled the speed of the tracks from the centerline of the road as a Gaussian distribution, and then extracted the directionality and turning restriction attributes for road maps such as OSM. Reference [32] used a probabilistic method to derive the number of traffic lanes from GNSS tracks by fitting a Gaussian mixture model (GMM) to the intersections between the GNSS traces and a sampling line perpendicular to the road's centerline. Reference [33] used data-mining techniques to extract the name and class of the road by integrating movement trajectories and geotagged data from social media with a support vector machine (SVM) method. Reference [34] used a fully connected deep neural network (DNN) to automatically extract deep features and classify trajectories based on transportation mode. Further, Reference [35] proposed a framework using feature engineering and noise removal to classify movement trajectories into typical transportation modes such as taxi, car, train, subway, walking, airplane, boat, bike, running, motorcycle, and bus.

The detected transportation modes of each movement trajectory can be used to update the attributes of the road map. Reference [19] adapted a robust map-matching algorithm to assure that each point was assigned to the current road map. Then, the missing intersections, turn restrictions, and road closures were detected and updated. Considering that OSM has become a common way for volunteers to draw maps, Reference [36] obtained newly drawn road data by analyzing OSM data. They then adopted a progressive buffering method to update the latest roads in the OSM data with roads from other data sources, including both geometry and attributes.

3. Methodology

3.1. General Description of the Proposed Method

The proposed method can be divided into three main parts (Figure 1). The first part is data filtering. In this step, two kinds of data, including navigation points located on the existing urban roads and records generated by pedestrians, were filtered out. The remaining navigation data were regarded as navigation points relating to driving on missing roads in city blocks. This part is introduced in Section 3.2.

Then, a clustering-based algorithm named RSC was proposed to extract road centerlines using the reserved navigation points. This is the second part of the proposed method. After that, the centerlines of the missing roads were determined (please see Section 3.3).

Finally, the topologies of missing roads and their relationship with the existing roads were established. The procedures are given in Section 3.4.

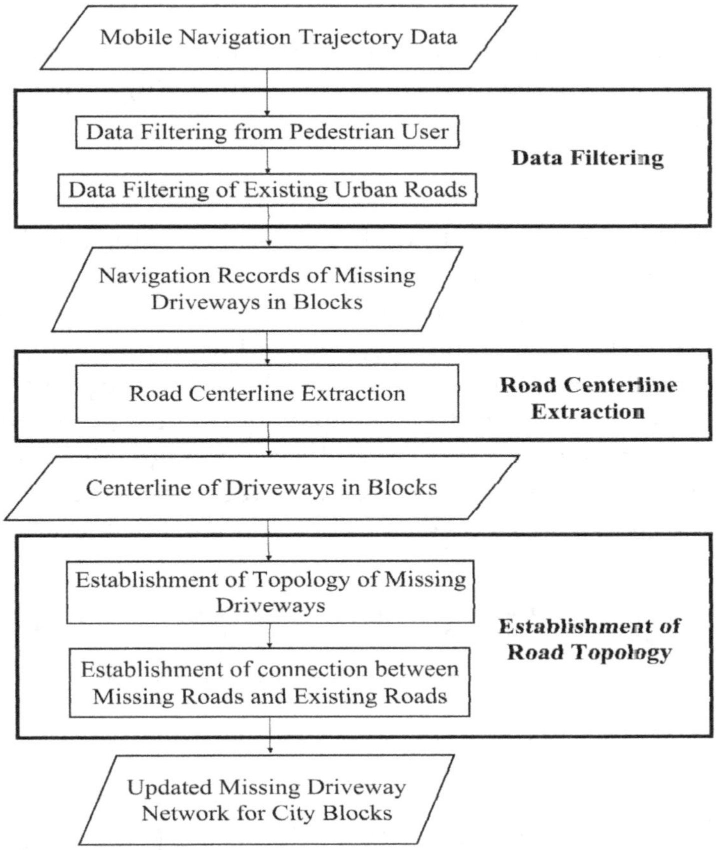

Figure 1. Flow chart of the proposed method.

3.2. Data Filtering

Mobile navigation data are collected as long as the navigation software is active. In order to reduce the duration of the calculation, data points located on the existing roads and generated by pedestrians should be filtered.

In this subsection, two main steps were adopted to filter the original data. The first step was to filter data from pedestrian users by the speed indicated by the record. Then, filtering was performed to segregate GNSS points based on existing urban roads via overlay analysis.

First, GNSS point speed information is the most appropriate indicator for distinguishing between pedestrian and vehicle users. According to previous studies, the average speeds of walking and driving are approximately 4.2 and 30 km/h, respectively. Therefore, in this research, 5.0 km/h was used as the threshold to separate pedestrian users from other navigation users. In other words, those GNSS points with speeds under 5.0 km/h were regarded as pedestrian users and were then removed.

Navigation data generated by the user's motion are partly located on existing roads and the rest on missing roads. According to our current research, navigation data points on existing roads account for more than 90% of all points in the used dataset. That is, the navigation data points on the roads inside city blocks only account for 10% of the total data volume. Therefore, filtering out the data points from the original data on existing roads will greatly improve the efficiency of the method as they account for the vast majority of the data.

Previous studies have been performed to filter the positioning points on existing urban roads, such as overlay analysis between the navigation-point layer and the existing urban road layer. Vector–raster analysis can also be used for filtering. In this paper, we adopted the method described in [37] to convert both existing roads and the navigation points into a raster format, so that filtering computing could be greatly accelerated.

3.3. Road Centerline Extraction

After filtering the raw data, purer navigation points of missing roads in city blocks were finally reserved. To extract the road network from these points, a clustering-based algorithm called RSC was proposed in this subsection.

3.3.1. Centerline Extraction via RSC

In this subsection, RSC was used to extract the centerlines of the missing roads. Figure 2 shows the four main RSC steps. The black dots indicate the navigation points, which were reserved after data filtering, as described in Section 3.2. Considering that the reserved GNSS points were located on the missing road, an initial point (P_{ini}) was randomly selected from the reserved points (see the yellow point in Figure 2a). Then, a radius parameter (R) was used to select the points that were covered by the ring with a radius of 2R (S_{2R}), which included the red circle and the green points of Figure 2b. R was the ring–radius parameter, which was approximately the average width of the roads in this region. After that, the highest density position could be calculated using the green points and could then be regarded as the node of the centerline (see the red point in Figure 2b).

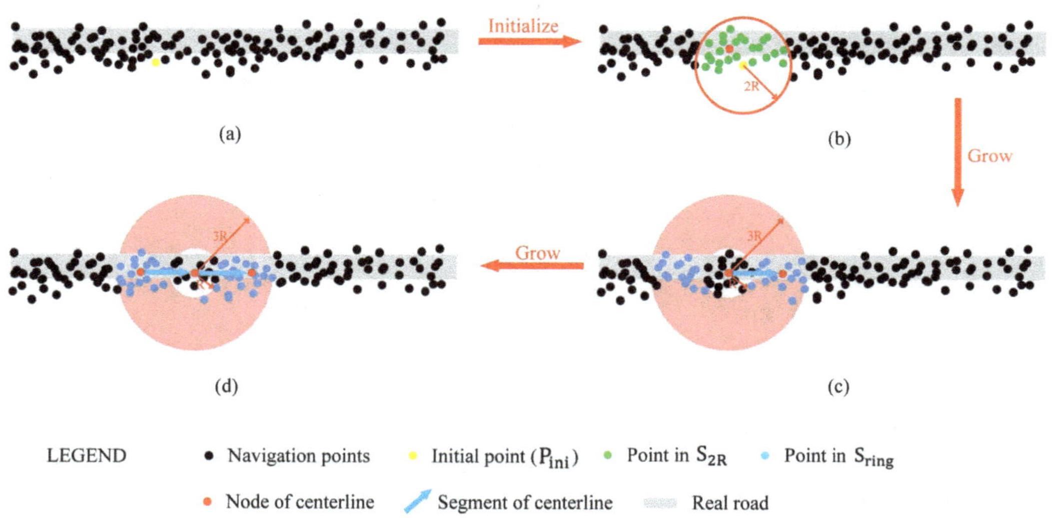

Figure 2. The main steps of RSC (ring-stepping clustering) are as follows: (**a**) a random GNSS (global navigation satellite system) point is chosen as the initial point, (**b**) the initial node is selected from the point set S_{2R} of the initial point, and (**c**,**d**) the next node is selected from the point set S_{ring} of the current node.

After the initial centerline node was found, a ring with an inner diameter R and an outer radius 3R was defined to find the next road centerline node. The points that fell on the ring were picked up to form the point set S_{ring} (see the blue points shown in Figure 2c).

The density of each point in S_{ring} was also calculated to find the highest density point. However, before deciding on the next node, each point in S_{ring} was calculated to determine whether it satisfied a U-turn. As shown in Figure 3 and Equation (1), value V was calculated from the potential segment (V_b) and the existing segment centerline (V_a).

$$V = V_a \cdot V_b \tag{1}$$

If V < 0, the potential node will not be selected as the next node.

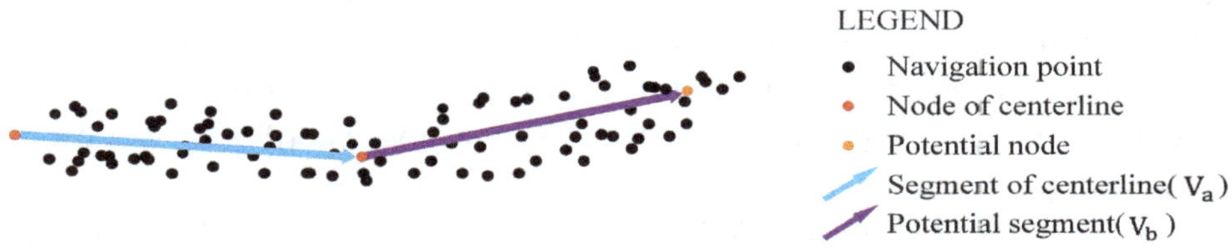

Figure 3. Determining a U-turn.

The pseudocode of the proposed RSC algorithm is presented in Algorithm 1. During data processing, the k–d tree [38] was introduced as an index of the point set for acceleration. The computational complexity of the proposed RSC algorithm is $\Theta(n^2)$.

Algorithm 1. RSC (ring-stepping clustering) Algorithm

Inputs: (1) a set of GNSS points **PointSet** = $\{P_1, P_2, P_3, \ldots, P_n\}$; (2) One parameter **R**.
Outputs: A set of centerlines **CenterLineSet** = $\{CL_1, CL_2, CL_3, \ldots, CL_m\}$, and each centerline is composed of a set of points NodeSet = $\{N_1, N_2, N_3, \ldots, N_k\}$

```
k–dtree ← Build k–d tree for PointSet
For i ← 1 to n do
  PointSet1 ← GetPointSetInRing(PointSet[i], 0, 2*R, PointSet)
  for j ← 1 to length(PointSet1) do
    PointSet1[j]'s PointSet2 = GetPointSetInRing(PointSet1[j], 0, R, PointSet)
    node ← the point whose PointSet2 has the most points in PointSet1
    numpts ← length of node's PointSet2
    If numpts > 0 then
      while numpts > 0 do
        add node to NodeSet
        PointSet3 ← GetPointSetInRing(Node, R, 3*R, PointSet)
        delete those points in PointSet3 that would make the centerline turn around
        for k ← 1 to length(PointSet3) do
          PointSet3[k]'s PointSet4 ← GetPointSetInRing(PointSet3[k], 0, R, PointSet3)
        end
        node ← the point whose PointSet4 has the most points in PointSet3
        numpts ← length of node's PointSet4
      end
    end
    add NodeSet to CenterLineSet
  end
end
return CenterLineSet
function GetPointSetInRing(Point, InsideRadius, OutsideRadius, PointSet)
  PointSetResult ← points in PointSet whose distance to Point is larger than InsideRadius and smaller than OutsideRadius (using kdtree)
return PointSetResult
```

3.3.2. Duplication Avoidance and Improvement

Using the above proposed algorithm, centerlines derived from GNSS points were obtained and organized by node sequence. However, after the centerline extraction for the missing road, the GNSS points on that road were still reserved. This led to centerline duplication.

Therefore, a rectangular buffer with width 3R for every single centerline segment was performed to remove the covered points (see the blue rectangle in Figure 4). All points that fell in the rectangular buffer were removed from the PointSet and were subject to further calculations. Moreover, the centerline extraction algorithm was terminated when PointSet was empty.

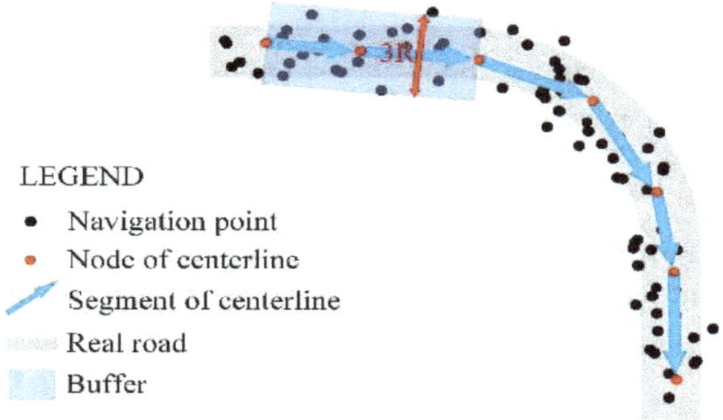

Figure 4. Centerline duplication avoidance and improvement.

3.4. Establishment of Road Topology

After extracting the centerlines of missing roadways, the road topology, which is important for the road map, remained missing. In this subsection, the road topology was rebuilt for the extracted roads. Two main parts were included: the topology establishment for the extracted missing roads and the relationship establishment for the existing roads.

3.4.1. Topology Establishment for Missing Roads

Establishing the topological relationship of the extracted centerlines consisted of three main steps. First, the topology relationships between different centerlines were built according to the end node (EN) of the centerlines. The distance between the end node (EN) and normal node (Ni) was calculated, and the minimum distance (D_{min}) was determined. If $D_{min} < 3R$, the connection between EN and Ni was regarded as a centerline, see Figure 5.

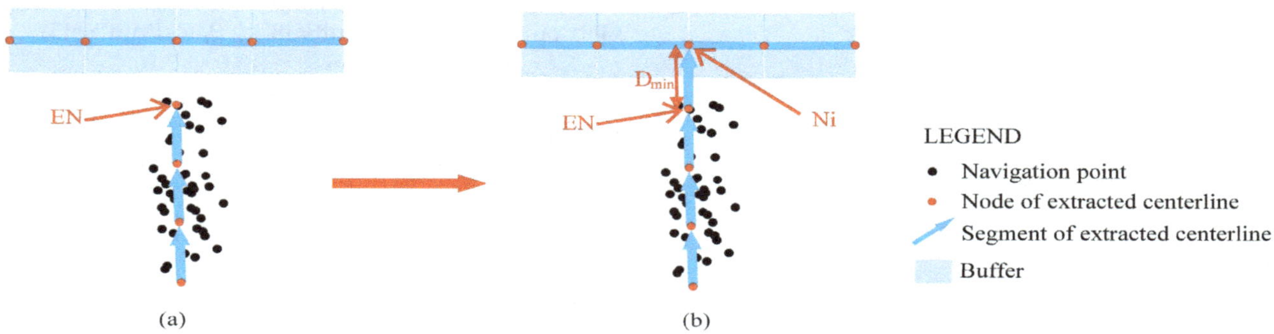

Figure 5. Relationship built between extracted centerlines. (**a**) Before building and (**b**) after building. EN is end node, D_{min} is minimum distance, Ni is normal mode.

Then, a cluster analysis was used to combine the adjacent topological nodes according to the distance between them. In this paper, the mean shift clustering algorithm [39] was adopted. Those topological nodes within a distance R were converted into a single topological node. This greatly improved the topological relationship of the missing roads.

Finally, a node classification procedure was performed to classify the nodes into two groups: the topological and normal nodes. The difference between the normal and topological nodes was the connected centerline segments. Once the connected centerline segments of a node were >2, the node was regarded as a topological node, and the centerlines were divided into two child centerlines (see the red nodes of Figure 6b).

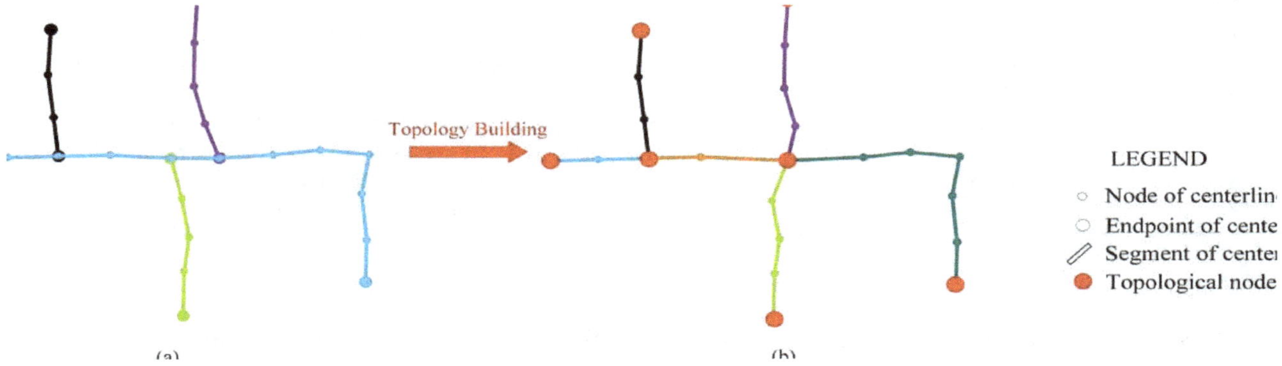

Figure 6. Topological rebuild of the extracted centerlines. The centerlines are shown (**a**) before rebuilding and (**b**) after rebuilding.

3.4.2. Connections between Missing and Existing Roads

After the topology of the missing roads was established, the connections between missing and existing roads were also established.

First, potential connections were built between the nodes and existing roads (see yellow lines of Figure 7). Then, potential connection verification was performed using the azimuth distribution of GNSS points around the potential connection; once this parameter was uniform with the main angular direction of the connection, the potential connection was regarded as valid. Otherwise, it was treated as an invalid connection.

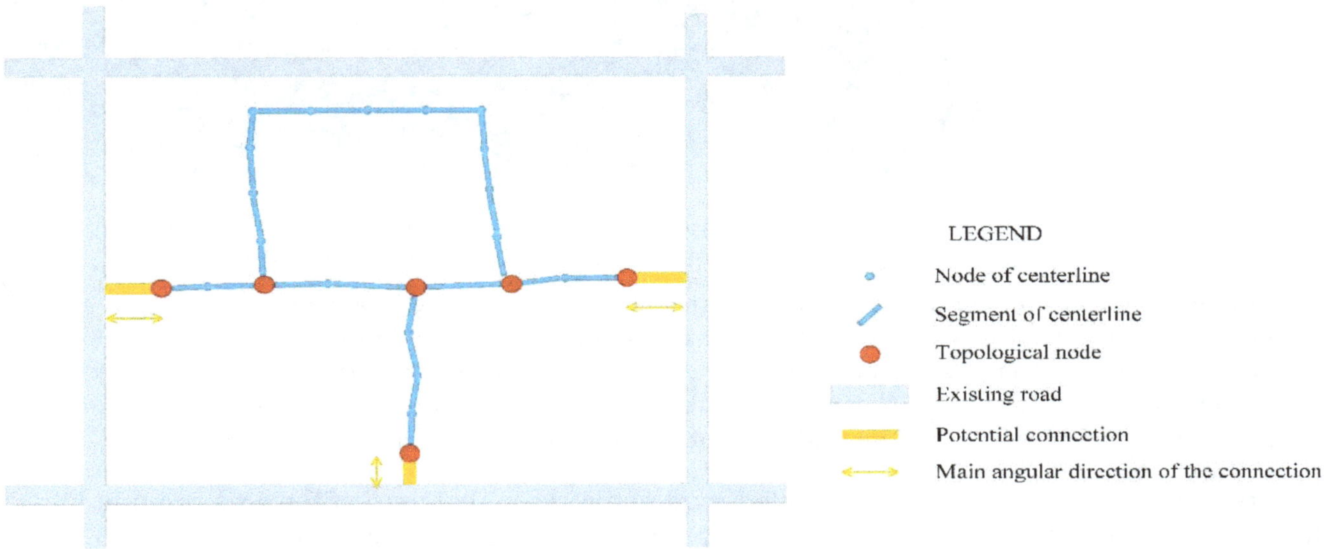

Figure 7. An illustrative example of the connections between missing and existing roads.

4. Case Study

4.1. Mobile Navigation Data

In this study mobile navigation data were used to extract missing roads. These data were generated and collected by mobile phones. These data were generated when the navigation application in the mobile phone was open, regardless of whether the user was using it for navigation. The sampling rate was one second per record, and each record contained seven main fields: day, time, ID, longitude, latitude, speed, and azimuth. A description of each field is given in Table 1.

Table 1. Description of the data fields.

Index	Field Name	Description
1	Day	The day the record was generated, including year, month, and day
2	Time	The time the record was generated, including hour, minute, and second
3	ID	ID of the user
4	Longitude	Longitude of the position of the user
5	Latitude	Latitude of the position of the user
6	Speed	Speed of the user
7	Azimuth	Azimuth of the user

4.2. Case Dataset

A region located in Shanghai was selected as the case area. It is 3.6 km long and 1.6 km wide, and it covers an area of 5.76 square kilometers. The case area covers two university campuses, and most of the roads in the case area are two-lane bidirectional roads. Around the case area, some municipal roads are available, such as Hujin Expressway, Jianchuan Road, Dongchuan Road, and Lianhua South Road. The location, imagery, and the existing municipal roads (marked with green color) of the case area are shown in Figure 8.

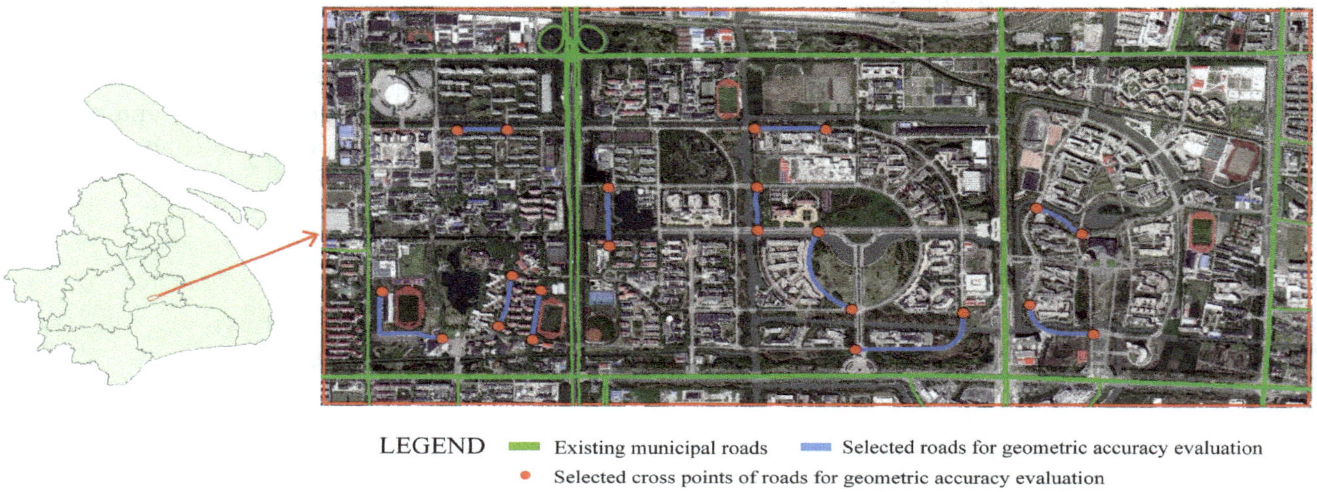

Figure 8. The location, imagery, existing municipal roads of the case area, and selected roads and cross points for quality evaluation. The imagery was collected on 13 April 2017 by DigitalGlobe's GeoEye-1. The spatial resolution of the imagery was 1 m. The imagery was matched to real map by the georeferencing toolbar of ArcGIS software. Relative to the real map, the 13 control points of the image had a positional accuracy of 1.97 m RMSE (error magnitudes range from 1.35 to 2.93 m).

To evaluate the quality of the extracted road networks, a real map of the area provided by the Shanghai Municipal Institute of Surveying and Mapping was used. The map was measured by total stations manually, and the precision of the map reached the centimeter level. Among all the real roads, 11 roads, including straight and curved roads, and 22 cross points (endpoints of the 11 selected roads) of the roads were selected to evaluate the performance of the proposed method. In order to compare the quality of satellite-derived method, the 11 selected roads were manually mapped from satellite imagery using the ArcGIS software—the mapping was done by three operators with good training in remote sensing image object extraction, respectively. Selected roads and cross points are shown in Figure 8. The manually mapped roads were compared with the real map of the area in order to evaluate the spatial accuracy, which is shown in Table 2. According to the evaluation results, the positioning precision of the digitalization was about 1.00 m. Compared with the real road data, the true difference was about 3.95 m.

Table 2. Result of spatial accuracy of the 22 manually mapped road points. Trueness is the distance between the average position and the corresponding real position (m). Precision is the mean square error of the points by different operators (m).

Point ID	Trueness (m)	Precision (m)	Point ID	Trueness (m)	Precision (m)
1	4.19	0.64	12	6.23	1.13
2	3.26	1.04	13	5.18	1.91
3	1.15	1.43	14	0.70	1.65
4	6.84	1.52	15	3.63	0.97
5	5.88	0.84	16	2.21	1.77
6	6.00	1.12	17	2.78	1.08
7	5.78	0.19	18	5.01	0.04
8	2.76	0.35	19	3.41	0.06
9	3.51	0.05	20	2.51	1.70
10	5.35	1.04	21	3.25	0.23
11	2.95	1.86	22	4.27	1.43
Average	3.95	1.00	–	–	–

Data used for analysis were collected in December 2017 (11–15 December) and consisted of 9,944,710 GNSS data points in total, belonging to 198,241 unique vehicle IDs. The data were provided by 1RenData (ShangHai) Technology Co., Ltd (Shanghai, China). Figure 9 shows the raw GNSS data distribution.

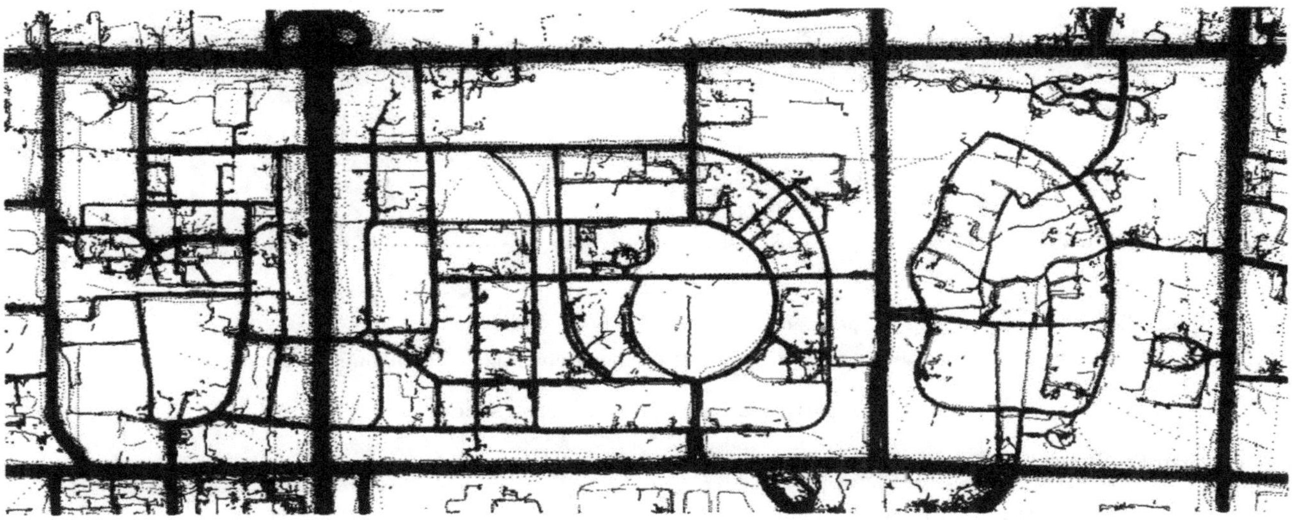

Figure 9. Raw research data. The black dots represent GNSS (Global Navigation Satellite System) points.

4.3. Data Filtering

Using the data filtering methods described in Section 3.2, 392,128 records belonging to 8676 unique user IDs were finally reserved. The reserved GNSS points took ~4.0% of the raw data. Components of the raw case data are shown in Table 3.

Table 3. Components of the case data.

Location		Pedestrian	Vehicle	Total
Existing Municipal Road	Value	3,054,797	6,179,641	9,234,438
	Percentage	30.7%	62.1%	92.8%
Missing Road	Value	318,144	392,128	710,272
	Percentage	3.2%	4.0%	7.2%
Total	Value	3,372,941	6,571,769	9,944,710
	Percentage	33.9%	62.1%	100.0%

4.4. Road Centerlines

4.4.1. Road Centerline Extraction

The proposed RSC method was used to extract road centerlines from the data after filtering. The method prototype was implemented in C-Sharp programming language. The experiment was carried out on a server with Intel Xeon CPU Platinum 8163@ 2.5 GHz and 16GB memory. As the roads' average width in the experimental area was ~7 m, the parameter R was set to 7.0 in this paper. The obtained centerlines are shown in Figure 10. There were 603 centerlines in total, and the average length of these centerlines was 116.09 m. It took 557 s to extract the centerlines of the 392,128 points.

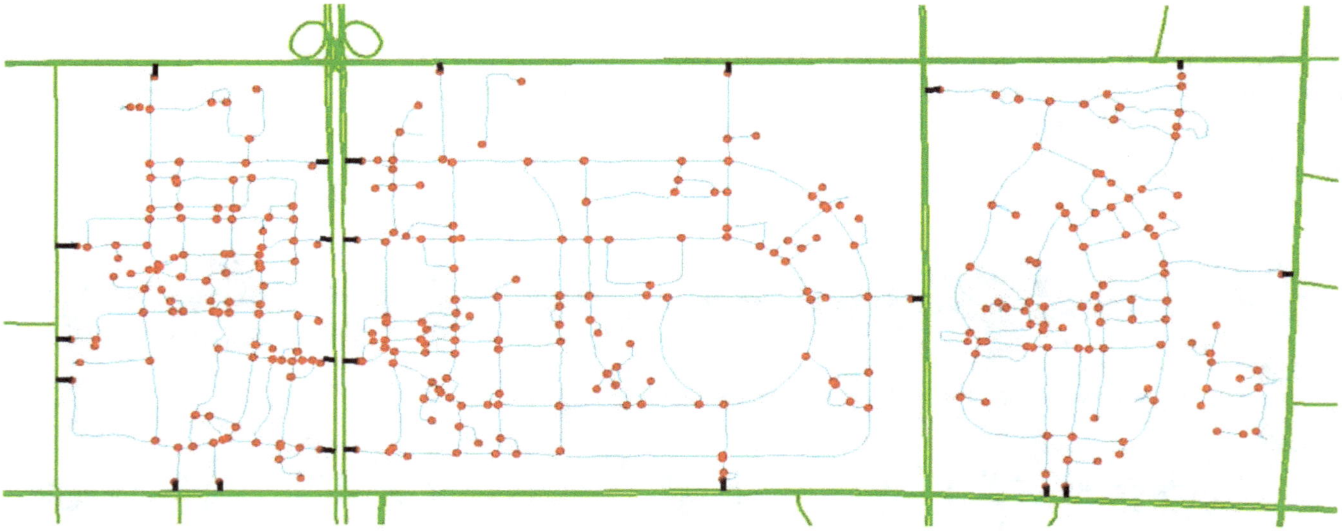

Figure 10. Results of the case study.

4.4.2. Geometric Quality Evaluation of Road Centerlines

■ Geometric Evaluation of Generated Centerlines Compared with Real Map

In this subsection, two indices were selected as quality measures of the extracted roads. The first index was the road's length [40,41]. Assume Lr is the length of the real road and Lg is the length of the road used for quality evaluation. Then, absolute error between the different road lengths was calculated by Equation (2):

$$Eg = |Lr - Lg| \tag{2}$$

where Eg is the absolute error between Lr and Lg.

The second index is the distance between each road centerline and the corresponding real road [42,43]. First, the region area between the real and extracted roads was calculated (see the blue area of Figure 11). Then, the quality values of the extracted centerlines Tg were calculated using Equation (3):

$$\text{Tg} = \frac{\text{Ag}}{\text{Lr}} \tag{3}$$

where Ag is the area of the region between the generated and real roads.

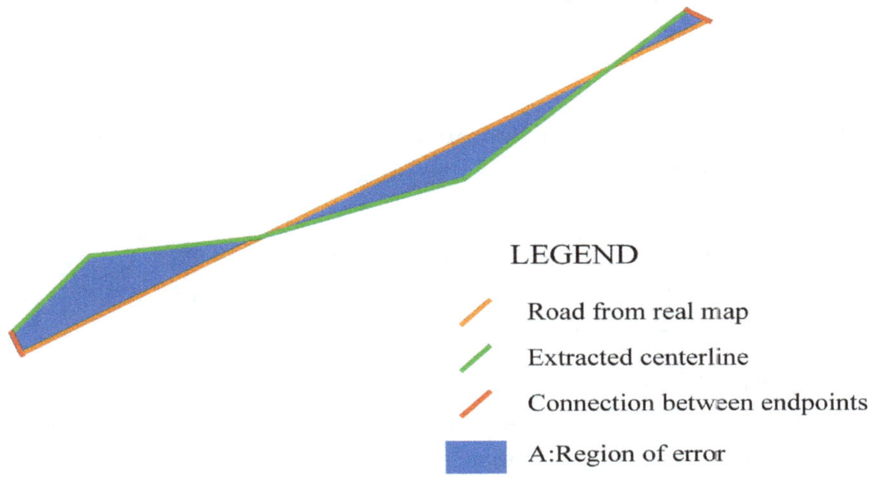

Figure 11. Distance calculation between real and extracted roads.

Using parameters in Equations (2) and (3), we listed the evaluation results of quality of the 11 selected roads in Table 4. According to Table 4, the average of Eg was 1.84 m and the average of Tg was 1.62 m, which demonstrates that the proposed method can achieve excellent results.

Table 4. Results of the geometric evaluation of generated centerlines (m).

Road ID	Lr	Lg	Eg	Tg
1	206.95	207.93	0.98	0.37
2	403.05	401.05	2.00	1.21
3	183.80	181.47	2.34	2.51
4	199.36	196.90	2.46	1.70
5	304.27	304.96	0.69	3.01
6	511.84	511.84	0.00	1.80
7	240.78	238.18	2.60	1.89
8	219.78	223.12	3.34	0.34
9	180.94	182.75	1.81	1.94
10	410.14	406.24	3.90	2.11
11	265.42	265.55	0.13	0.96
Average	284.21	283.64	1.84	1.62

- Geometric Comparison between Digitalization and Proposed Method

In order to compare the geometric accuracy between the proposed method and digitalization by remote sensing image, the true difference between selected cross points of the real roads and corresponding cross points of the roads, which were named as Dg, was used. In this paper, 22 road cross points were selected to compute the true difference with the real map. Then, the minimum, maximum, and the average Dg of two methods are listed in Table 5.

Table 5. Minimum, maximum, and average Dg of digitalization and proposed method (m).

Digitalization from Remote Sensing Image			Proposed Method		
Minimum	**Maximum**	**Average**	**Minimum**	**Maximum**	**Average**
0.23	6.00	3.95	0.81	6.02	3.43

According to Table 5, the average of Dg of the mapped roads was 3.95 m, which was larger than the proposed method. It shows that the accuracy of the proposed method was slightly better than the digitalization method.

■ F-Score Evaluation

In addition to the evaluation method mentioned above for the generated centerline, the F − score proposed in [11] was also adopted to evaluate the proposed method. The F − score was computed as follows:

$$\text{spurious} = \frac{\text{spurious marbles}}{(\text{spurious marbles} + \text{matched marbles})}$$
$$\text{missing} = \frac{\text{empty holes}}{(\text{empty holes} + \text{matched holes})}$$
$$\text{F} - \text{score} = 2 \times \frac{(1-\text{spurious})(1-\text{missing})}{(1-\text{spurious})+(1-\text{missing})}.$$

Starting from a random location, the roads were explored by placing point samples on each graph during a traversal outward within a maximum radius. Sample points on the roads that required evaluation were considered as "marbles" and on the real roads as "holes". In this case, spurious marbles represent the number of points on the evaluated roads that do not get a match, matched marbles represent the number of points on the evaluated roads that get a match, empty holes represent the number of points on the real roads that do not get a match and matched holes represent the number of points on the evaluated roads that get a match.

The F − score of the proposed method was compared with that of the other methods [11–14]. The comparison is shown in Figure 12. Obviously, the proposed method offered a significant improvement over the previous methods.

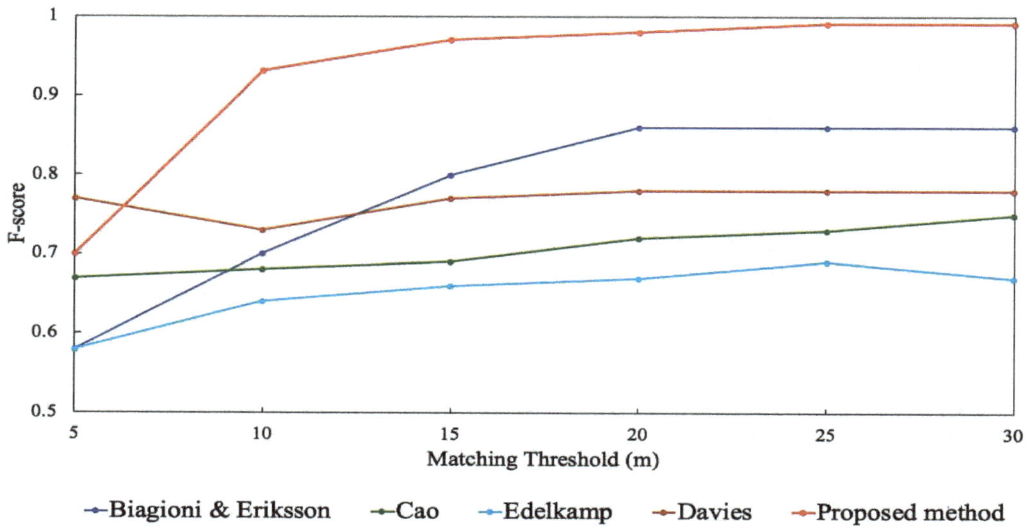

Figure 12. F − score of the proposed and existing methods.

4.5. Topology Evaluation

4.5.1. Topology of Missing Roads

The topology was built after centerline extraction. There were 371 topological nodes in total. After manual verification, 296 were real intersections on the roads. The correctness was ~79.8%. The wrongly extracted topological nodes, usually located at houses, were too densely localized or were located on roads that were too complex. This was probably caused by the low positioning accuracy and multipath effects of GNSS devices and can be improved after enhancing the dataset.

4.5.2. Connections between Missing and Existing Roads

Twenty-six potential connections between topological nodes and existing roads were found. After azimuth verification, 23 connections were reserved; when compared to the corresponding remote sensing images, they were correct. These correct connections represent both entrances and exits.

5. Discussion and Conclusions

This paper proposes a new method for generating missing road networks in city blocks using big mobile navigation trajectory data. An algorithm named RSC was designed according to high frequency GNSS data. After extracting centerlines, methods of building road topologies in city blocks and establishing connections between missing and existing roads were proposed. A case area (5.76 km^2) was used to verify the feasibility and validity of the proposed method. The results showed that compared to real roads, the average length difference is approximately 1.84 m and the average distance is approximately 1.64 m, indicating that the proposed method can achieve meter-level missing road extraction results. Data from satellite-extracted roads indicate that the proposed method achieves greater results than image-derived method. Meanwhile, using the $F-$ score index, the proposed method can achieve the best results compared to previous studies.

The novelty of the proposed method is the higher geometric quality of the extracted missing roads. The length difference and the distance between extracted roads with real roads are approximately 1.84 m and 1.64 m, respectively. This allows the possibility of generating complicated road networks. Meanwhile, the $F-$ score performance of the proposed method offered a large improvement over previous methods, meaning that the road networks generated by the proposed method are far more meticulous.

However, the complexity of the proposed method is $\Theta(n^2)$. This metric indicates that when the amount of input of GNSS data increases, resource and time consumption by the algorithm will also increase geometrically. Though the method introduced in the paper worked well in most areas, the result and quality will be affected by a few factors.

First, due to the chaotic trajectories in car parks, there will be a jumble of GNSS points in the corresponding region. Therefore, roads through car parks cannot be extracted using the proposed method. Further, a bridge present above the road, an underground car park, or an underpass under the road will result in failure to generate the road's centerline at ground level.

Second, when the road is between two high buildings or trees, a very thick canopy is formed, and the shadow and multipath effects may cause the coordinates of the GNSS points to be burdened by significant errors; thus, the quality of the centerline extracted via the proposed method will be poor.

Moreover, when two roads are close to each other and in parallel and when they are both not wide enough, their related GNSS points will be difficult to distinguish, so they will probably be extracted as one road.

Finally, during data filtering, some GNSS data with the speeds < 5.0 km/h is removed to filter pedestrian and vehicle users. Such an approach would result in excluding some vehicles traveling at a lower speed; thus, the GNSS data involved in the calculation will also be reduced.

Author Contributions: Conceptualization, Hangbin Wu and Guangjun Wu; methodology, Zeran Xu and Hangbin Wu; software, Zeran Xu; validation, Zeran Xu; formal analysis, Hangbin Wu and Zeran Xu; investigation, Zeran Xu; resources, Guangjun Wu; writing—original draft preparation, Hangbin Wu and Zeran Xu; writing—review and editing, Hangbin Wu and Zeran Xu; visualization, Zeran Xu and Hangbin Wu; supervision, Hangbin Wu; project administration, Hangbin Wu; funding acquisition, Hangbin Wu.

Acknowledgments: The authors would like to appreciate Quan Yuan and Rui Jia for their manual operations in the digitalization part. Also, authors appreciate the contributions made by anonymous reviewers.

References

1. Lee, S.; Lee, D.; Lee, S. Network-oriented road map generation for unknown roads using visual images and gps-based location information. *IEEE Trans. Consum. Electron.* **2009**, *55*, 1233–1240. [CrossRef]
2. Wu, T.; Xiang, L.; Gong, J. Updating road networks by local renewal from gps trajectories. *ISPRS Int. J. Geo-Inf.* **2016**, *5*, 163. [CrossRef]
3. Biagioni, J.; Eriksson, J. Inferring road maps from global positioning system traces. *Transp. Res. Rec. J. Transp. Res. Board* **2015**, *2291*, 61–71. [CrossRef]
4. Shan, Z.; Wu, H.; Sun, W.; Zheng, B. COBWEB: A robust map update system using GPS trajectories. In Proceedings of the ACM International Joint Conference on Pervasive and Ubiquitous Computing, Osaka, Japan, 7–11 September 2015; pp. 927–937.
5. Costa, G.H.R.; Baldo, F. Generation of road maps from trajectories collected with smartphone—A method based on genetic algorithm. *Appl. Soft Comput.* **2015**, *37*, 799–808. [CrossRef]
6. Liu, X.; Liu, X.; Wei, H.; Forman, G.; Zhu, Y. CrowdAtlas: Self-updating maps for cloud and personal use. In Proceedings of the International Conference on Mobile Systems, Applications, and Services, Taipei, Taiwan, 25–28 June 2013; pp. 469–470.
7. Park, S.; Bang, Y.; Yu, K. Techniques for updating pedestrian network data including facilities and obstructions information for transportation of vulnerable people. *Sensors* **2015**, *15*, 24466–24486. [CrossRef]
8. El-Sheimy, N.; Schwarz, K.P. Navigating urban areas by VISAT—A mobile mapping system integrating GPS/INS/digital cameras for GIS applications. *Navigation* **1998**, *45*, 275–285. [CrossRef]
9. Zhang, Y.; Liu, J.; Qian, X.; Qiu, A.; Zhang, F. An Automatic Road Network Construction Method Using Massive GPS Trajectory Data. *ISPRS Int. J. Geo-Inf.* **2017**, *6*, 400. [CrossRef]
10. Sghaier, M.O.; Lepage, R. Road Extraction From Very High Resolution Remote Sensing Optical Images Based on Texture Analysis and Beamlet Transform. *IEEE J. Sel. Top. Appl. Earth Obs. Remote Sens.* **2016**, *9*, 1946–1958. [CrossRef]
11. Biagioni, J.; Eriksson, J. Map inference in the face of noise and disparity. In Proceedings of the International Conference on Advances in Geographic Information Systems, Redondo Beach, CA, USA, 6–9 November 2012.
12. Cao, L.; Krumm, J. From GPS traces to a routable road map. In Proceedings of the Workshop on Advances in Geographic Information Systems, New York, NY, USA, 4–6 November 2009; pp. 3–12.
13. Edelkamp, S.; Schrödl, S. Route Planning and Map Inference with Global Positioning Traces. In *Computer Science in Perspective, Essays Dedicated to Thomas Ottmann*; Springer: Berlin, Heidelberg, 2003; pp. 128–151.
14. Davies, J.J.; Beresford, A.R.; Hopper, A. Scalable, distributed, real-time map generation. *IEEE Pervasive Comput.* **2006**, *5*, 47–54. [CrossRef]
15. Zhao, Y.; Liu, J.; Chen, R.; Li, J.; Xie, C.; Niu, W. A new method of road network updating based on floating car data. *Geosci. Remote Sens. Symp.* **2011**, *24*, 1878–1881.
16. Freitas, T.R.M.; Coelho, A.; Rossetti, R.J.F. Improving digital maps through GPS data processing. In Proceedings of the International IEEE Conference on Intelligent Transportation Systems, St. Louis, MO, USA, 12–15 October 2009; pp. 1–6.
17. Freitas, T.R.M.; Coelho, A.; Rossetti, R.J.F. Correcting routing information through GPS data processing. In Proceedings of the International IEEE Conference on Intelligent Transportation Systems, Piscataway, NJ, USA, 19–22 September 2010; pp. 706–711.
18. Boucher, C.; Noyer, J.C. Automatic detection of topological changes for digital road map updating. *IEEE Trans. Instrum. Meas.* **2012**, *61*, 3094–3102. [CrossRef]
19. Wang, Y.; Wei, H.; Forman, G. Mining large-scale gps streams for connectivity refinement of road maps. *Comput. J.* **2018**, *58*, 2109–2119.
20. Agamennoni, G.; Nieto, J.I.; Nebot, E.M. Robust inference of principal road paths for intelligent transportation systems. *IEEE Trans. Intell. Transp. Syst.* **2011**, *12*, 298–308. [CrossRef]
21. Schroedl, S.; Wagstaff, K.; Rogers, S.; Langley, P.; Wilson, C. Mining gps traces for map refinement. *Data Min. Knowl. Discov.* **2004**, *9*, 59–87. [CrossRef]
22. Guo, T.; Iwamura, K.; Koga, M. Towards high accuracy road maps generation from massive GPS Traces data.

In Proceedings of the 2007 IEEE International Geoscience and Remote Sensing Symposium, Barcelona, Spain, 23–28 July 2007; pp. 667–670.
23. Jang, S.; Kim, T.; Lee, E. Map generation system with lightweight GPS trace data. In Proceedings of the International Conference on Advanced Communication Technology, Miyazaki, Japan, 23–25 June 2010; Volume 2, pp. 1489–1493.
24. Niehöfer, B.; Burda, R.; Wietfeld, C.; Bauer, F.; Lueert, O. GPS Community Map Generation for Enhanced Routing Methods Based on Trace-Collection by Mobile Phones. In Proceedings of the International Conference on Advances in Satellite and Space Communications, Colmar, France, 20–25 July 2009; pp. 156–161.
25. Shi, W.; Shen, S.; Liu, Y. Automatic generation of road network map from massive GPS, vehicle trajectories. In Proceedings of the International IEEE Conference on Intelligent Transportation Systems, St. Louis, MO, USA, 4–7 October 2009; pp. 1–6.
26. Chen, C.; Cheng, Y. Roads Digital Map Generation with Multi-track GPS Data. *IEEE Comput. Soc.* **2008**, *12*. [CrossRef]
27. Li, L.; Li, D.; Xing, X.; Yang, F.; Rong, W.; Zhu, H. Extraction of road intersections from gps traces based on the dominant orientations of roads. *Int. J. Geo-Inf.* **2017**, *6*, 403. [CrossRef]
28. Xie, X.; Bingyungwong, K.; Aghajan, H.; Veelaert, P.; Philips, W. Inferring directed road networks from gps traces by track alignment. *ISPRS Int. J. Geo-Inf.* **2015**, *4*, 2446–2471. [CrossRef]
29. Qiu, J.; Wang, R. Automatic extraction of road networks from gps traces. *Photogramm. Eng. Remote Sens.* **2016**, *82*, 593–604. [CrossRef]
30. Winden, K.V.; Biljecki, F.; Spek, S.V.D. Automatic update of road attributes by mining gps tracks. *Trans. GIS* **2016**, *20*, 664–683. [CrossRef]
31. Zhang, L.; Thiemann, F.; Sester, M. Integration of GPS traces with road map. In Proceedings of the Second International Workshop on Computational Transportation Science, Seattle, WA, USA, 3 November 2009; pp. 17–22.
32. Chen, Y.; Krumm, J. Probabilistic modeling of traffic lanes from GPS traces. In Proceedings of the 18th SIGSPATIAL International Conference on Advances in Geographic Information Systems, San Jose, CA, USA, 2–5 November 2010; pp. 81–88.
33. Li, J.; Qin, Q.; Han, J.; Tang, L.-A.; Lei, K.H. Mining trajectory data and geotagged data in social media for road map inference. *Trans. GIS* **2014**, *19*, 18. [CrossRef]
34. Endo, Y.; Toda, H.; Nishida, K.; Kawanobe, A. Deep feature extraction from trajectories for transportation mode estimation. In *Pacific-Asia Conference on Knowledge Discovery and Data Mining*; Springer: Berlin, Germany, 2016; pp. 54–66.
35. Etemad, M.; Soares Júnior, A.; Matwin, S. Predicting Transportation Modes of GPS Trajectories using Feature Engineering and Noise Removal. In *Advances in Artificial Intelligence: 31st Canadian Conference on Artificial Intelligence, Canadian AI 2018, Toronto, ON, Canada, May 8–11, 2018, Proceedings 31*; Springer International Publishing: New York, NY, USA, 2018; pp. 259–264.
36. Liu, C.; Xiong, L.; Hu, X.; Shan, J. A progressive buffering method for road map update using openstreetmap data. *ISPRS Int. J. Geo-Inf.* **2015**, *4*, 1246–1264. [CrossRef]
37. Bresenham, J.E. Algorithm for computer control of a digital plotter. *IBM Syst. J.* **1999**, *4*, 25–30. [CrossRef]
38. Chen, Y.; Zhou, L.; Tang, Y.; Singh, J.P.; Bouguila, N.; Wang, C.; Wang, H.; Du, J. Fast neighbor search by using revised kd tree. *Inf. Sci.* **2019**, *472*, 145–162. [CrossRef]
39. Cheng, Y. Mean shift, mode seeking, and clustering. *IEEE Trans. Pattern Anal. Mach. Intell.* **1995**, *17*, 790–799. [CrossRef]
40. Šuba, R.; Meijers, M.; Oosterom, P.V. Continuous road network generalization throughout all scales. *ISPRS Int. J. Geo-Inf.* **2016**, *5*, 145. [CrossRef]
41. Zhang, J.; Wang, Y.; Zhao, W. An Improved Hybrid Method for Enhanced Road Feature Selection in Map Generalization. *ISPRS Int. J. Geo-Inf.* **2017**, *6*, 196. [CrossRef]
42. Ahmed, M.; Fasy, B.T.; Hickmann, K.S.; Wenk, C. A path-based distance for street map comparison. *ACM Trans. Spat. Algorithms Syst.* **2015**, *1*, 1–28. [CrossRef]
43. Liu, X.; Biagioni, J.; Eriksson, J.; Wang, Y.; Forman, G.; Zhu, Y. Mining large-scale, sparse GPS traces for map inference: Comparison of approaches. In Proceedings of the 18th ACM SIGKDD International Conference on Knowledge Discovery and Data Mining, Beijing, China, 12–16 August 2012; pp. 669–677.

5

MapReduce-Based D_ELT Framework to Address the Challenges of Geospatial Big Data

Junghee Jo [1,*] and Kang-Woo Lee [2]

1 Busan National University of Education, Busan 46241, Korea
2 Electronics and Telecommunications Research Institute (ETRI), Daejeon 34129, Korea; kwlee@etri.re.kr
* Correspondence: dreamer@bnue.ac.kr

Abstract: The conventional extracting–transforming–loading (ETL) system is typically operated on a single machine not capable of handling huge volumes of geospatial big data. To deal with the considerable amount of big data in the ETL process, we propose D_ELT (delayed extracting–loading –transforming) by utilizing MapReduce-based parallelization. Among various kinds of big data, we concentrate on geospatial big data generated via sensors using Internet of Things (IoT) technology. In the IoT environment, update latency for sensor big data is typically short and old data are not worth further analysis, so the speed of data preparation is even more significant. We conducted several experiments measuring the overall performance of D_ELT and compared it with both traditional ETL and extracting–loading– transforming (ELT) systems, using different sizes of data and complexity levels for analysis. The experimental results show that D_ELT outperforms the other two approaches, ETL and ELT. In addition, the larger the amount of data or the higher the complexity of the analysis, the greater the parallelization effect of transform in D_ELT, leading to better performance over the traditional ETL and ELT approaches.

Keywords: ETL; ELT; big data; sensor data; IoT; geospatial big data; MapReduce

1. Introduction

In recent years, numerous types of sensors have been connected to the Internet of Things (IoT) and have produced huge volumes of data with high velocity. A large percentage of these sensor big data is geospatial data, describing information about physical things in relation to geographic space that can be represented in a coordinate system [1–4]. With the advance of IoT technologies, more diverse data have now become available, thereby greatly increasing the amount of geospatial big data.

Given the general properties of big data, the unique characteristics of geospatial data create an innovative challenge in data preparation [5]. Geospatial data typically include position data. These coordinate data differ from normal string or integer data, requiring the data pre-processing process to include a lot of floating-point arithmetic computations. Examples include transformation in geometry, converting coordination reference systems, and evaluating spatial relationships. Among these, the most well-known aspect of geospatial data is spatial relationship, describing the relationship of some objects in a specific location to other objects in neighboring locations. The calculation of spatial relationship is mostly included in spatial analysis and has been generally regarded as a sophisticated problem [6]. Moreover, processing temporal elements also complicates the handling of geospatial data.

To deal with the challenges in processing and analyzing geospatial big data, several systems have emerged. Systems designed for big data have existed for years (e.g., Hadoop [7] and Spark [8]); however, they are uninformed about spatial properties. This has led to a number of geospatial systems (e.g., SpatialHadoop [9] and GeoSpark [10]) being developed, mostly by injecting spatial data types or functions inside existing big data systems. Hadoop, especially, has proven to be a mature big data platform and so several geospatial big data systems have been constructed by inserting spatial data awareness into Hadoop. However, it is still not easy for big data software developers to create geospatial applications. Typically, to generate a MapReduce job for a required operation in Hadoop, developers need to program a map and reduce functions. Spatial analysis usually requires handling more than one MapReduce step, where the output of the data from a previous MapReduce step becomes the input to the next MapReduce step. As the complexity level of spatial analysis is increased, the number of MapReduce steps is also increased, resulting in augmented difficulties for the developers to write iterative code to define the increasingly more complicated MapReduce steps.

To resolve this issue, in our previous work [11], we found a way to represent spatial analysis as a sequence of one or more units of spatial or non-spatial operators. This allows developers of geospatial big data applications to create spatial applications by simply combining built-in spatial or non-spatial operators, without having any detailed knowledge of MapReduce. Once the sequence of operators has been incorporated, it is automatically transformed to the map and reduces jobs in our Hadoop-based geospatial big data system. During this conversion process, our system controls the number of MapReduce steps in such a way as to achieve better performance by decreasing the overhead of mapping and reducing. The challenges for geospatial big data, however, lie in confronting not only how to store and analyze the data, but also how to transform the data while achieving good performance.

Currently, a large amount of geospatial data is continuously provided from many spatial sensors. It is important to analyze this geospatial big data as soon as possible to extract useful insights. However, the time required to transform massive amounts of geospatial data into the Hadoop platform has gradually increased. That is, it takes a lot of time to prepare the data required for geospatial analysis, thereby delaying obtaining the results of spatial analysis results. For example, we found that it took about 13 hours and 30 minutes to load 821 GB of digital tachograph (DTG) data using the traditional ETL method. In the ETL process, data are extracted from data sources, then transformed, involving normalization and cleansing, and loaded into the target data base. The conventional ETL system is typically operated on a single machine that cannot effectively handle huge volumes of big data [12]. To deal with the considerable quantity of big data in the ETL process, there have been several attempts in recent years to utilize a parallelized data processing concept [13–15].

One study [14] proposed ETLMR using a MapReduce framework to parallelize ETL processes. ETLMR is designed by integrating with Python-based MapReduce. This study conducted an experimental evaluation assessing system scalability based on different scales of jobs and data to compare with other MapReduce-based tools. Another study [15] compared Hadoop-based ETL solutions with commercial ETL solutions in terms of cost and performance. They concluded that Hadoop-based ETL solutions are better in comparison to existing commercial ETL solutions. The study in [16] implemented P-ETL (parallel-ETL), which is developed on Hadoop. Instead of the traditional three steps of extracting, transforming, and loading, P-ETL involves five steps of extracting, partitioning, transforming, reducing, and loading. This study has shown that P-ETL outperforms the classical ETL scheme. Many studies, however, have focused on big data analysis, but there have been insufficient studies attempting to increase the speed of preparing the data required for big data analysis.

In this paper, we continue our previous study on storing and managing geospatial big data and explain our approach to enhance the performance of ETL processes. Specifically, we propose a method to start geospatial big data analysis in a short time by reducing the time required for data transformation under the Hadoop platform. A transformation is defined as data processing achieved by converting source data into a consistent storage format aiming to query and analyze. Due to the complex nature of transformations, performance of the ETL processes depend mostly on how efficiently the transformations are conducted, which is the rate-limiting step in the ETL process. Our approach allows MapReduce-based parallelization of the transformation in the ETL process. Among the various sources of geospatial big data, we concentrate on sensor big data. With the increasing number of IoT sensing devices, the amount of sensor data is expected to grow significantly over time for a wide range of fields and applications. IoT-based sensor data are, however, essentially loosely structured and typically incomplete, much of it being directly unusable. In addition, in the IoT environment, the update period—the time between the arrival of raw data and when meaningful data are made available—occurs more frequently than for typical batch data. These difficulties require that considerable resources are used for transformation in the ETL process.

This paper extends our research work presented in [11] and suggests a way to increase performance of the transformation functionality in the ETL process by taking advantage of the MapReduce framework. First, in Section 2 we briefly explain our previous work on constructing a geospatial big data processing system by extending the original Hadoop to support spatial properties. We focus particularly on explaining automatically converting a user-specified sequence of operators for spatial analysis to MapReduce steps. Section 3 describes up-to-date ETL research followed by our approach on improving performance of transformation in the ETL processes based on MapReduce. Our conducted experimental settings and results are described in Sections 4 and 5, respectively. Section 6 concludes our work and presents our plans for future research.

2. Geospatial Big Data Platform

In our previous study [11], we developed a high performance geospatial big data processing system based on Hadoop/MapReduce, named Marmot [17]. In Marmot, spatial analysis is defined as a sequence of RecordSetOperators, where a RecordSet is a collection of records and a RecordSetOperator is a processing element using a RecordSet, similar to a relational operator in Relational Database Management System (RDBMS). A sequence of RecordSetOperators is defined as a Plan, as shown in Figure 1·

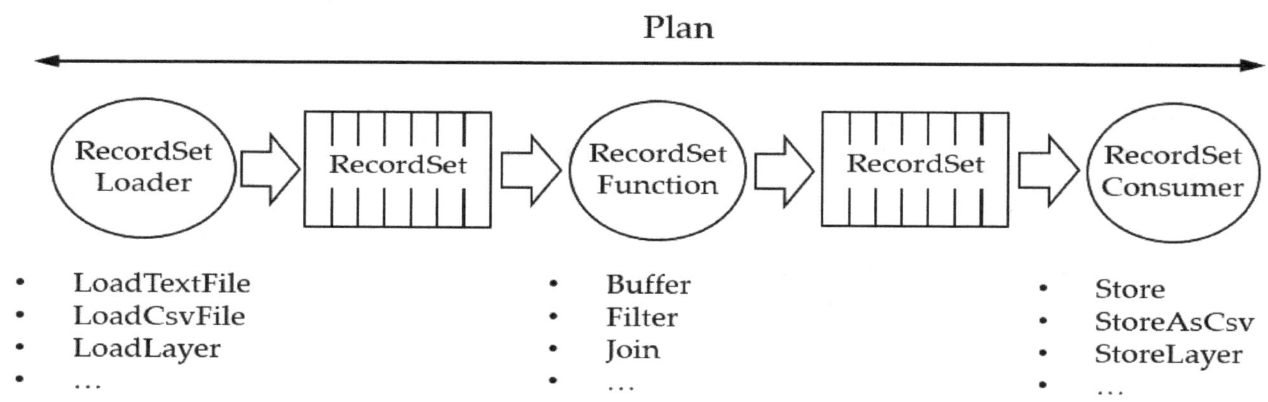

Figure 1. Representation of spatial analysis in Marmot: A sequence of one or more units of spatial or non-spatial operators.

In Marmot, a RecordSetOperator is classified as three possible types: RecordSetLoader, RecordSetFunction, or RecordSetConsumer. RecordSetLoader is a non-spatial operator loading source data and transforming it to a RecordSet; RecordSetFunction is a spatial or non-spatial operator taking a RecordSet as source data and producing a new RecordSet as output data; RecordSetConsumer is a non-spatial operator storing a finally created RecordSet as a result of a given spatial analysis outside of Marmot.

To process a given spatial analysis, a developer creates a corresponding Plan by combining spatial operators and non-spatial operators and injects the Plan into Marmot. Marmot processes each RecordSetOperator one by one and automatically transforms the given Plan to map and reduce jobs, as shown in Figure 2.

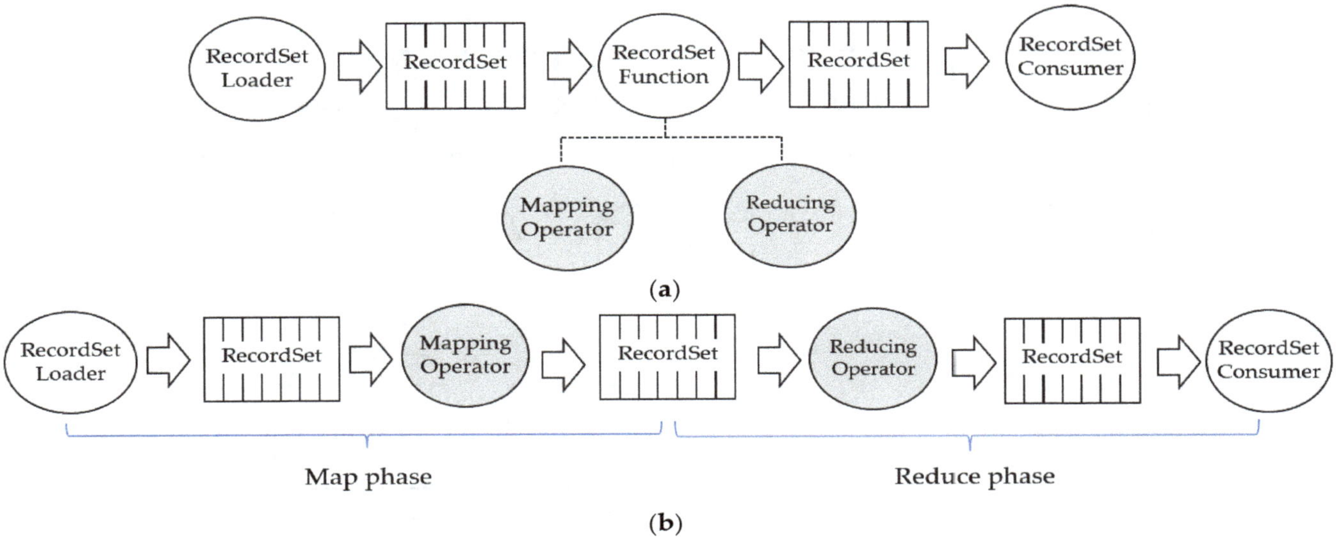

Figure 2. Automatic transformation of a Plan into MapReduce jobs. (**a**) A Plan having a RecordSetFunction divided into mapping and reducing operators; (**b**) An automatically transformed Plan.

While parsing a given Plan, when Marmot meets a RecordSetFunction that can be separated into mapping and reducing operators (e.g., ReduceByGroupKey), as shown in Figure 2a, Marmot decomposes the RecordSetFunction into the mapping operator and reducing operator, and eventually transforms the Plan into MapReduce jobs consisting of map and reduce phases, as shown in Figure 2b. During this transformation, Marmot controls the number of MapReduce phases in a way to achieve better performance by decreasing the overhead of mapping and reducing. To describe how Marmot handles such processes in detail, an example of spatial analysis to retrieve subway stations in a city is shown in Figures 3 and 4.

```
Plan plan;
plan = marmot.planBuilder("Subway stations per city")
            .load("logs/subway stations")
            .update("the_geom=ST_Centroid(the_geom)")
            .spatialJoin("the_geom", "region/cadastral", "the_geom",
                     INTERSECTS, "*, param.sig_cd")
            .reduceByGroupKey("sig_cd")
                .aggregate(COUNT())
            .storeAsCsv("result")
            .build();
```

Figure 3. An example code for searching subway stations per city.

Figure 3 is a Marmot code for an example of spatial analysis. The analysis is represented as a Plan consisting of five RecordSetOperators: Load, Update, SpatialJoin, ReduceByGroupKey, and StoreAsCsv. As shown in Figure 4, using the Load operator, Marmot reads the boundaries of each subway station and computes their center coordinates. The calculated center points are then utilized as the representative locations of each subway station via the Update operator. For each subway station, using the SpatialJoin operator, Marmot identifies the city that is the center point of the subway station. Finally, the number of subway stations per city is calculated via the ReduceByGroupKey operator and the results are stored in a CSV file named "result" via the StoreAsCsv operator.

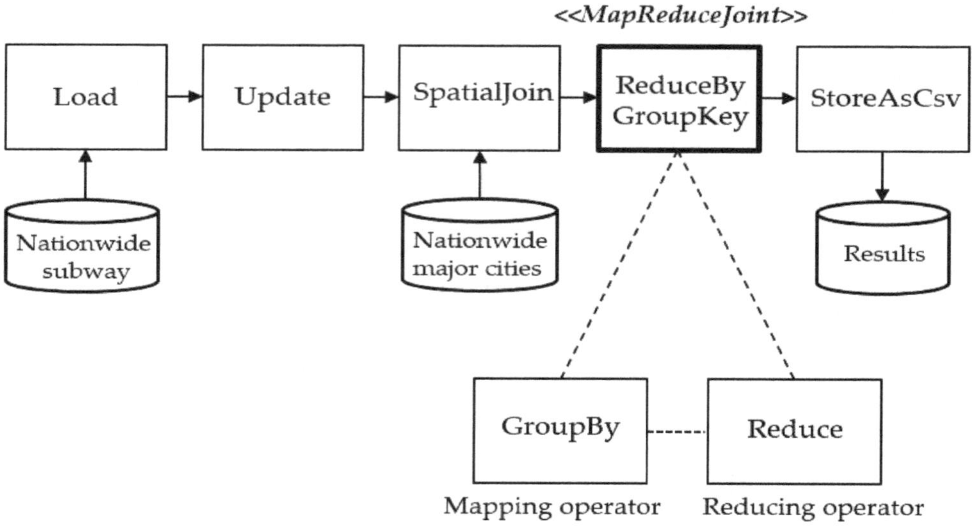

Figure 4. An example Plan for searching subway stations per city.

During the process of transforming the Plan to a sequence of MapReduce jobs, ReduceByGroupKey is decomposed into GroupBy and Reduce as a mapping operator and a reducing operator, respectively. Accordingly, Load, Update, SpatialJoin, and GroupBy are executed during the Map phase; Reduce and StoreAsCsv, during the Reduce phase.

3. Our MapReduce-Based D_ELT Framework

As mentioned in the previous section, we constructed the Marmot, high-performance data management system that enables developers with no specific knowledge of big data technologies to implement improved performance spatial analysis applications to geospatial big data. The issues concerning geospatial big data, however, lie not only in how to efficiently manage the data for fast analysis, but also in how to efficiently transform the data for fast data preparation.

DTG data, for example, have been used to analyze the status of transportation operations to identify improvement points and to identify disadvantaged areas in terms of public transportation. Transportation authorities, e.g., the Korea Transportation Safety Authority, collect DTG data from commercial vehicles and apply analytics to such big data to extract insights and facilitate decision making. Often, the results of data analysis must be derived periodically within a specific time, e.g., every single day, to be prepared for emergent cases. In this situation, to complete the given analysis in time, not only the data analysis speed, but also the data preparation speed is a critical factor affecting the overall performance. In the IoT environment, update latency for sensor big data, the focus of this paper among various sources of geospatial big data, is typically short and old data are not worth further analysis, making data preparation speed even more important. Moreover, sensor big data is machine-generated; therefore, the source data contains more noise or errors compared to human-generated data, complicating data preparation even more.

Traditional ETL [18–20] can no longer accommodate such situations. The ETL is designed for light-weight computations on small data sets, but is not capable of efficiently handling massive amounts of data. Figure 5a describes the data preparation and analysis in the ETL process. In this approach, data are extracted from various sources and then transformed on an ETL server, which is typically one machine, and loaded into a Hadoop distributed file system (HDFS). The loaded data are finally analyzed in a big data platform for decision-making. In this approach, an *analysis* operation is processed in a parallel/distributed way using MapReduce [21,22], which guarantees reasonable performance, but bottlenecks can occur during a *transform* operation. In fact, *transform* is the most time consuming phase in ETL because this operation includes filtering or aggregation of source data to fit the structure of the target database. Data cleaning should also be completed for any duplicated data, missing data, or different data formats. Moreover, in big data environments, due to heterogeneous sources of big data, the traditional *transform* operation will create even more computational burdens. The overall performance of the ETL processes, therefore, depends mainly on how efficiently the *transform* operation is conducted.

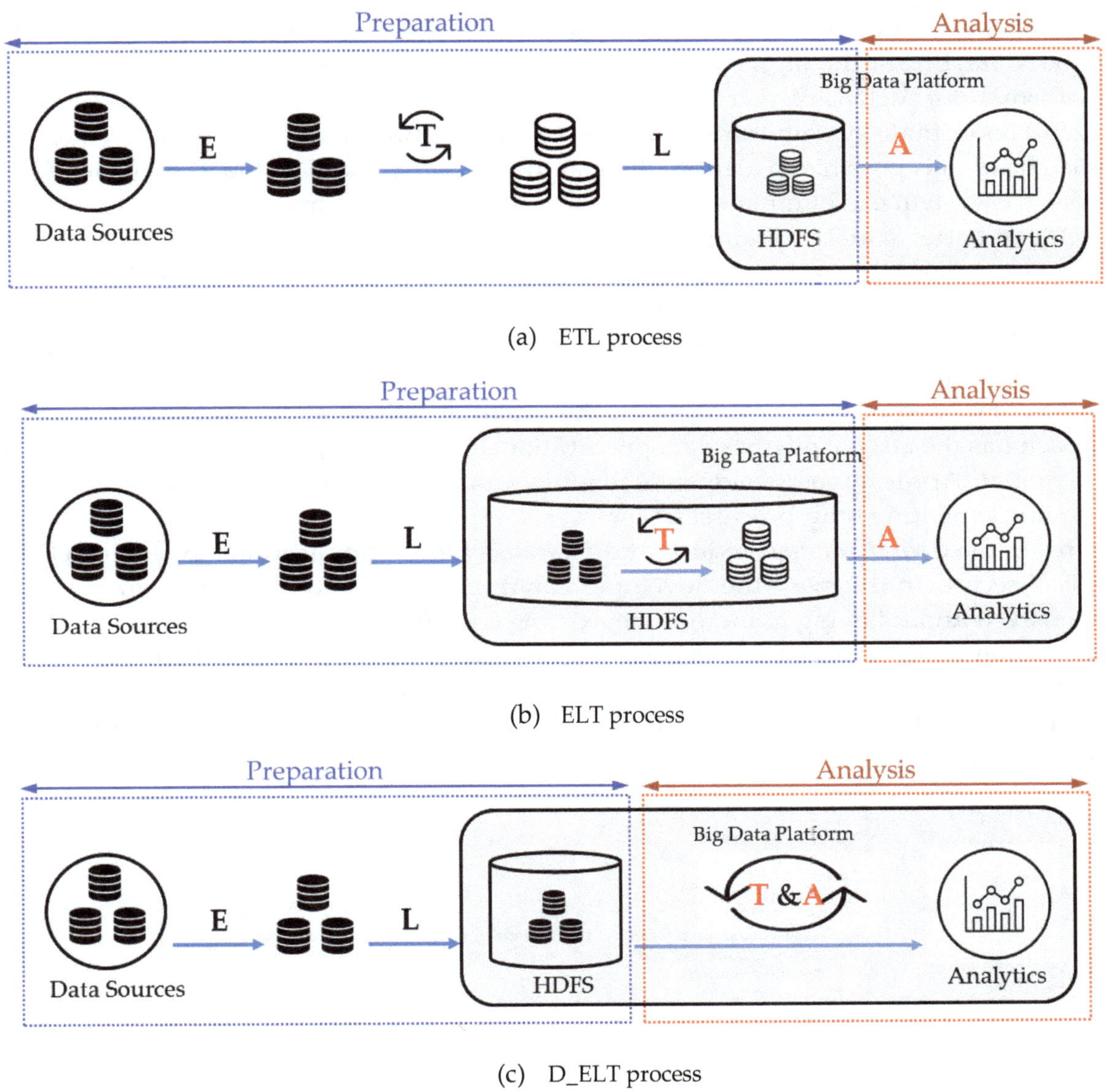

(a) ETL process

(b) ELT process

(c) D_ELT process

Figure 5. Illustration of geospatial big data preparation and analysis processes comparing three cases: (**a**) ETL; (**b**) ELT; (**c**) D_ELT. In the figure, "E" stands for extract, "T" stands for transform, "L" stands for load, and "A" stands for analysis.

To overcome the drawbacks of traditional ETL and to speed up the data preparation process, the processes of ELT was devised [23–25]. The nature of traditional ETL is to perform *transform* immediately after the *extract* operation and then start the *load* operation. In contrast, the basic idea of ELT is to conduct the *load* operation immediately after the *extract* operation, and perform the *transform* after storing the data in the HDFS, as shown in Figure 5b. This approach has several advantages over ETL. The *transform* operation can be done at the run time when needed and it is possible to use *transform* even multiple times to handle changing requirements for data. In addition, this approach eliminates a separate transformation engine, the ETL server, between the source and target and makes the overall system less costly. Above all, ELT allows raw source data to be loaded directly into the target and also leverages the target system to perform the *transform* operation. In that sense, ELT can speed up *transform* using parallelization/distribution supported in the Hadoop-based big data platform.

Despite these advantages, ELT still has limitations in handling big data. The ELT framework can speed up *transform* using MapReduce, but *analysis* is initiated only after the *transform* has been completed. In this approach, it is difficult to optimize *transform* in conjunction with *analysis* because the *transform* is performed in a batch regardless of the context of *analysis*. For example, in the case of geospatial data, one of the high computational overheads in conducting *transform* occurs during type transformation, such as converting the x–axis and y–axis of plain-text into (x,y) coordinates of the point and coordinate system transformation for conducting spatial analysis. If *analysis* does not require such tasks, it is possible to identify them at the *transform* phase and load only the required data. By doing so, the system can eliminate unnecessary transformations and speed up performance.

To achieve better scalability and performance in conducing *transform* on geospatial big data, this paper offers a new approach for data preparation called D_ETL—in the sense that the decision of how to perform *transform* is delayed until the context of *analysis* is understood. As shown in Figure 5c, in our approach, *transform* is executed in parallel/distributed with *analysis* within our geospatial big data platform, Marmot. In Marmot, the operators for *transform* are considered a type of RecordSetOperator and are also composed of a Plan, along with the existing RecordSetOperator designed for *analysis*. This approach has the advantage that data preparation and analysis processes are described using the same data model. Application developers, therefore, can be free from the inconvenience of having to get used to implementing both processes.

Regarding the operators required to conduct *transform*, the application developer specifies them in the D_ELT script. In this way, the developer can implement both data preparation and analysis simultaneously, without having to modify the existing code for conducting *analysis*. The D_ELT script consists of the names of operators and a list of the key-values of the parameters, as shown in Figure 6. For convenience, if a developer needs a new operator for conducting *transform*, the operator can be separately implemented as a form of plug-in and can be used in Marmot, in the same way as for existing operators.

```
{
 "name": "import_plan",
 "operator": [{
  "parseCsv": {
   "delimiter": ",",
   "options": {
    "headerColumn": ["car_no", "ts", "month", "sid_cd", "besselX", "besselY",
    "status", "company", "driver_id", "xpos", "ypos"],
    "commentMarker": "#"
   }
  }
 }, {
  "expand": {
   "column": [{
    "name": "status"
   }]
  }
 }, {
  "toPoint": {
   "xColumn": "xpos",
   "yColumn": "ypos",
   "outColumn": "the_geom"
  }
 }, {
  "transformCrs": {
   "geometryColumn": "the_geom",
   "sourceSrid": "EPSG:4326",
   "targetSrid": "EPSG:5186"
  }
 }, {
  "project": {
   "columnExpr": "the_geom,*-
{the_geom,xpos,ypos,besselX,besselY,month,sid_cd}"
  }
 }]
}
```

Figure 6. An example of D_ELT (delayed extracting–loading –transforming) script describing operators required for data transformation.

To perform a spatial analysis, Marmot first loads the D_ELT script to determine what operators need to be executed for *transform*. Then, Marmot (1) examines the operators needed to be executed for *analysis*, (2) loads only the required data based on the need of *analysis*, and (3) executes both *transform* and *analysis* in a parallel distributed way. At this time, part of the transformed data can be used for *analysis* and not have to wait for all the data to finish being transformed. Figure 7 shows the sequence of operators executed for *transform* and *analysis* and their composition as a form of a plan. In this example Plan, "ParseCSV" is the operator for *transform* and "Filter" is the operator for *analysis*. They are allocated in the Map phase and executed in a parallel distributed way. The outputs from the Map phase are combined during the Reduce phase and the results are written in the output file.

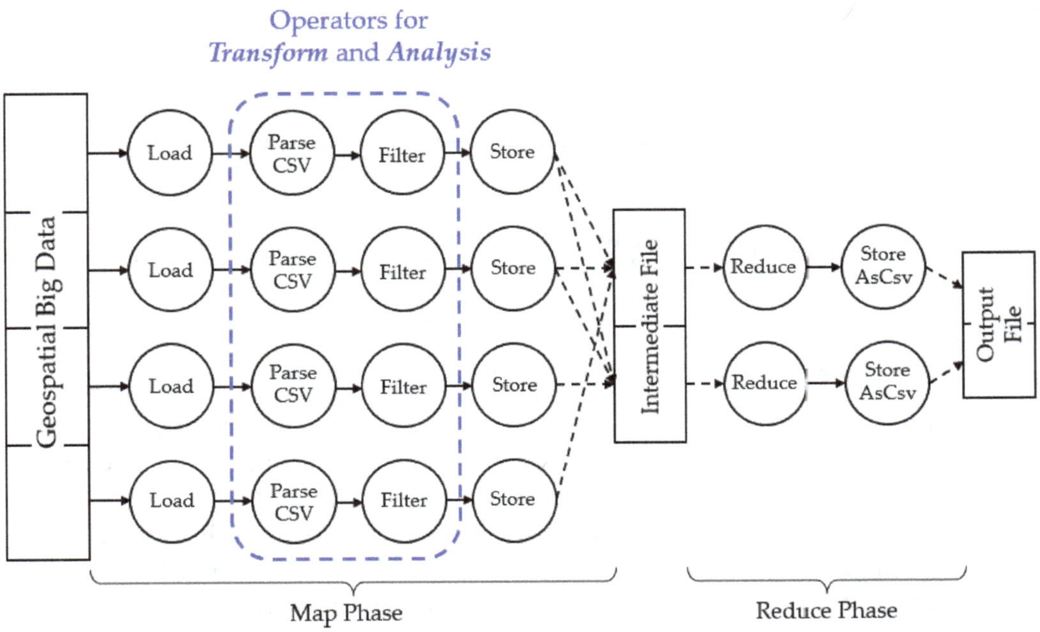

Figure 7. Illustration of the Map and Reduce phases during the D_ELT process.

The reason why we implemented D_ELT using MapReduce instead of Spark, another well-known engine for big data processing, is that our previously developed geospatial big platform is based on Hadoop and we had the goal of improving the data transformation time in that environment. In addition, the data we are currently handling is a large amount of DTG data, generating 20–30 TB every month. Using Spark, when running spatial analysis based on this large size of data, we anticipated that unexpected problems may occur (e.g., disk swapping), but to our knowledge, concrete solutions have not yet been proposed.

It is also important to note that ELT and D_ELT are identical in terms of performing data transformation during the MapReduce phase in Hadoop. The difference between ELT and D_ELT is as follows. In ELT, once raw data are uploaded to Hadoop, the data are transformed using MapReduce. After completely finishing the transformation, analysis is then started using another MapReduce. In D_ELT, however, data transformation is not conducted, although all of the raw data are uploaded to Hadoop but delayed until the time of conducting the analysis. That is, the transformation task is piggybacked onto the analysis task and both tasks are performed together using the same MapReduce. In this way, part of the transformed data can be used for analysis immediately without having to wait for all the data to be transformed.

4. Experimental Evaluation

This section explains our evaluation of the improvement in performance achieved by our proposed approach, D_ELT. In addition, the scalability of three different approaches (traditional ETL, ELT, and D_ELT) were measured and compared by varying data size and levels of analysis complexity.

4.1. Experimental Setup

Our experiments were conducted on the four nodes of a Hadoop cluster. Each node was a desktop computer with a 4.0 GHZ Intel 4 core i7 CPU, a 32 GB main memory, and a 4 TB disk. The operating system was CentOS 6.9 and the Hadoop version was Hortonworks HDP 2.6.1.0 with Ambari 2.5.0.3. PostgreSQL 9.5 was used for the database management system along with Oracle JDK 1.8. The 2.7.3 version of MapReduce2 was used.

The test data used in the experiment were DTG data installed in vehicles, which record the driving record in real time. The structure of the data consisted of timestamp, vehicle number, daily mileage, accumulated mileage, speed, acceleration, RPM, brake, x_position, y_position, and angle. The data were classified into three different sizes: small, 9.9GB; medium, 19.8 GB; and large, 29.8 GB, as shown in Table 1. For the geospatial big data platform, we used our developed system Marmot, as explained in Section 2.

Table 1. Data size: small, medium, and large.

	Data Size
Small	9.9 GB
Medium	19.8 GB
Large	29.8 GB

4.2. Experiment 1: Measurement of Data Preparation Time

In this experiment, we compared the data preparation time of ETL and ELT to our proposed D_ELT, and the scalability of each approach based on the different data size. The overall results from this experiment are presented in Table 2.

As shown in Figure 5 in Section 3, the total time for data preparation in the ETL process includes time for extracting, transforming, and loading. In the case of ELT, the total time spent on data preparation is the summation of times for extracting, loading, and transforming. While in the ETL process, *transform* was conducted on a single machine, which is based on non-MapReduce and *transform* in the ELT was performed in a parallel distributed way based on MapReduce.

Table 2. Data preparation time (in seconds): ETL, ELT, and D_ELT.

	ETL [1]	**ELT** [2]	**D_ELT** [3]
Small	579	413	116
Medium	1158	808	231
Large	1727	1175	345

[1] Data preparation time in ETL: E+T+L, where E for extract, T for transform, L for load; [2] Data preparation time in ELT: E+L+T, where T is executed in a parallel distributed way; [3] Data preparation time in D_ELT: E+L.

In the case of D_ELT, the total time spent on data preparation is the summation of times only for extracting and loading, but does not include the time for transforming. *Transform* was simultaneously executed along with *analysis* in the data analysis phase, and so that the data preparation in D_ELT does not include *transform* but only *extract* and *load*.

4.3. Experiment 2: Measurement of Data Analysis Time

In this experiment, we compared the data analysis time of ETL, ELT, and our proposed D_ELT, and the scalability of each approach based on the different data size. Additionally, this experiment also included optimized D_ELT, D_ELT_Opt, conducting data analysis by filtering and using only the required data. In order to see the variation in performance according to the different complexity levels of analysis, we utilized three analyses—Count, GroupBy, and SpatialJoin—for low, middle, and high-level complex analysis, respectively. The overall results from this experiment are presented in Table 3.

Table 3. Data analysis time (in seconds): ETL(or ELT), D_ELT, and optimized D_ELT.

		ETL(or ELT) [1]	D_ELT [2]	D_ELT_Opt [3]
Count	Small	68	96	57
	Medium	126	179	104
	Large	181	257	143
GroupBy	Small	76	98	87
	Medium	139	179	162
	Large	203	256	233
SpatialJoin	Small	391	406	406
	Medium	772	806	806
	Large	1087	1190	1148

[1] Data analysis time in ETL or ELT: A, where A is executed in parallel; [2] Data analysis time in D_ELT: T+A, T, where T, A are executed in parallel; [3] Data analysis time in optimized D_ELT: T+A, where T, A are executed in parallel using only the required data.

As shown in Figure 5 in Section 3, the total time for data analysis in ETL includes only the *analysis*, which is conducted in a parallel distributed way using MapReduce. In the case of ELT, once the data preparation is completed, the *analysis* will be conducted in the same way as for ETL. In the cases of D_ELT and optimized D_ELT, the total time spent on data analysis is the time required to execute *transform* and *analysis* in a parallel distributed way using MapReduce.

5. Results and Discussion

5.1. Results

The first experiment for measuring the data preparation time for each approach reveals the following points. As shown in Figure 8, D_ELT is about 5 times faster than ETL (116 sec vs. 579 sec for small data; 231 sec vs.1158 sec for medium data; 345 sec vs. 1727 sec for large data) and about 3 times faster than ELT (116 sec vs. 413 sec for small data; 231 sec vs. 808 sec for medium data; 345 sec vs. 1175 sec for large data), regardless of the data size. The ELT approach is about 1.4 times faster than ETL. This is because of the parallel distributed processing effect using Marmot's MapReduce when performing *transform* in ELT.

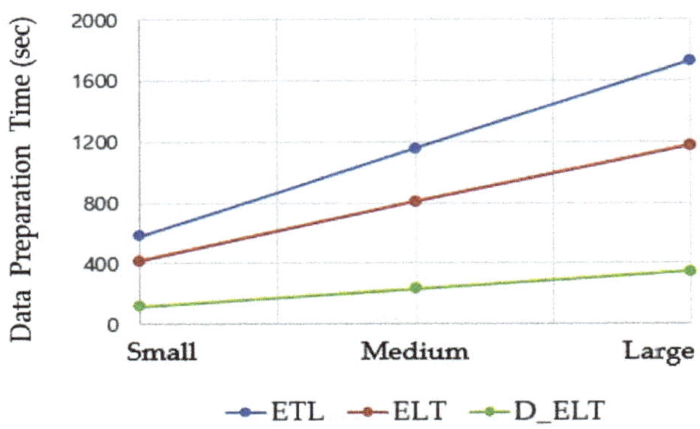

Figure 8. Data preparation time (in seconds): ETL, ELT, and D_ELT.

The data analysis time is measured by the second experiment and reveals the following points. Table 4 compares the performance between D_ELT and ETL(or ELT) and optimized D_ELT and ETL(or ELT). In both cases, the ratio of performance to analysis is almost identical regardless of the data size. An interesting point is that the data analysis time in the D_ELT process contains time for *transform*, while the ETL(or ELT) process does not include this time. Although D_ELT is slower than ETL(or ELT), there is little difference in performance—D_ELT is up to 1.4 times slower. In the case of optimized D_ELT, the process is only up to 1.2 times slower than the ETL(or ELT) approach. In D_ELT, in the case of simple analysis, the time involved in data transforming is relatively large compared to the analysis time and consumes a large part of the total execution time. However, in the case of complex spatial analysis, the time involved in data transforming is relatively small compared to data analysis, and so the transformation overhead incurred is relatively small.

Table 4. Performance comparison: D_ELT/ETL(or ELT) and optimized D_ELT/ETL(or ELT).

		D_ELT/ETL(or ELT)	D_ELT_Opt/ETL(or ELT)
Count	Small	1.41	0.84
	Medium	1.42	0.83
	Large	1.42	0.79
GroupBy	Small	1.29	1.14
	Medium	1.29	1.17
	Large	1.26	1.15
SpatialJoin	Small	1.04	1.04
	Medium	1.04	1.04
	Large	1.09	1.06

It is important to note that these have no effect on overall performance degradation, considering that D_ELT is about 3–5 times faster than ETL(or ELT) during data preparation, as shown in Table 2 in Section 4 and Figure 8. Therefore, the performance of both D_ELT and optimized D_ELT is much greater than that of ETL or ELT.

Figure 9 compares the performance between D_ELT and ETL(or ELT) during data analysis according to the different data size and analysis type. As aforementioned, the analysis time in the D_ELT includes *transform* as opposed to ETL(or ELT), and so D_ELT is slower than ETL(or ELT), as shown in Table 3 in Section 4. However, the higher the complexity of the analysis (Count < GroupBy < SpatialJoin), the smaller the difference between the D_ELT and the ETL(or ELT) performance. The reason is that the higher the complexity of analysis, the higher the effect of parallelizing the *transform* of D_ELT, thereby enhancing the performance of D_ELT.

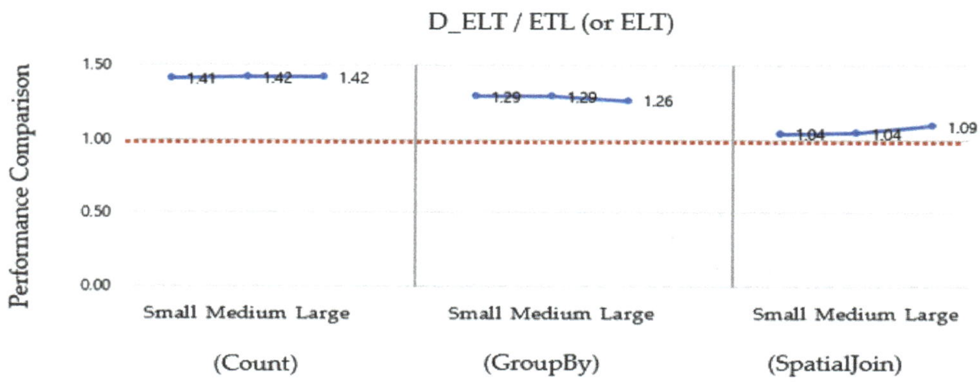

Figure 9. Performance comparison of D_ELT/ETL(or ELT) based on the small, medium, and large data size for each of three analyses: Count, Group-By, and SpatialJoin.

Similarly, Figure 10 compares the performance between optimized D_ELT and ETL(or ELT) during data analysis, according to different data size and analysis type. Compared to Figure 9, in the case of two simple analysis cases, Count and GroupBy, optimized D_ELT is faster than D_ELT. This is because for simple analysis, large amounts of data are often unrelated to the analysis, and so more data can be included in the optimization target, resulting in an incremental increase in performance of optimized D_ELT.

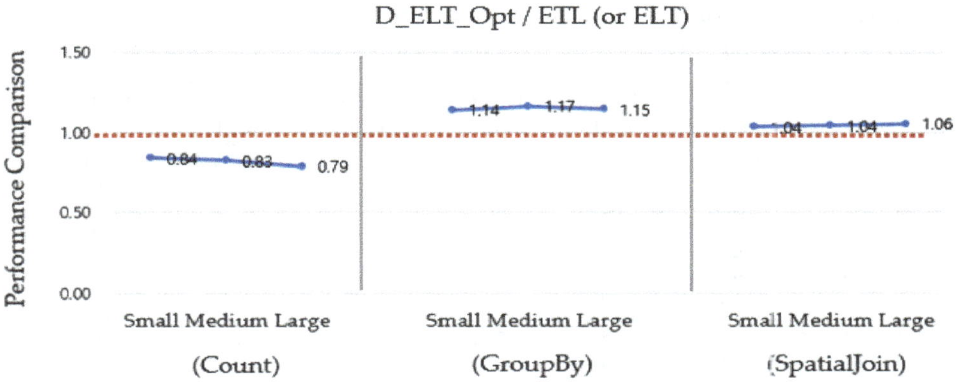

Figure 10. Performance comparison of optimized D_ELT/ETL(or ELT) based on the small, medium, and large data size for each of three analyses: Count, Group-By, and SpatialJoin.

In both cases, the performance ratio of the analysis is very similar regardless of data size. Thus, we chose only the small data size to compare the performance ratio, as shown in Figure 11. This shows that the higher the complexity of the analysis, the smaller the performance difference between D_ELT and optimized D_ELT. This is because the more complex the analysis, the more data is involved in the analysis, which reduces the scope of optimization. In the case of SpatialJoin, which has the highest complexity among the three analyses, the two values in Figure 11 converge to almost 1.0, showing that there is almost no performance difference between D_ELT and optimized D_ELT.

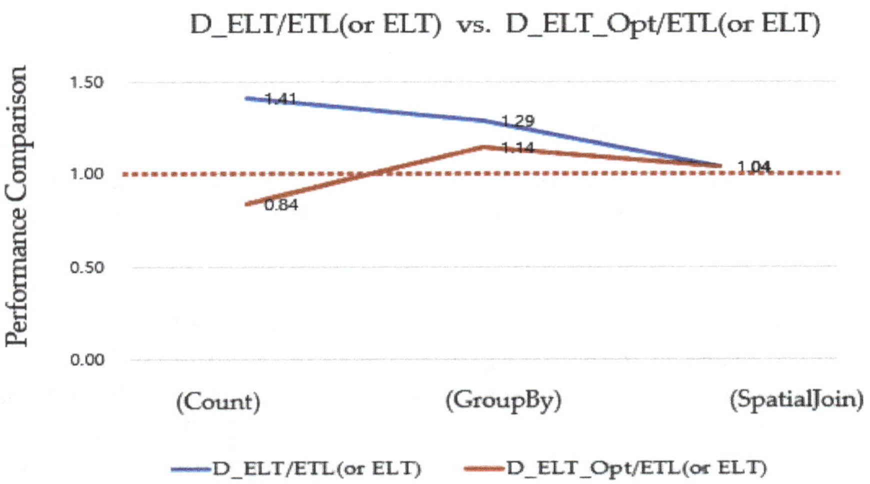

Figure 11. Performance comparison of D_ELT/ETL(or ELT) vs. optimized D_ELT/ETL(or ELT) based on the small data size for each of three analyses: Count, Group-By, and SpatialJoin.

The overall performance of ETL, ELT, D_ELT, and optimized D_ELT is derived by summing the data preparation and analysis times. Table 5 shows that the overall performance of D_ELT is much faster than that of the ETL or ELT approaches. D_ELT is up to 3 times faster than ETL and 2 times faster than ELT. Optimzed D_ELT is up to 4 times faster than ETL and 3 times faster than ELT. The results are derived from the two simple analysis cases, Count and GroupBy, but not SpatialJoin. In the case of SpatialJoin, both D_ELT and optimized D_ELT still perform better than ETL or ELT, but there is almost no difference between the overall performance of D_ELT and optimized D_ELT. Figure 12 shows that as the complexity of the analysis is increased, the gap between D_ELT and optimized D_ELT is decreased.

Table 5. Overall performance of ETL, ELT, D_ELT, optimized D_ELT (in seconds), and performance comparison among ELT vs. ETL, D_ELT vs. ETL, and optimized D_ELT vs. ETL.

		ETL (sec)	ELT (sec)	D_ELT (sec)	D_ELT_Opt (sec)	ELT/ ETL	D_ELT/ ETL	D_ELT_Opt/ ETL
Count	Small	647	481	212	173	0.74	0.33	0.27
	Medium	1284	934	410	335	0.73	0.32	0.26
	Large	1908	1356	602	488	0.71	0.32	0.26
GroupBy	Small	655	489	214	203	0.75	0.33	0.31
	Medium	1297	947	410	393	0.73	0.32	0.30
	Large	1930	1378	601	578	0.71	0.31	0.30
SpatialJoin	Small	970	804	522	522	0.83	0.54	0.54
	Medium	1930	1580	1037	1037	0.82	0.54	0.54
	Large	2814	2262	1535	1493	0.80	0.55	0.53

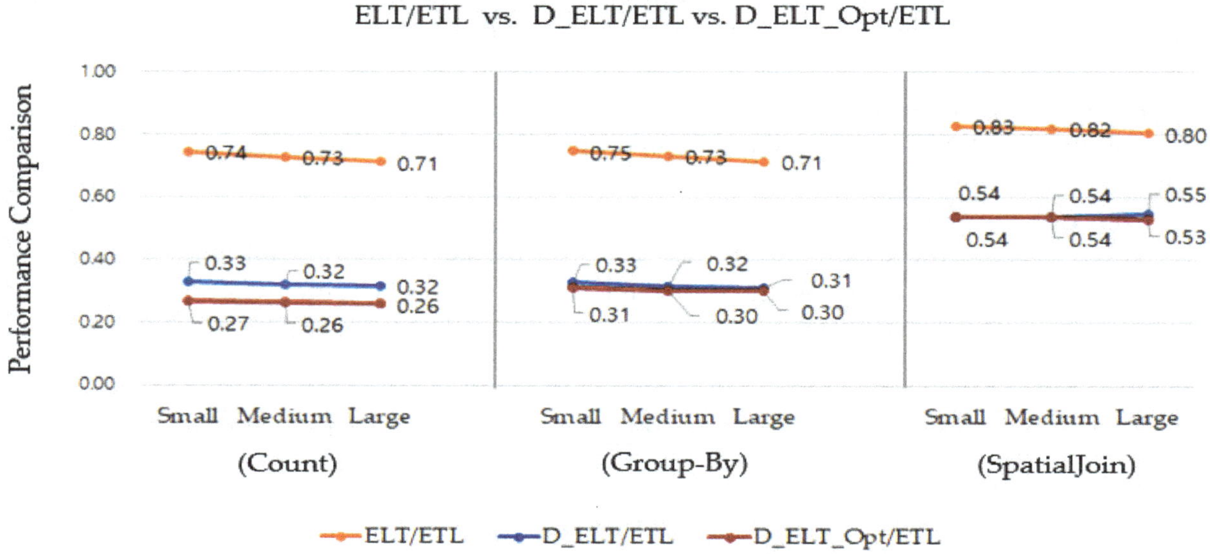

Figure 12. Overall performance comparison of ELT/ETL vs. D_ELT/ETL vs. optimized D_ELT/ETL based on a small, medium, and large data size for each of three analyses: Count, Group-By, and SpatialJoin.

5.2. Discussion

There are two conventional methods—ETL and ELT. The traditional ETL method does not use the distributed/parallel method during data pre-processing, causing problems especially when the volume of data to be pre-processed is large. The ELT method improves traditional ETL methods to speed up data pre-processing using the distributed/parallel method. Our proposed D_ELT method reduces overhead in data pre-processing. In D_ELT, the transformation task is piggybacked onto the analysis task and both tasks are performed together using the same MapReduce. This way allows one to conduct the analysis immediately without storing transform results and also excludes unnecessary transformations that are not utilized in the analysis.

Compared to existing methods, however, the D_ELT method significantly reduces the data preparation time, but has the disadvantage in the following cases. First, the case that the same kind of analysis must be conducted repetitively. For example, the D_ELT method results in a 1382-second reduction (large data, Table 2) in data preparation time compared to that of the conventional ETL method, but 103 seconds is added (large data, SpatialJoin, Table 3) every time an analysis is conducted. Therefore, the greater the number of conducting analyses, the more inefficient D_ELT is compared to traditional methods. In the example above, the D_ELT method is more inefficient than the existing method when the same analysis is conducted more than 14 times in succession. Second, in the case that a large amount of input data is invalid, a large amount of data can be removed as a result of the transform. In D_ELT, the transformation task is piggybacked every time an analysis task is executed, a large amount of invalid data is repeatedly read, resulting in unnecessary I/O and computation burden. Finally, the method proposed in this paper does not consider real-time applications. However, it provides the advantage that required analysis results can be obtained relatively more quickly than other conventional methods.

6. Conclusions

This paper presents our proposed D_ELT approach to efficiently transform and analyze data, thereby making it usable for a large amount of sensor big data, especially geospatial big data. Based on the experimental results, we made several observations as follows. First, D_ELT outperforms ETL and ELT during data preparation. Second, D_ELT shows performance degradation during data analysis. However, the higher the complexity of the analysis, the smaller the performance degradation, resulting in overall improved performance compared to ETL or ELT. Finally, in the case of simple analysis increasing the scope of optimization, optimized D_ELT outperforms ELT. In the future, we plan to further increase the overall performance of our developed system including D_ELT and Marmot by investigating the spatial index, to better support spatial queries in dealing with geospatial big data.

Author Contributions: K.-W.L. designed and implemented the D_ELT; K.-W.L. and J.J. conducted the testing of D_ELT and analyzed the experiment results; J.J. wrote and K.-W.L. revised the manuscript.

Acknowledgments: This research, 'Geospatial Big data Management, Analysis and Service Platform Technology Development', was supported by the MOLIT (The Ministry of Land, Infrastructure and Transport), Korea, under the national spatial information research program supervised by the KAIA(Korea Agency for Infrastructure Technology Advancement) (19NSIP-B081011-06).

References

1. Li, S.; Dragicevic, S.; Castro, F.A.; Sester, M.; Winter, S.; Coltekin, A.; Pettit, C.; Jiang, B.; Haworth, J.; Stein, A.; et al. Geospatial big data handling theory and methods: A review and research challenges. *ISPRS J. Photogramm. Remote Sens.* **2016**, *115*, 119–133. [CrossRef]
2. Morais, C.D. Where Is the Phrase "80% of Data is Geographic?". Available online: http://www.gislounge.com/80-percent-data-is-geographic (accessed on 4 April 2018).
3. Jeansoulin, R. Review of forty years of technological changes in geomatics toward the big data paradigm. *ISPRS Int. J. Geo-Inf.* **2016**, *5*, 155. [CrossRef]
4. He, Z.; Liu, Q.; Deng, M.; Xu, F. Handling multiple testing in local statistics of spatial association by controlling the false discovery rate: A comparative analysis. In Proceedings of the IEEE 2nd International Conference 2017Big data Analysis (ICBDA), Beijing, China, 10–12 March 2017; pp. 684–687.
5. Liu, P.; Di, L.; Du, Q.; Wang, L. Remote Sensing Big data: Theory, Methods and Applications. *Remote Sens.* **2018**, *10*, 711. [CrossRef]
6. Chen, P.; Shi, W. Measuring the Spatial Relationship Information of Multi-Layered Vector Data. *ISPRS Int. J. Geo-Inf.* **2018**, *7*, 88. [CrossRef]
7. White, T. *Hadoop: The Definitive Guide*, 3rd ed.; O'Reilly Media, Inc.: Sebastopol, CA, USA, 2012; ISBN 1449338771.
8. Zaharia, M.; Chowdhury, M.; Franklin, M.J.; Shenker, S.; Stoica, I. *Spark: Cluster Computing with Working Sets*; HotCloud: Boston, MA, USA, 22 June 2010.
9. Eldawy, A. SpatialHadoop: Towards flexible and scalable spatial processing using MapReduce. In Proceedings of the SIGMOD PhD symposium 2014, Snowbird, UT, USA, 22 June 2014; pp. 46–50.
10. Yu, J.; Wu, J.; Sarwat, M. Geospark: A cluster computing framework for processing large-scale spatial data. In Proceedings of the 23rd SIGSPATIAL International Conference on Advances in Geographic Information Systems, Bellevue, WA, USA, 3–6 November 2015; p. 70.
11. Jo, J.; Lee, K.W. High-Performance Geospatial Big data Processing System Based on MapReduce. *ISPRS Int. J. Geo-Inf.* **2018** *7*, 399. [CrossRef]
12. Sabtu, A.; Azmi, N.F.M.; Sjarif, N.N.A.; Ismail, S.A.; Yusop, O.M.; Sarkan, H.; Chuprat, S. The challenges of extract, transform and loading (ETL) system implementation for near real-time environment. In Proceedings

of the 2017 International Conference on Research and Innovation in Information Systems (ICRIIS) 2017, Langkawi, Malaysia, 16–17 July 2017; pp. 1–5.

13. Bala, M.; Boussaid, O.; Alimazighi, Z. A Fine Grained Distribution Approach for ETL Processes in Big data Environments. *Data Knowl. Eng.* **2017**, *111*, 114–136. [CrossRef]
14. Liu, X.; Thomsen, C.; Pedersen, T.B. ETLMR: A highly scalable dimensional ETL framework based on MapReduce. In *Transactions on Large-Scale Data-and Knowledge-Centered Systems VIII*; Springer: Berlin/Heidelberg, Germany, 2013; pp. 1–31.
15. Misra, S.; Saha, S.K.; Mazumdar, C. Performance Comparison of Hadoop Based Tools with Commercial ETL Tools-A Case Study. In Proceedings of the International Conference on Big Data Analytics, Mysore, India, 16–18 December 2013; pp. 176–184.
16. Bala, M.; Boussaid, O.; Alimazighi, Z. P-ETL: Parallel-ETL based on the MapReduce paradigm. In Proceedings of the 2014 IEEE/ACS 11th International Conference on Computer Systems and Applications (AICCSA), Doha, Qatar, 10–13 November 2014; pp. 42–49.
17. Marmot from GitHub. Available online: https://github.com/kwlee0220/marmot.server.dist (accessed on 23 September 2019).
18. Trujillo, J.; Lujan-Mora, S. A UML based approach for modeling ETL processes in data warehouses. In *Conceptual Modeling—ER 2003, Proceedings of the International Conference on Conceptual Modeling, Chicago, IL, USA, 13–16 October 2003*; Springer: Berlin/Heidelberg, Germany, 2003; pp. 307–320.
19. El Akkaoui, Z.; Zimanyi, E. Defining ETL worfklows using BPMN and BPEL. In *DOLAP '09, Proceedings of the ACM Twelfth International Workshop on Data Warehousing and OLAP, Hong Kong, China, 6 November 2009*; ACM: New York, NY, USA, 2009; pp. 41–48.
20. Thomsen, C.; Bach Pedersen, T. pygrametl: A powerful programming framework for extract-transform-load programmers. In *DOLAP '09, Proceedings of the ACM Twelfth International Workshop on Data Warehousing and OLAP, Hong Kong, China, 6 November 2009*; ACM: New York, NY, USA, 2009; pp. 49–56.
21. Zheng, L.; Sun, M.; Luo, Y.; Song, X.; Yang, C.; Hu, F.; Yu, M. Utilizing MapReduce to Improve Probe-Car Track Data Mining. *ISPRS Int. J. Geo-Inf.* **2018**, *7*, 287. [CrossRef]
22. Yao, X.; Mokbel, M.; Ye, S.; Li, G.; Alarabi, L.; Eldawy, A.; Zhao, Z.; Zhao, L.; Zhu, D. LandQv2: A MapReduce-based system for processing arable land quality big data. *ISPRS Int. J. Geo-Inf.* **2018**, *7*, 271. [CrossRef]
23. Cohen, J.; Dolan, B.; Dunlap, M.; Hellerstein, J.M.; Welton, C. MAD skills: New analysis practices for big data. *Proc. VLDB Endow.* **2009**, *2*, 1481–1492. [CrossRef]
24. Devi, P.S.; Rao, V.V.; Raghavender, K. Emerging Technology Big data-Hadoop over Datawarehousing ETL. In Proceedings of the International Conference (IRF), Pretoria, South Africa, 2–4 September 2014; pp. 30–34.
25. Storey, V.C.; Song, I.Y. Big data technologies and management: What conceptual modeling can do. *Data Knowl. Eng.* **2017**, *108*, 50–67. [CrossRef]

Geographic Knowledge Graph (GeoKG): A Formalized Geographic Knowledge Representation

Shu Wang [1,2,3], Xueying Zhang [1,2,3,*], Peng Ye [1,2,3], Mi Du [1,2,3], Yanxu Lu [1,2,3] and Haonan Xue [1,2,3]

1 Key Laboratory of Virtual Geographic Environment (Nanjing Normal University), Ministry of Education, Nanjing 210023, China; 141301020@stu.njnu.edu.cn (S.W.); 161301027@stu.njnu.edu.cn (P.Y.); 171301021@stu.njnu.edu.cn (M.D.); 171302083@stu.njnu.edu.cn (Y.L.); 181302106@stu.njnu.edu.cn (H.X.)

2 State Key Laboratory Cultivation Base of Geographical Environment Evolution (Jiangsu Province), Nanjing 210023, China

3 Jiangsu Center for Collaborative Innovation in Geographical Information Resource Development and Application, Nanjing 210023, China

* Correspondence: zhangxueying@njnu.edu.cn

Abstract: Formalized knowledge representation is the foundation of Big Data computing, mining and visualization. Current knowledge representations regard information as items linked to relevant objects or concepts by tree or graph structures. However, geographic knowledge differs from general knowledge, which is more focused on temporal, spatial, and changing knowledge. Thus, discrete knowledge items are difficult to represent geographic states, evolutions, and mechanisms, e.g., the processes of a storm "{9:30-60 mm-precipitation}-{12:00-80 mm-precipitation}- . . . ". The underlying problem is the constructors of the logic foundation (ALC description language) of current geographic knowledge representations, which cannot provide these descriptions. To address this issue, this study designed a formalized geographic knowledge representation called GeoKG and supplemented the constructors of the ALC description language. Then, an evolution case of administrative divisions of Nanjing was represented with the GeoKG. In order to evaluate the capabilities of our formalized model, two knowledge graphs were constructed by using the GeoKG and the YAGO by using the administrative division case. Then, a set of geographic questions were defined and translated into queries. The query results have shown that GeoKG results are more accurate and complete than the YAGO's with the enhancing state information. Additionally, the user evaluation verified these improvements, which indicates it is a promising powerful model for geographic knowledge representation.

Keywords: geographic knowledge representation; geographic knowledge graph; formalization; GeoKG

---

1. Introduction

Geographic knowledge consists of the product of geographic thinking and reasoning about the world's natural and human phenomena, which plays an important role in geographic studies and applications [1]. Nearly every geographer is trying to answer the question of "how to perceive, understand and organize geographic knowledge scientifically." [2] Generally, geographic knowledge representation is a type of human expression of the real world that is of great importance to storage and computation [3]. Especially in the era of Big Data, well-structured geographic knowledge is a benefit to all kinds of geospatial applications, because formalization is the foundation of geospatial big data computing, mining, and visualization.

At present, the most popular knowledge representation is the knowledge graph. It organizes knowledge with a set of concepts, relations, and facts, which are associated by two types {entity, relation, entity} and {entity, attribute, attribute value} [4]. There are only three basic elements in knowledge graphs: the entity, relation, and attribute. These three elements can explicitly represent general information, such as "when did the Beijing storm occur on 21 July—9:30, 21 July". However, geographic knowledge is more complicated than general knowledge. More processes and evolutions need to be answered, e.g., "what caused the 7·21 Beijing storm", "how did it develop", and "what were the effects of the 7·21 Beijing storm". Entities, relations, and attributes cannot easily and directly answer these mechanics questions. For example, the geographic knowledge graph representation of the 7·21 Beijing storm is shown in Figure 1.

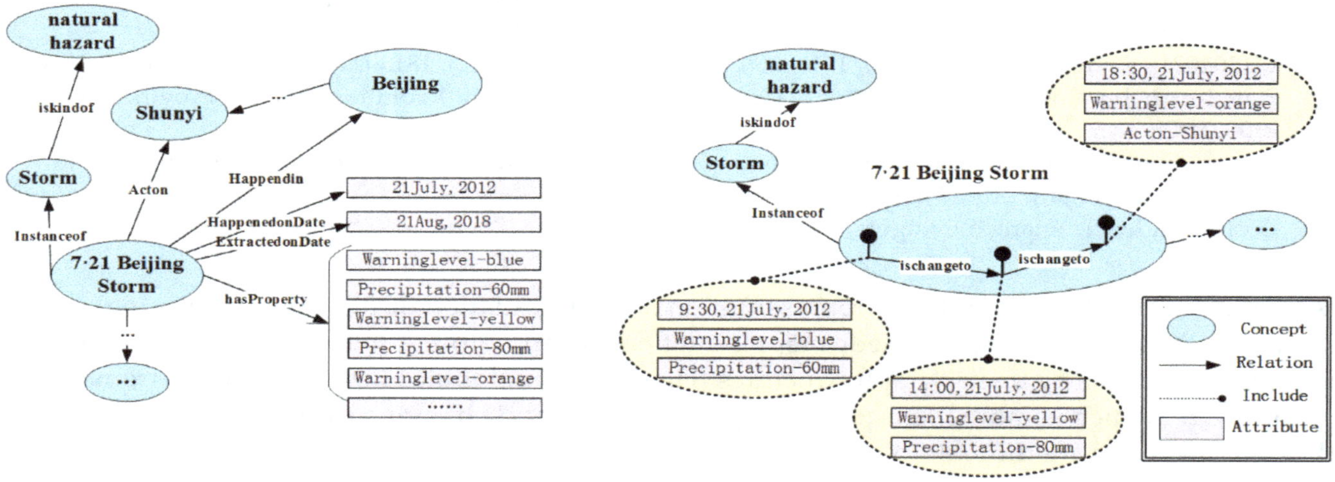

(a) (b)

Figure 1. Different geographic knowledge representations of the 7·21 Beijing Storm. (**a**) Knowledge graph data structure and (**b**) procedural knowledge data structure.

Figure 1a organizes geographic knowledge of the 7·21 Beijing storm using the data structure of the current knowledge graph. This knowledge representation model can explicitly represent each fact and its relations. However, it is not able to represent evolutions or mechanisms, which are key topics in geography. Moreover, this type of knowledge representation differs greatly from procedural knowledge data structure shown in Figure 1b. In general, humans perceive objects, events, and activities through the processing of declarative knowledge, procedural knowledge, and structural knowledge [5]. And procedural knowledge gives the framework to the declarative knowledge state by state, which is benefit for underlying mechanism understanding [6,7]. The 7·21 Beijing storm includes three main stages: 9:30, 14:00, and 18:30. Each stage has a list of attributes. This procedural knowledge data structure helps people acknowledge the evolution or mechanism more explicitly. For example, people cannot directly understand that all the attributes (warning level-blue, warning level-yellow, etc.) link to "7·21 Beijing storm", whereas people could know that the storm has different warning level on different stages.

The purpose of this paper is to improve the declarative discrete facts of knowledge graph to procedural aggregated knowledge. To address this issue, this paper presents a formalized model for geographic knowledge representation from a geography perspective, called GeoKG, and supplements the constructors of the ALC descriptive language.

The remainder of this paper is organized as follows: Section 2 reviews the related works on geographic ontology and geographic knowledge graph. Section 3 describes the methodology by stating the basic ideas from the six core geographical questions and proposes a formalized model of geographic knowledge representation called GeoKG. Section 4 gives an evolution case study of administrative divisions of Nanjing with the formalized GeoKG model. Section 5 constructs the administrative division case by using the GeoKG model and the YAGO model, sets a series of questions, and analyzes the results. Finally, Section 6 presents the conclusions.

2. Related Works

There are two main representations of geographic knowledge for geospatial big data computing and reasoning: geographic ontology, and geographic knowledge graphs.

2.1. Geographic Ontology

Geographic ontology originates from ontology, which represents the most basic philosophical theories that represent the nature and characteristics of the real world [8]. In the 1960s, "ontology" was introduced in information science for categorization, representation, knowledge sharing, and reuse [9]. Geographic ontology is a domain concept, which explicitly and formally defines the geographic concepts and their relations within geography by hierarchical relations [10–12]. These hierarchical relations between concepts are significant to geographic knowledge representation, information integration, knowledge interoperation, and information retrieval. Thus, geographic ontology is an important geographic knowledge representation method that is widely implemented in various geographical information applications [13]. However, computer simulations not only require the standard hierarchical concept logic but also massive amounts of instance information in geographic knowledge representation. There are two types of typical geographic ontologies that are limited by the representations of geographic knowledge.

First, geographic ontology focuses on the structure of a conceptual system, which is built by strict hyponymy information [9]. These relationships are well suited for categorization, disambiguation, identification and inference but not for describing the states and phenomena of changing geographic objects [6,14]. The descriptions of these changing states and phenomena require copious information, which is lacking in geographic ontology [15]. Although hyponymy is strictly defined by hierarchical tree structures in geographic ontology, this structure cannot directly represent the relationships between multiple concepts that are important to represent evolutions and mechanisms in geography [11]. In addition, the relationships between vertices in a tree are not bi-directional, which limits the representation of the interactions between geographical objects. The cause of these problems is related to the tree structure limiting the representation of geographic knowledge [16].

Second, the logic foundation of geographic ontology is description logic (DL) of attributive concept language with complements (ALC) [17]. DL is an object-based formal knowledge representation language. It contains four components: a construction set represents concepts and roles (e.g., a river is a concept; disjunction is a role), assertion about a concept of terminology (terminology box, Tbox, e.g., each river has its own length), assertion about an individual item (assertion box, Abox, e.g., the length of the Changjiang River is 6300 km) and the reasoning mechanism of Tbox and Abox. DL can construct complicated concepts and roles with simple concepts and roles by constructors. According to the different constructors, DL can be classified as ALC, ALCN, S, SH, SHIQ, etc. ALC is the basic DL that contains intersections ($\sqcap$), unions ($\sqcup$), complements ($\neg$), universal restrictions ($\forall$), and existential restrictions ($\exists$). ALCN consists of basic ALC operators and number restrictions (Q; $\geq$ n and $\leq$ n); ALC+R^+, short for S, consists of basic descriptions and enhancing relationship operators (R^+; role or concept transitions); the SH language, with concept inclusions and role inclusions ($\sqsubseteq$); and SHIQ includes inverse roles (I) and role transitions (R^+). At present, description logic SHIQ has been certified to represent changes in the field of logical theory [18]. Note that "change" is an absolutely essential element for geographic knowledge representation and means that the ALC constructors cannot represent all logical relations of geographic knowledge, especially in quantity expressions and state changes. For example, number restriction constructors are required to represent the geographic knowledge of "the Yangzi River has at least three branches" ($\exists$ has a branch; the Yangzi River $\geq$ 3), and transition constructors are required to represent the geographic knowledge of "Beiping was renamed

Beijing on 27 September 1949" (Beiping ≡ Trans(Beijing)). Meanwhile, many studies theoretically demonstrated and proved the decidability, soundness and completeness of the operators of a series of DL (from ALC, S, SI, SHI to SHIQ, etc.) on the Tableau algorithm [19–22], and the complexity of SI (role or concept transitions) is PSPACE complete and the following SHI and SHIQ are EXPTIME complete [22,23].

2.2. *Geographic Knowledge Graph*

A knowledge graph is a graph-formed knowledge representation model with strict logic, different concepts, various relations, and massive instances [10]. It was first presented by Google in 2012, containing over 5.7 billion entities and 0.18 billion facts [24]. With this wealth of information, the real world can be explicitly described. Graph-based storage has properties of connection, direction, and multi-vertices that are suitable for representing the interactions between concepts. Thus, knowledge graphs are promising models to represent knowledge and have been widely built, e.g., YAGO [25], Freebase [26], Probase [27], and DBpedia [28]. A geographic knowledge graph is a domain knowledge graph that is in the exploratory stage.

At present, most geographic knowledge graphs are organized as universal knowledge graphs, e.g., CSGKB [4], NCGKB [29], and CrowdGeoKG [15]. The common sense geographic knowledge base (CSGKB) uses a data structure that links the concepts of geographic features, geographic locations, spatial relationships and administrators for geographic information retrieval (GIR) instead of traditional gazetteers. Moreover, the naive Chinese geographic knowledge base (NCGKB) constructs a GIR-oriented geographic knowledge base based on Chinese Wikipedia based on given concept relations and their instances. CrowdGeoKG uses a crowdsourced geographic knowledge graph that extracts different types of geo-entities from OpenStreetMap and enriches them with human geography information from Wikidata. All of the concepts of these geographic knowledge graphs are developed based on geographic ontologies that follow the ALC descriptive language, resulting in the same problem as geographic ontology.

More importantly, three current bases organize the geographic knowledge as a set of concepts, relations, and facts, which are associated by two kinds of types {entity, relation, entity} and {entity, attribute, attribute value} [4]. Actually, there are only three basic elements in knowledge graph: entity, relation and attribute. These three elements can explicitly represent general information as "when did 7·21 Beijing storm— 9:30, 21 July". However, geographic knowledge is more complicated than general knowledge. More processes and evolutions need to be answered, e.g., "what causes the 7·21 Beijing storm", "how did it develop", and "what are the effects of the 7·21 Beijing storm". Entities, relations, and attributes cannot easily and directly answer these mechanics questions.

Scholars indicated that more elements are required. PLUTO supplemented the element of time with "before" and "after" in the knowledge graph model to describe the change trajectories of geographic objects [30]. Geological knowledge graphs have been applied with the evolution element for stating changes between different geological objects [31]. YAGO also explored anchoring spatial and temporal dimensions to the knowledge base, called YAGO2 [7]. YAGO2 let time points and time intervals with standard format to describe the temporal information and set geographical coordinates associate to entities to complete their spatial information. In fact, these spatial and temporal knowledge stored in YAGO system are just regarded as general attributes by adding the predicates like "wasBornOnDate", "occursSince", "hasGeoCoordinates", etc., whereas declarative discrete information cannot directly answer the proceeding questions, evolutions and mechanisms. Additionally, ten core concepts of geographic information sciences were proposed for transdisciplinary research: location, neighbourhood, field, object, network, event, granularity, accuracy, meaning, and value [23]. These concepts can cover every corner of geoscience, but they were extremely difficult to relate to one conceptualized model. More recently, six factors (geographic semantics, location, shape, evolutionary process, relationship between elements, and attribute) were proposed to describe information from geographic element, object, or phenomenon [22]. Though these factors were designed for information representation of the geographic objects, they can also provide guidance for geographic knowledge representation. And all the above studies indicated that geographic knowledge can be represented more effectively by supplementing elements, whereas it also brings a foundation question: "how to organize geographic knowledge scientifically and cognitively?" Therefore, a conceptualized model of geographic knowledge graph from the geography perspective warrants further study.

3. Methodology

3.1. Basic Idea

3.1.1. Guiding Ideology

To address the aforementioned issues, the core question of the GeoKG model is to define the types of geographic knowledge that need to be stored. Geography (from the Greek γεωγραφία, geographia, literally "earth description") is a field of science devoted to the study of the lands, features, inhabitants, and phenomena of Earth [32]. As a type of human understanding of the geographic environment, geographic knowledge should answer questions about geography. The questions about geography have been separated into six core questions by the International Geographical Union (IGU), which is a part of the international charter on geographical education [2]. Therefore, GeoKG begins to define the basic elements and the conceptualized model using the six core questions in geography. Each question corresponds to one core issue:

- Where is it? →**space**
- What is it like? →**state**
- Why is it there? →**evolution**
- When and how did it happen? →**change**
- What impacts does it have? →**interaction**
- How should it be managed for the mutual benefit of humanity and the natural environment? →**usage**

3.1.2. Main Elements

These aforementioned questions can be used to describe six core aspects of geographic knowledge that should be represented by GeoKG. Each aspect requires some elements to describe them, and we try to find the basic elements among all aspects:

- Space →{**object**, location, time, relation, ... }
- State →{**object**, time, location, attribute ... }
- Evolution →{**object**, state, change, time, location, attribute, ... }
- Change →{**object**, time, location, attribute, relation, ... }
- Interaction →{**object**, relation, change, ... }
- Usage →{**object**, change, state, ... }

There are seven types of elements among the description of these six aspects: *object, location, time, attribute, relation, state,* and *change*. Three typical characteristics of these seven elements in describing the six aspects are as follows:

- **Object-centered representation**. All descriptions of the six aspects require geographic objects. Without **objects**, other elements are meaningless. Therefore, the six basic elements are formed around the object element.
- **Combined representation**. A description of a single basic element is just a statement. To represent these aspects in geography, the basic elements should be combined. Thus, all of the basic elements can be linked.

- **Stepped representation**. Note that the six aspects from the core geographical questions are not equal. Space and state focus on the static conditions of objects. Evolution and change pay more attentions to the dynamic conditions of objects. Moreover, interaction and usage rely on relationships and mechanisms between geographic objects. Thus, the basic elements cannot be treated as equals.

According to the three typical characteristics of the basic elements, we discovered that a geographic object is a type of media used to represent geographic knowledge. There are six basic elements used to describe geographic knowledge (see Figure 2). *Location, time, attribute, state, change,* and *relation* can co-efficiently represent geographic objects from different aspects. Note that these basic elements are not equivalent. *Location, time,* and *attribute* belong to the first level and represent a single static state of a geographic object. *State, change,* and *relation* describe the dynamic evolutions and relations to geographic objects.

- A geographic object is the core of geographic knowledge representation and is the minimum unit to perceive the world. The six basic elements (*location, time, attribute, state, change and relation*) represent geographic knowledge from different perspectives, which are linked to geographic objects.
- Static independent geographic objects can be described by elements of *location, time,* and *attribute*. Location shows the spatial patterns of geographic objects. Time gives the temporal dimension of geographic objects for human cognition. Attribute describes the static features of geographic objects.
- Any geographic object has an entire life cycle, including stages of generation, change, evolution and extinction. Different stages in the life cycle represent different *states*. States are represented by sets of attributes of geographic objects under a particular spatial-temporal dimension.
- Geographic objects are not always static. Any change in other elements of a geographic object will turn a state to another state or a relation to another relation. Thus, *change* is an essential part of geographic knowledge representation.
- Geographic objects are not isolated. Any scene, phenomenon, and environment consists of many geographic objects and complex relations between them. Thus, *relation* is the key descriptor of the interactions among complex geographic objects.

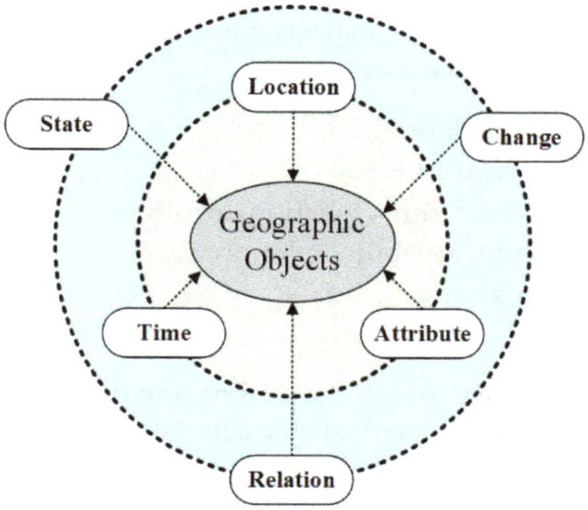

Figure 2. The six basic elements to represent a geographic object.

3.1.3. GeoKG Model

A conceptualized model of GeoKG is shown in Figure 3, which is based on the ideas mentioned above. The six core elements represent geographic objects and their information together. In this model, geographic objects consist of a series of states. Any state of a geographic object is represented by attributes under a specific spatial-temporal condition. Any two continuous states or different states between two geographic objects could result in a change element. The change element can be categorized into time changes, location changes or attribute changes. If the essential attribute is changed, the geographic object will become another geographic object. The relation element exists between any time, location, and attribute of different states, regardless of whether they are the same geographic objects or not.

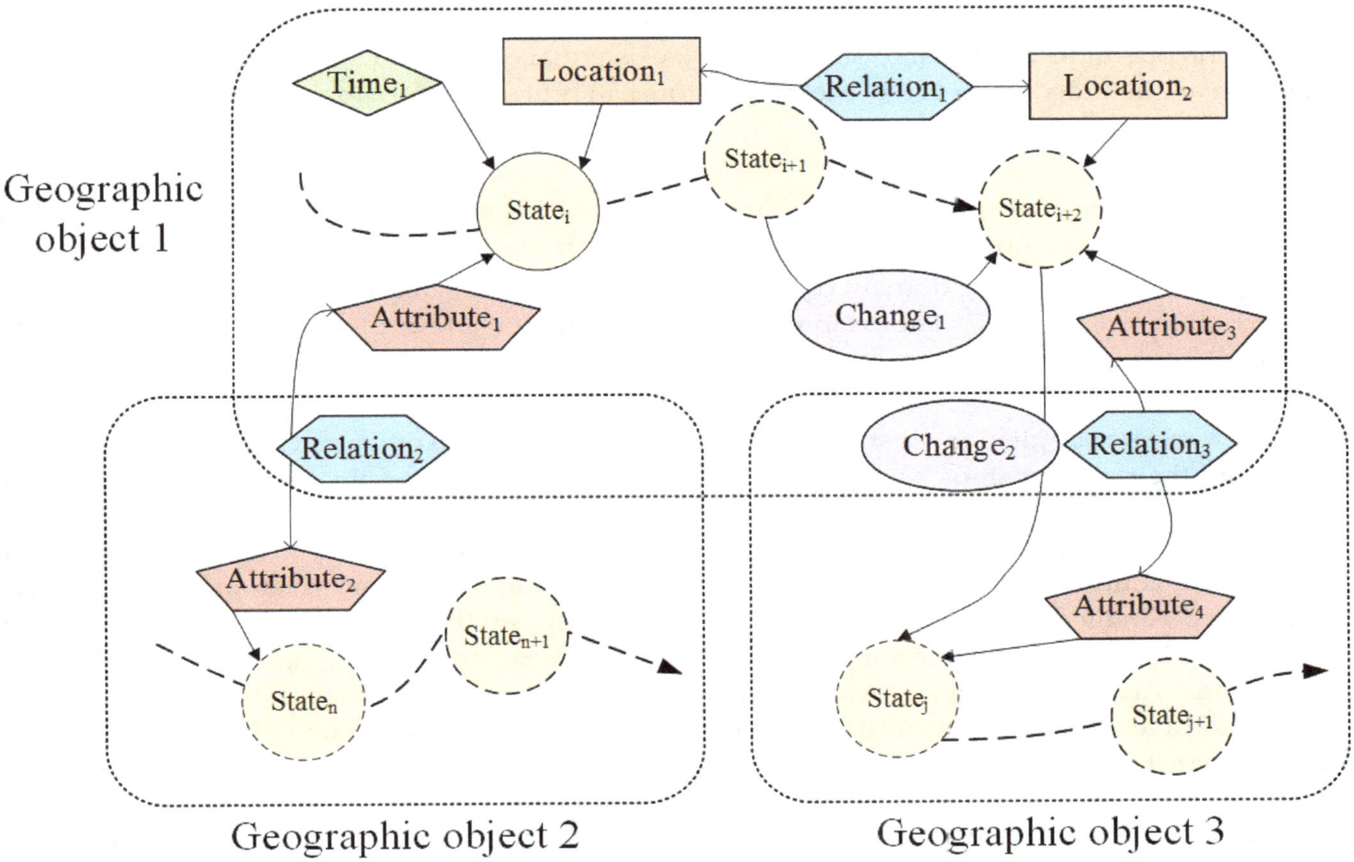

Figure 3. A conceptualized model of GeoKG based on the six basic elements.

3.2. Model Formalization

To organize geographic knowledge in consideration of the basic ideas, GeoKG must be based on a thorough and formalized model. This section provides the model semantics of GeoKG by using description logic (DL), which is not limited to only the attribute language complement (ALC) level. Using description logic, a user can create a conceptual description for the representation and computation of geographic knowledge that is clear and formal.

3.2.1. DL and Construction Operators

DL is comprised of three basic components: concepts, individuals (instances), and roles. Concepts describe the common features about individual sets, e.g., all land mass that projects well above its surroundings forms the concept of "mountain". Individuals are the instances of concepts, e.g., a geographic entity, such as the "Rocky Mountains". Roles can be explained as the binary relation between individuals as properties, e.g., spatial relation (conjunction, disjunction). A description logic system contains four parts. These parts include a construction set, which represent concepts and roles, an assertion about concept terminology (terminology box, Tbox), an assertion about an individual (assertion box, Abox) and the reasoning mechanism of Tbox and Abox. Tbox are sets containing the definitions of the relationship of concepts and the axiom of relationships, which contain explanations of the concepts and roles. Abox includes axiom sets describing specific situations, which contain the instance information of Tbox. Abox include two forms. One is the concept assertion, which expresses whether an object belongs to a concept. The other one is the relation assertion, which express whether two objects satisfy a certain relation. Description logic can represent complicated concepts and relations on atomic concepts and atomic relations based on the given construction operator. The basic construction operators are and ($\sqcap$), or ($\sqcup$), not ($\neg$), existential quantifier ($\exists$), and universal quantifier ($\forall$), which are included in ALC DL. More operators can represent more logic, which form different types of DL.

Let C and D be concepts; a, b, and c, individuals; and R is a role between individuals. S is a simple role, and n is a nonnegative integer. As usual, an interpretation $\mathcal{I} = \left(\Delta^{\mathcal{I}}, \cdot^{\mathcal{I}}\right)$ consists of a non-empty set $\Delta^{\mathcal{I}}$, called the domain of $\mathcal{I}$, and a valuation $\cdot^{\mathcal{I}}$, which associates, with each role R, a binary relation $R^{\mathcal{I}} \subseteq \Delta^{\mathcal{I}} \times \Delta^{\mathcal{I}}$. For comprehensive background reading, please refer to the referenced paper [20]. The primary operators that differ from DL are shown in Table 1.

Diagrams are supplied to illustrate the graphic meanings of the operators related to geographic objects and their relationships. A top concept indicates all concepts or objects, e.g., $\top$ River means all the rivers. A bottom concept indicates no concepts or objects in the set, e.g., $\bot$River means there are no rivers in the set. An atomic concept indicates the minimize concept, e.g., Ac could be the river, ocean, city, or country. An atomic role indicates the relationships between two atomic concepts, e.g., R $\subseteq$ river $\times$ ocean means that there exists a relationship between the river and ocean. A conjunction indicates two individuals that are joint or connected, e.g., Yangzi River $\sqcap$ Nanjing indicates a joint part of the Yangzi River and Nanjing. A disjunction indicates the logic disjunction of two individuals, e.g., Yangzi River $\sqcup$ Nanjing means the combination set of the Yangzi River and Nanjing. A negation indicates the set of all individuals not in the target individual, e.g., $\neg$Yangzi River means all individuals except the Yangzi River. An exist restriction indicates the existence of an individual or a role, e.g., $\exists$Yangzi River means there exists a Yangzi River and $\exists$R $\subseteq$ Yangzi River $\times$ Zhong Mountains means there exists a role between the Yangzi River and Zhong Mountains. A value restriction indicates all individuals or roles, e.g., $\forall$ River means all rivers and $\forall$ R $\subseteq$ Yangzi River $\times$ Zhong Mountains means all roles between the Yangzi River and Zhong Mountains. A concept inclusion indicates a concept belonging to another concept, e.g., rain $\sqsubseteq$ precipitation means rain is a kind of precipitation. A role inclusion indicates a role belonging to a role set, e.g., $R_{location\_Yangzi\ River-Zhong\ Mountain} \sqsubseteq$ Yangzi River $\times$ Zhong Mountains indicates that the location relation between the Yangzi River and Zhong Mountains is one of the roles of the entire role set of the Yangzi River and Zhong Mountains. An inverse role indicates that a role has reversibility. A trans role indicates that a role has transmissibility. A qualifying at least/at most restriction indicates there exists at least or at most, e.g., ($\exists \geq$ 3rivers) $\subseteq$ Yangzi River means the Yangzi River has at least three branches.

Table 1. Syntax and semantics of the main construction operators of the description logic.

Category (Symbol)	Construction Operators	Syntax	Semantics	Diagrams
ALC	Top concept	$\top$	$\Delta^{\mathcal{I}}$	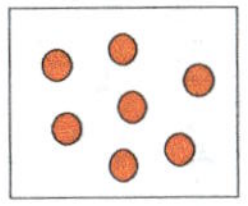
	Bottom concept	$\perp$	$\varnothing$	
	Atomic concept	Ac	$Ac^{\mathcal{I}} \subseteq \Delta^{\mathcal{I}}$	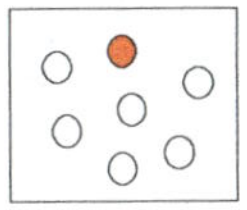
	Atomic role	R	$R^{\mathcal{I}} \subseteq \Delta^{\mathcal{I}} \times \Delta^{\mathcal{I}}$	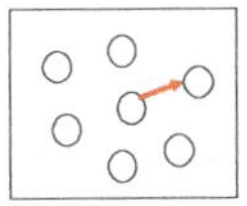
	Conjunction	$C \sqcap D$	$C_1^{\mathcal{I}} \cap C_2^{\mathcal{I}}$	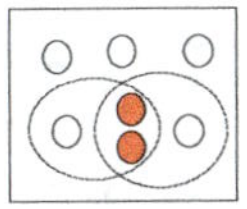
	Disjunction	$C \sqcup D$	$C_1^{\mathcal{I}} \cup C_2^{\mathcal{I}}$	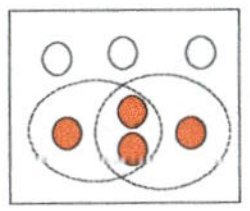
	Negation	$\neg C$	$\Delta^{\mathcal{I}} \backslash C^{\mathcal{I}}$	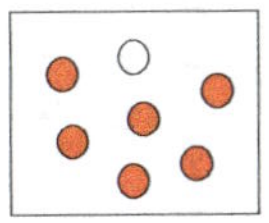
	Exist restriction	$\exists R.C$	$\{a \in C^{\mathcal{I}} \mid \exists y, (a,b) \in R^{\mathcal{I}} \;\&\; b \in C^{\mathcal{I}}\}$	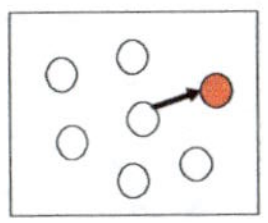
ALC	Value restriction	$\forall R.C$	$\{a \in C^{\mathcal{I}} \mid \forall y, (a,b) \in R^{\mathcal{I}} \;\&\; b \in C^{\mathcal{I}}\}$	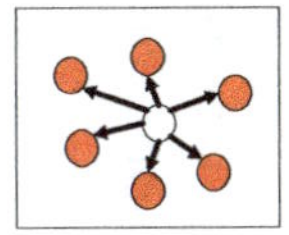
H	Concept inclusion	$C_1 \sqsubseteq C_2$	$C_1^{\mathcal{I}} \subseteq C_2^{\mathcal{I}}$	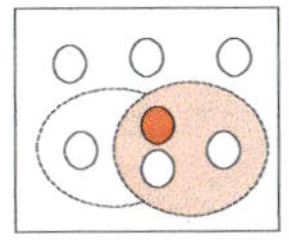
	Role inclusion	$R \sqsubseteq S$	$R^{\mathcal{I}} \subseteq S^{\mathcal{I}}$	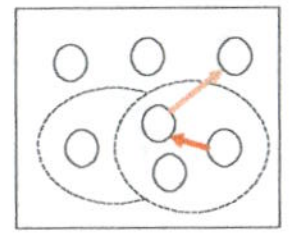
I	Inverse role	R^-	$\{(a,b) \in R^{\mathcal{I}} \mid (b,a) \in R^{\mathcal{I}}\}$	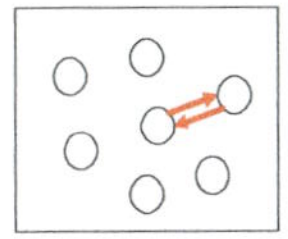
R^+	Trans role	Trans(R)	$\{(a,c) \in R^{\mathcal{I}} \mid \exists (a,b) \in R^{\mathcal{I}} \wedge (b,c) \in R^{\mathcal{I}}\}$	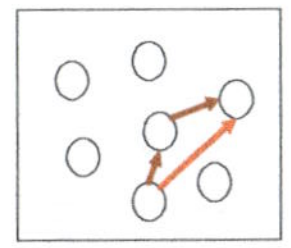
Q	Qualifying at least restriction	$R.C \geq n$	$\{a \in C^{\mathcal{I}} \mid \#(\{b \mid (a,b) \in R^{\mathcal{I}} \;\&\; b \in C^{\mathcal{I}}\}) \geq n\}$	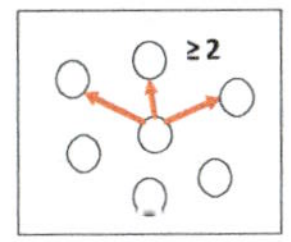
	Qualifying at most restriction	$R.C \leq n$	$\{a \in C^{\mathcal{I}} \mid \#(\{b \mid (a,b) \in R^{\mathcal{I}} \;\&\; b \in C^{\mathcal{I}}\}) \leq n\}$	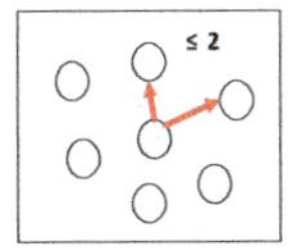

means the number of. Note the decidability, soundness, and completeness of all these operators have been demonstrated [6,23].

3.2.2. Formalization Representation

In this section, the semantics of the GeoKG model are defined. First, we prescribe the set of geographic knowledge GK sourced from the entire world's natural and human phenomena W. GeoGK is a set of GK that can be defined as follows:

$$\text{GeoGK} = \{\langle GK\rangle | GK \in W\}$$

GK is a tuple that consists of geographic object O and its basic elements E:

$$GK = \{\langle O,\ E\rangle | \exists O \neq \varnothing,\ \exists E \neq \varnothing\}$$

The basic element set E contains six different elements: *location L*, *time T*, *attribute A*, *state St*, *change Ch* and *relation Re*. Thus, E is a six-tuple:

$$E = \{\langle L,\ T,\ A,\ St,\ Ch, Re\rangle | \exists L \parallel T \parallel A \parallel St \parallel Ch \parallel Re \neq \varnothing\}$$

Each element is identified as follows:

(1) Time

Time describes the temporal information of the state of a geographic object. Let St_i indicate a specific state of geographic object O_i; the basic element *time T* can be defined as follows:

$$T = \{\exists T \in St_i | \forall O_i \neq \varnothing, St_i \in O_i\}$$

Time should be described by both the basic types and reference time information. The basic types are point time, interval time and reference time. Point time T_{poi} records the moment of the state of a geographic object. Interval time T_{int} indicates the time interval between two point times. Reference time T_{ref} indicates the time of other elements of a geographic object, e.g., "2018 World Cup" is an event with a unique time period that could reference the specific time accurately. Time reference knowledge $tref$ indicates the additional knowledge of time descriptions. Let tw indicate the time word. A time word indicates a point time that could contain several time descriptive parts, e.g., 12-July-2018, ten past nine and tomorrow morning. The point time T_{poi}, the interval time T_{int} and the reference time T_{ref} are defined as follows:

$$T_{poi} = \{\langle tw, tref\rangle \big| \forall ! tw \in T\}$$

$$T_{int} = \{\langle tw,\ tref\rangle \big| \forall tw \in T, \# tw \geq 2, \forall \mathrm{R} \sqsubseteq tw\}$$

$$T_{ref} = \{\langle E,\ tref\rangle \big| \forall E\ \&\ \forall ! T \sqsubseteq St_i\}$$

where R is the interval relation of two time words. Time reference knowledge $tref$ is a set of reference knowledge consisting of commonality, relativity, fuzziness, continuity, and periodicity, namely, $tref = \{\langle com,\ rel,\ fuz, con, per\rangle\}$. There are some examples for each reference time word. For example, "12-July-2018" is a common time, and the Late Jurassic is a domain time description. Relativity indicates whether time is relative, e.g., "two days ago" is a relative time that refers to the absolute time "today", "9 o'clock" is an accurate time, "around 9" is a fuzzy time, "12-July" is an instance time, and "until 12, July" is a continuous time. Periodicity can be easily understood, such as "every weekend", "every month", and "annually".

(2) Location

Location describes the spatial information of the state of a geographic object. Let St_i indicate a specific state of geographic object O_i; the basic element *location L* can be identified as follows:

$$L = \{\exists L \in St_i | \forall O_i \neq \varnothing, St_i \in O_i\}$$

According to the complexity of location descriptions, a location can be set into basic types and reference location information. The basic types include toponym, address, coordinates, and reference location. Toponym L_{top} describes a location with a common name. Address L_{add} indicates a location with orderly numbers and streets named by administrators. Coordinate L_{coo} records the location with a series of numbers organized mathematically. Reference location L_{ref} indicates the location of other elements in a geographic object. Location reference knowledge $lref$ indicates the additional knowledge of location descriptions. Let tp, ad, co indicate toponym, address, and coordinate, respectively. Toponym L_{top}, address L_{add}, coordinates L_{coo}, and reference location L_{ref} are identified as follows:

$$L_{top} = \{\langle tp, lref\rangle \big| \forall! tp \in L\}$$

$$L_{add} = \{\langle ad, lref\rangle \big| \forall ad \in L\}$$

$$L_{coo} = \{\langle co, lref\rangle \big| \forall co \in L\}$$

$$L_{ref} = \{\langle E,\ lref\rangle \big| \forall E \ \&\ \forall! L \sqsubseteq St_i\}$$

Location reference knowledge $lref$ is a set of reference knowledge consisting of the space type, spatial reference, commonality, relativity, and fuzziness, namely, $lref = \{\langle typ, ref, com,\ rel,\ fuz\rangle\}$. Space type describes what types of space, such as reality, virtual or a specific domain location. For example, Pandora is a toponym of the virtual world of the movie Avatar. Spatial reference illustrates the system of a location description, e.g., WGS84 and Mercator projection. Commonality stores whether a location is a domain location, e.g., Beijing is a common toponym that could be coded as "-.-..--...-.---/-..---.-.-.--.." in a Morse code system. Relativity indicates whether a location is relative, e.g., "20 km south of Beijing" is a relative location description that refers to the absolute location "Beijing". Fuzziness states whether a location description is accurate or not, e.g., "near Times Square" is a fuzzy location description.

(3) Attribute

An attribute describes the feature information of the state of a geographic object. Let St_i indicate a specific state of geographic object O_i; the basic element *attribute A* can be identified as follows:

$$A = \{\exists A \in St_i | \forall O_i \neq \varnothing, St_i \in O_i\}$$

All the feature descriptions of a geographic object belong to an attribute, e.g., shape, color, speed, etc. To organize the attributes of a geographic object, identifying what is an attribute is key. An attribute is a single feature description of one geographic object. For example, "a typhoon is a mature tropical cyclone that develops between 180° and 100° E in the Northern Hemisphere, with peak months from August to October" describes three attributes: the typical attribute of "mature tropical cyclone", the location attribute of "develops between 180° and 100° E in the Northern Hemisphere" and the frequency attribute of "peak months from August to October". It is noted that attribute can be divided into two types: essential attribute A_{es} and non-essential attribute A_{ne}:

$$A_{es} = \{\exists A_{es} \in St_i | \forall O_i \neq \varnothing, St_i \in O_i,\ \#A_{es} \geq 1\}$$

$$A_{ne} = \{A_{ne} \in A' \big| \forall O_i \neq \varnothing, St_i \in O_i, A' = A_{es}^{-}\}$$

An essential attribute is a mark attribute that identifies a geographic object from others. When an essential attribute changes, a geographic object could change to another object. For example, when a mature tropical cyclone develops in the Atlantic Ocean, it cannot be a typhoon. A non-essential attribute is another feature description of a geographic object, e.g., the frequency attribute of a typhoon is "peak months from August to October". These attributes cannot determine the nature of a geographic object.

(4) State

The state illustrates the different stages of a geographic object. It can be seen that the above three basic elements work together to express the state. Thus, the element *state St* can be identified as follows:

$$St = \{\exists St_i \in O | \exists! L \sqsubseteq St_i, \exists! T \sqsubseteq St_i, \exists A \sqsubseteq St_i, \#A \geq 0\}$$

where ∃! means the unique existence. The formulation means that the state is a part of a geographic object. As the element *state St* is represented by sets of attributes of geographic objects under a particular spatial-temporal dimension, it must depend on the element *location L* and the element *Time*. Note that the element *location L* and the element *Time T* exist uniquely, because of time and space are two dimensions to represent the stage in Euclidean space. For example, the state of a typhoon includes all features for a specific spatial-temporal reference frame, e.g., "Typhoon Maria, 23:00/10July-2018, E123.40°/N25.60°, central pressure 945 hpa, max speed 30 km/h". The state cannot be defined without the temporal and spatial information. By contrast, the element *state St* does not depend on the element *Attribute A*. The attributes are the descriptive records that cannot affect whether the state exists. For example, "Typhoon Maria, 23:00/10July-2018, E123.40°/N25.60°" also defines a state of Typhoon Maria. Thus, the attribute element is defined different from location element and time element.

(5) Change

A change describes the changes in a geographic object from one state to another. Thus, change *Ch* must contain at least one difference between two states, which can be a location change, time change or attribute change. A change contains four main components:

$$Ch = \{\langle St, act, CE, type\rangle \in O \big| \exists St, \#St = 2, CE \in \{T, L, A\}, type \in (Ch_d, Ch_e)\}$$

where *St* indicates the state (including two different ones), *act* indicate the action of the change, *CE* indicate change elements and *type* indicates the type of the change. It is noted that there are two types of changes: a developing change and an evolving change. A developing change shows the changes from one geographic object, and an evolving change describes the changes between two different geographic objects. Let Ch_d indicate a developing change and Ch_e indicate an evolving change; the formalized definitions are as follows:

$$Ch_d = \{\exists Ch_d = St_i \times St_{i+1} \big| \exists St_i \& St_{i+1} \in O_m, St_i \neq St_{i+1}\}$$

$$Ch_e = \{\exists Ch_e = St_{end} \times St_i | \exists St_{end} \in O_m, \exists St_i \in O_n, \exists! St_{end}.A_{es} \neq St_i.A_{es}\}$$

where O, O_m, *and* O_n are geographic objects, St_i and St_{i+1} indicate the continuous states of the geographic objects, St_{end} indicates the last state of the geographic objects, and A_{es} indicate the essential attribute of the geographic objects.

(6) Relation

A relation expresses the differences between the elements of geographic objects, which includes three typical types: location relation, time relation, and attribute relation. These three types describe the spatial difference, temporal difference and feature difference, respectively. A relation contains three main components: the elements of two states *E*, the semantic of the relation *Sem*, and the type of the relation *type*:

$$Re = \{\langle E, Sem, type\rangle \in O \big| \exists E \& \#E \geq 2, type \in (Re_l,\ Re_t,\ Re_a)\}$$

Let Re_l, Re_t, *and* Re_a indicate location relation, time relation, and attribute relation, respectively, L_i and L_j indicate the locations of different states, T_i and T_j indicate the times of different states, and A_i and A_j indicate the attributes of different states. The different types of relations are identified as follows:

$$Re_l = \left\{\exists Re_l = L_i \times L_j \big| \exists St_i \& St_j, St_i \neq St_{i+1}\right\}$$

$$Re_t = \left\{\exists Re_t = T_i \times T_j \big| \exists St_i \& St_j, St_i \neq St_{i+1}\right\}$$

$$Re_a = \left\{\exists Re_a = A_i \times A_j \big| \exists St_i \& St_j, St_i \neq St_{i+1}\right\}$$

A location relation describes the spatial relationships between different states, e.g., the location relations between the different states of a typhoon or the location relations between two different city centres under development. A time relation illustrates the temporal relationships between different states, i.e., the time span between two states, e.g., the time span of river diversion. An attribute relation describes the feature relationships between different states, i.e., the differences between two states of a typhoon, e.g., the max wind speed, central pressure, etc.

4. Case Study

In this section, a full example is shown to illustrate the geographic knowledge representation using the GeoKG model. To describe the geographic knowledge representation clearly, an evolution case of administrative divisions of Nanjing was selected. The given example includes the basic geographic objects (e.g., Yangzi River, Zhongshan Mountain), the changing area of Nanjing, and several affiliated districts in different eras.

4.1. Research Area

Nanjing, formerly romanized as Nanking and Nankin, is the capital of Jiangsu province of the People's Republic of China and the second largest city in the East China region, with an administrative area over 6000 km^2. The inner area of Nanjing enclosed by the city wall is Nanjing Centre District, with an area of 55 km^2, while the Nanjing Metropolitan Region includes surrounding cities and areas. Three representative stages were chosen to represent the revolution of Nanjing: 1368, 1949, and 2018. The sketch maps were shown in Figure 4.

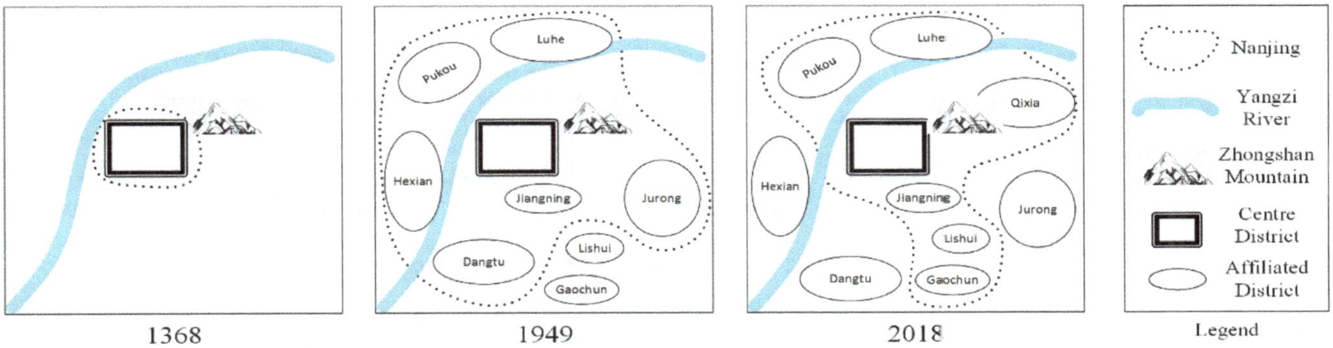

Figure 4. The sketch maps of administrative divisions evolution of Nanjing in 1368, 1949, and 2018.

The first stage is Ming dynasty, which firstly named this city in the word of "Nanjing". The first emperor of the Ming dynasty, Zhu Yuanzhang, who overthrew the Yuan dynasty, renamed the city of Nanjing, rebuilt it, and made it the dynastic capital in 1368. He constructed a 48 km long city wall around Nanjing. That is the centre district of Nanjing, which is situated in the south of the Yangzi River and to the west of the Zhongshan Mountain.

The second stage is the founding of the People's Republic of China. The government set Nanjing as a province unit, which directly controlled by the government. At that stage, Nanjing administrated the centre district and several affiliated districts. The centre district included district 1–10 and affiliated districts involved Jiangning, Jurong, Dangtu, Hexian, Pukou, and Luhe. In 1949, Nanjing had been expended through Yangzi River and Zhongshan Mountain.

The third stage is 2018, which refers to the current administrative boundaries of Nanjing. After a series of administrative division adjustments, Gaochun and Lishui was supplemented into Nanjing and Jurong, Dangtu, and Hexian was removed from the boundaries.

During over 600 years development of Nanjing, numerous elements were changed including the boundaries, affiliated districts, the relations between Nanjing and other geographic objects (e.g., Yangzi River and Zhongshan Mountain). Different relations happened in different stages among these geographic objects. Thus, the GeoKG model was used to represent these changing geographic knowledge. The formalization is introduced in the next section.

4.2. Formalization

In this example, administrative division evolution was organized by using the GeoKG model. A geographic object is the key to represent geographic knowledge. First, this case identifies six relevant geographic objects: Nanjing O_{nj}, Yangzi River O_{yr}, Zhongshan Mountain O_{zm}, Centre District O_{cd}, Jiangning O_{jn}, and Gaochun O_{gc}. Jiangning and Gaochun are representative affiliated districts which were selected in this case. Jiangning is always been part of Nanjing in 1949 and 2018 and Gaochun has an administrative division adjustment. Each geographic object consists of a series of states, changes and relations. For example, Nanjing O_{nj} contains three states $S_{nj} = \{S_{nj1}, S_{nj2}, S_{nj3}\}$, six changes $C_{nj} = \{C_{nj11}, C_{nj12}, C_{nj13}, C_{nj21}, C_{nj22}, C_{nj23}\}$, and 12 relations $R_{nj} = \{R_{nj11}, R_{nj12}, R_{nj13}, R_{nj21}, R_{nj22}, R_{nj23}, R_{nj24}, R_{nj31}, R_{nj32}, R_{nj33}, R_{nj34}, R_{nj35}\}$. Thus, Nanjing O_{nj} can be defined as follow and the corresponding diagram is shown in Figure 5.

$$O_{nj} = \left\{ \begin{array}{c} S_{nj} \sqsubseteq O_{nj}, C_{nj} \sqsubseteq O_{nj}, R_{nj} \sqsubseteq O_{nj} \big| \\ S_{nj} = \left\{ S_{nj1}, S_{nj2}, S_{nj3} \middle| S_{nj}.number \leq 3, S_{nj}.number \geq 3 \right\}, \\ C_{nj} = \left\{ C_{nj11}, C_{nj12}, C_{nj13}, C_{nj21}, C_{nj22}, C_{nj23} \middle| C_{nj}.number \leq 6, C_{nj}.number \geq 6 \right\} \\ R_{nj} = \left\{ \begin{array}{c} R_{nj11}, R_{nj12}, R_{nj13}, R_{nj21}, R_{nj22}, R_{nj23}, R_{nj24}, R_{nj31}, R_{nj32}, R_{nj33}, R_{nj34}, R_{nj35} \big| \\ R_{nj}.number \leq 12, R_{nj}.number \geq 12 \end{array} \right\} \end{array} \right\}$$

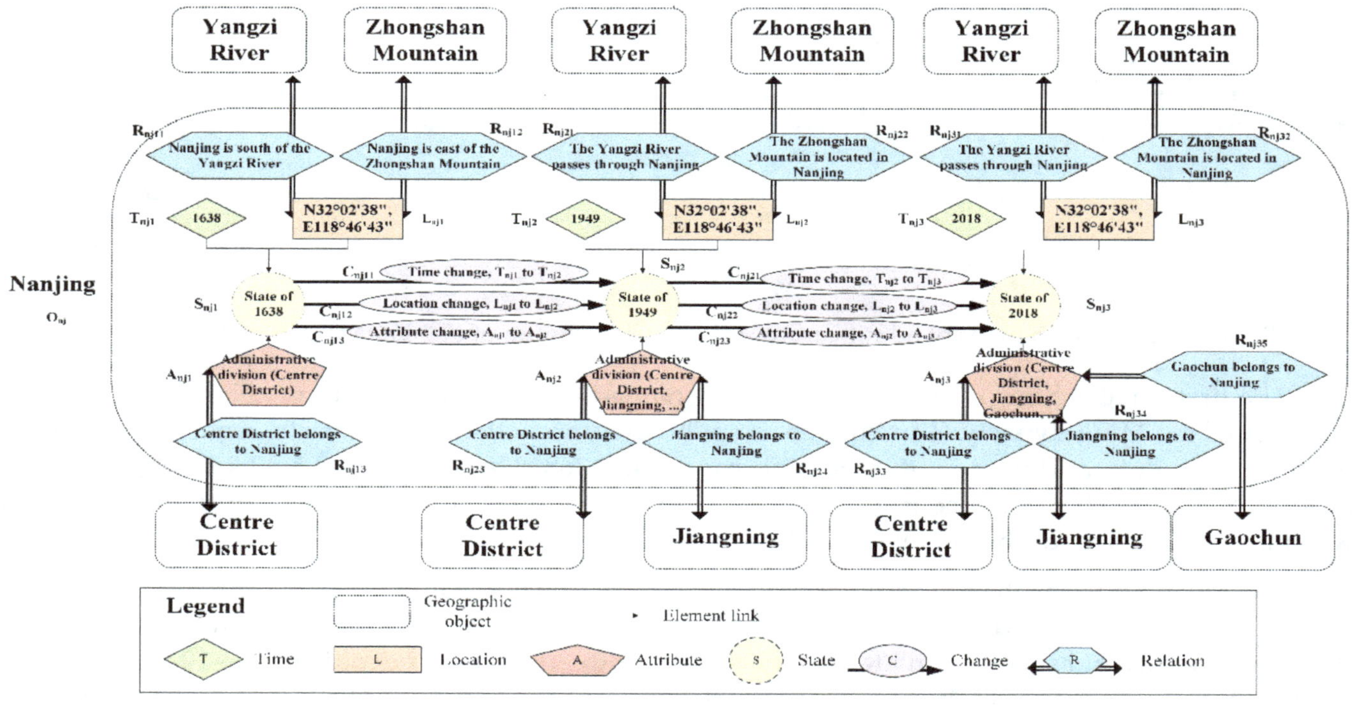

Figure 5. The diagram of different elements of Nanjing by using the GeoKG model.

Actually, different states of Nanjing $\{S_{nj1}, S_{nj2}, S_{nj3}\}$ indicate three different stages of 1368, 1949, and 2018. Each state contains different time, location, and attribute elements. For example, the state S_{nj1} of Nanjing contains time element T_{nj} of "1368", location element L_{nj} of "location descriptions in 1368" and attribute element A_{nj} of "administrative region". The state S_{nj1} of Nanjing can be defined as follows:

$$S_{nj1} = \left\{ \begin{array}{c} T_{nj} \sqsubseteq S_{nj1}, L_{nj} \sqsubseteq S_{nj1}, A_{nj} \sqsubseteq S_{nj1} | \\ T_{nj} = \{T_{nj1} | T_{nj}.number \leq 1, T_{nj}.number \geq 1\}, \\ L_{nj} = \{L_{nj1} | L_{nj}.number \leq 1, L_{nj}.number \geq 1\}, \\ A_{nj} = \{A_{nj1} | A_{nj}.number \leq 1, A_{nj}.number \geq 1\} \end{array} \right\}$$

Different states could contain changes indicating different kinds of changes from one state to another one. For example, there are three main changes $\{C_{nj11}, C_{nj12}. C_{nj13}\}$ from the state S_{nj1} of Nanjing in 1368 to the state S_{nj2} of Nanjing in 1949: the change C_{nj11} between time elements, the change C_{nj12} between location elements and the change C_{nj13} between the attribute elements of "administrative region". Note that all these changes belong to developing change type which indicates the change do not create a new geographic object. The changes can be defined as follows:

$$C_{nj11} = \left\{ \begin{array}{c} St, act,\ CE, type \sqsubseteq C_{nj11} | \\ St = \{S_{nj1}, S_{nj2}\}, act = \{"time\ change"\}, CE = \{T_{nj1}, T_{nj2}\}, type = Ch_d \end{array} \right\} \sqsubseteq O_{nj}$$

$$C_{nj12} = \left\{ \begin{array}{c} St, act,\ CE, type \sqsubseteq C_{nj12} | \\ St = \{S_{nj1}, S_{nj2}\}, act = \{"location\ change"\}, CE = \{L_{nj1}, L_{nj2}\}, type = Ch_d \end{array} \right\} \sqsubseteq O_{nj}$$

$$C_{nj13} = \left\{ \begin{array}{c} St, act,\ CE, type \sqsubseteq C_{nj13} | \\ St = \{S_{nj1}, S_{nj2}\}, act = \{"attribute\ change"\}, CE = \{A_{nj1}, A_{nj2}\}, type = Ch_d \end{array} \right\} \sqsubseteq O_{nj}$$

Relation is an indispensable element which exists in geographic objects referring to the relationships between different elements. In this example, there are three relations $\{R_{nj11}, R_{nj12}, R_{nj13}\}$ relate to Nanjing in 1368: the spatial relation R_{nj11} between Nanjing O_{nj} and Yangzi River O_{yz}, the spatial relation R_{nj12} between Nanjing O_{nj} and Zhongshan Mountain O_{zm}, and the attribute relation R_{nj13} between Nanjing O_{nj} and Centre District O_{cd}, where L_{yz1} is the location of Yangzi River O_{yz} in 1368, L_{zm1} is the location of Zhongshan Mountain O_{zm} in 1368 and A_{cd1} is the "administrative region" attribute of Centre District O_{cd} in 1368. The relations can be defined as follows and the diagram of these relations was shown in Figure 6.

$$R_{nj11} = \left\{ \begin{array}{c} E, Sem, type \sqsubseteq R_{nj11} | \\ E = \{L_{nj1}, L_{yz1}\}, Sem = \{"Nanjing\ is\ south\ of\ the\ Yangzi\ River"\}, type = Re_l \end{array} \right\} \sqsubseteq O_{nj}$$

$$R_{nj12} = \left\{ \begin{array}{c} E, Sem, type \sqsubseteq R_{nj12} | \\ E = \{L_{nj1}, L_{zm1}\}, Sem = \{"Nanjing\ is\ east\ of\ the\ Zhongshan\ Mountain"\}, type = Re_l \end{array} \right\} \sqsubseteq O_{nj}$$

$$R_{nj13} = \left\{ \begin{array}{c} E, Sem, type \sqsubseteq R_{nj13} | \\ E = \{A_{nj1}, A_{cd1}\}, Sem = \{"Centre\ District\ is\ part\ of\ Nanjing"\}, type = Re_a \end{array} \right\} \sqsubseteq O_{nj}$$

Correspondingly, Yangzi River contains the relation $R_{yz1} = R_{nj11}^{-}$, Zhongshan Mountain contains the relation $R_{zm1} = R_{nj12}^{-}$, and Centre District contains the relation $R_{cd1} = R_{nj13}^{-}$:

$$R_{yz1} = \left\{ \begin{array}{c} E, Sem, type \sqsubseteq R_{yz1} | \\ E = \{L_{nj1}, L_{yz1}\}, Sem = \{"Nanjing\ is\ south\ of\ the\ Yangzi\ River"\}, type = Re_l \end{array} \right\} \sqsubseteq O_{yz}$$

$$R_{zm1} = \left\{ \begin{array}{c} E, Sem, type \sqsubseteq R_{zm1} | \\ E = \{L_{nj1}, L_{zm1}\}, Sem = \{"Nanjing\ is\ east\ of\ the\ Zhongshan\ Mountain"\}, type = Re_l \end{array} \right\} \sqsubseteq O_{zm}$$

$$R_{cd1} = \left\{ \begin{array}{c} E, Sem, type \sqsubseteq R_{cd1} | \\ E = \{A_{nj1}, A_{cd1}\}, Sem = \{"Centre\ District\ is\ part\ of\ Nanjing"\}, type = Re_a \end{array} \right\} \sqsubseteq O_{cd}$$

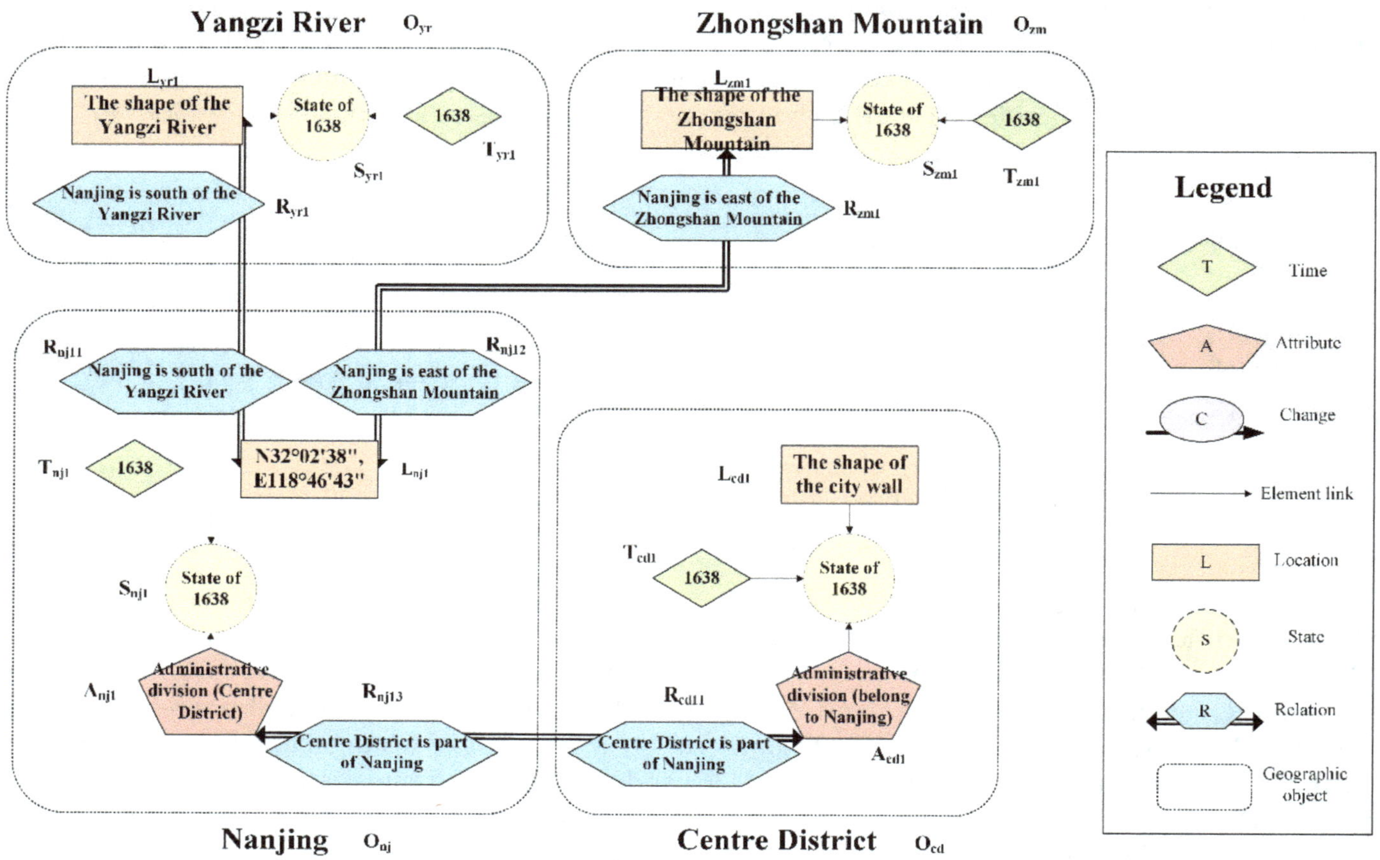

Figure 6. The diagram of relation elements of Nanjing in 1368.

The whole evolution case of administrative divisions of Nanjing can be shown in Figure 7. Corresponding to Figure 4, each geographic object contains one to three states. For instance, Yangzi River and Zhongshan Mountain have three stages of 1368, 1949, and 2018 and Jiangning and Gaochun have two stages of 1949 and 2018. As inner changes are not considered, Centre District only represented one stage. Between different stages, different kinds of changes were considered. For example, different stages of Yangzi River and Zhongshan Mountain include time change $\{C_{yz11}, C_{yz21}, C_{zm11}, C_{zm21}\}$, different stages of Nanjing include time change $\{C_{nj11}, C_{nj21}\}$, location change $\{C_{nj12}, C_{nj22}\}$, and attribute change $\{C_{nj13}, C_{nj23}\}$, and different stages of Jiangning and Gaochun include time change $\{C_{jn11}, C_{gc11}\}$ and attribute change $\{C_{jn12}, C_{gc12}\}$. Additionally, relations link different elements among both different geographic objects and same geographic object. For example, Nanjing in 1368 has relations to Yangzi River R_{nj11}, Zhongshan Mountain R_{nj12}, and Centre District R_{nj13}. Then, Nanjing in 1949 has relations to Yangzi River R_{nj21}, Zhongshan Mountain R_{nj22}, Centre District R_{nj23}, and Jiangning R_{nj24}. In 2018, Nanjing has relations to Yangzi River R_{nj31}, Zhongshan Mountain R_{nj32}, Centre District R_{nj33}, Jiangning R_{nj34}, and Gaochun R_{nj35}.

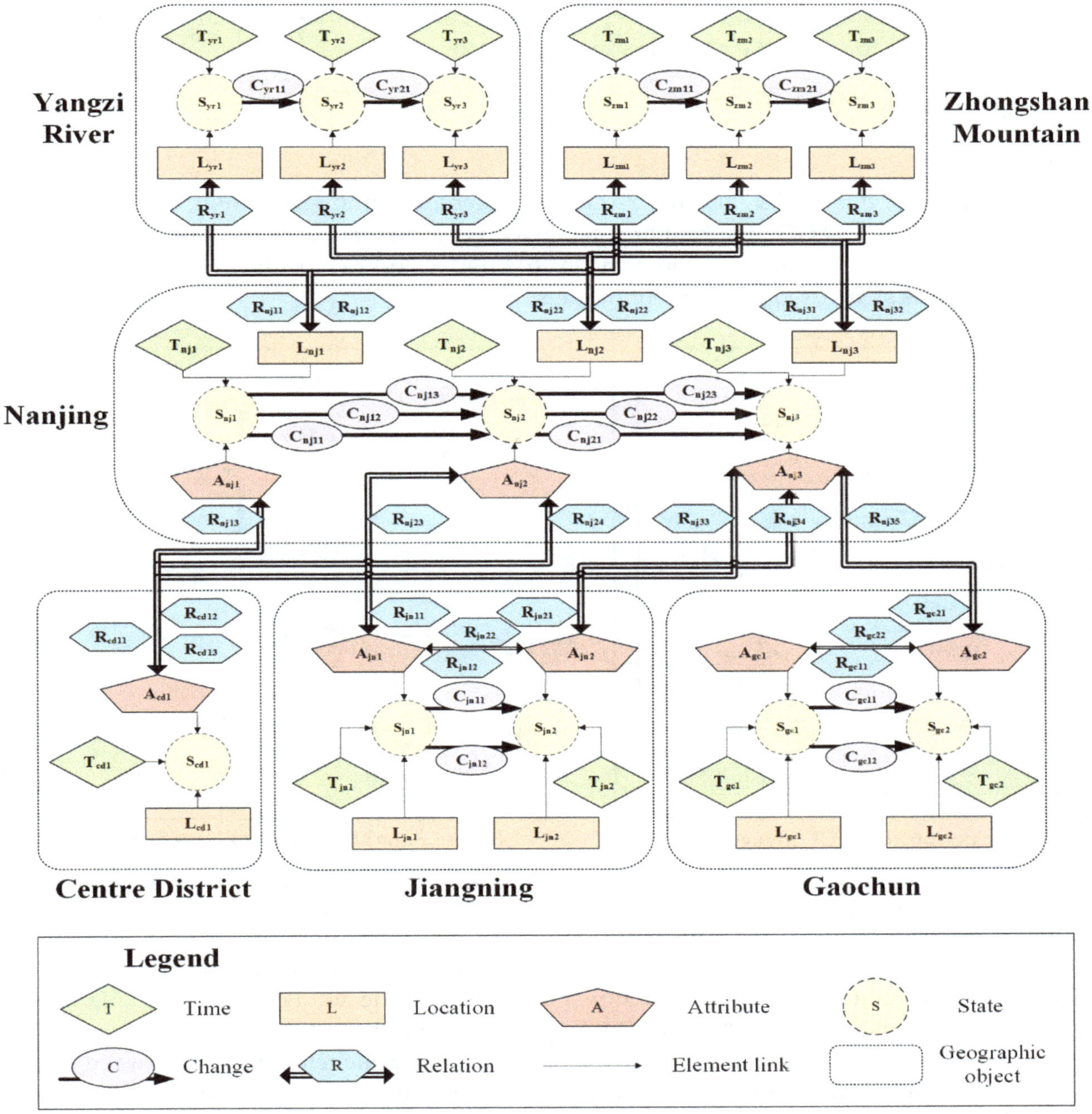

Figure 7. An overview of evolution case of administrative divisions of Nanjing and relevant geographic objects.

Note that there are also inner relations between elements. In this case, the administrative division of Jiangning in 1949 has an attribute relation R_{jn12} of "inheritance relationship" to the administrative division of Jiangning in 2018. Gaocun has the same attribute relation R_{gc11}. All these relations have the inverse relations in the opposite sides.

5. Discussion

In this section, the case study of administrative division evolution of Nanjing was constructed by using the GeoKG model and the YAGO model. YAGO is a representative open source knowledge graph with different versions. Note that we compared our model with YAGO2, a spatially and temporally enhanced version from https://www.mpi-inf.mpg.de/departments/databases-and-information-systems/research/yago-naga/yago/. Then, three kinds of core geographic questions were posted and the results were analyzed to evaluate the knowledge representation ability of these two models. Finally, a user evaluation was given to verify the comparisons objectively.

5.1. The GeoKG and the YAGO

5.1.1. Structures

The structures of the GeoKG and the YAGO are different. Although Section 2 briefly introduced the characteristics of the YAGO, the comparison between two different structures needs to be analyzed in order to understand the following comparisons of queries and the results in next section. Figure 8 shows the examples structured by different models.

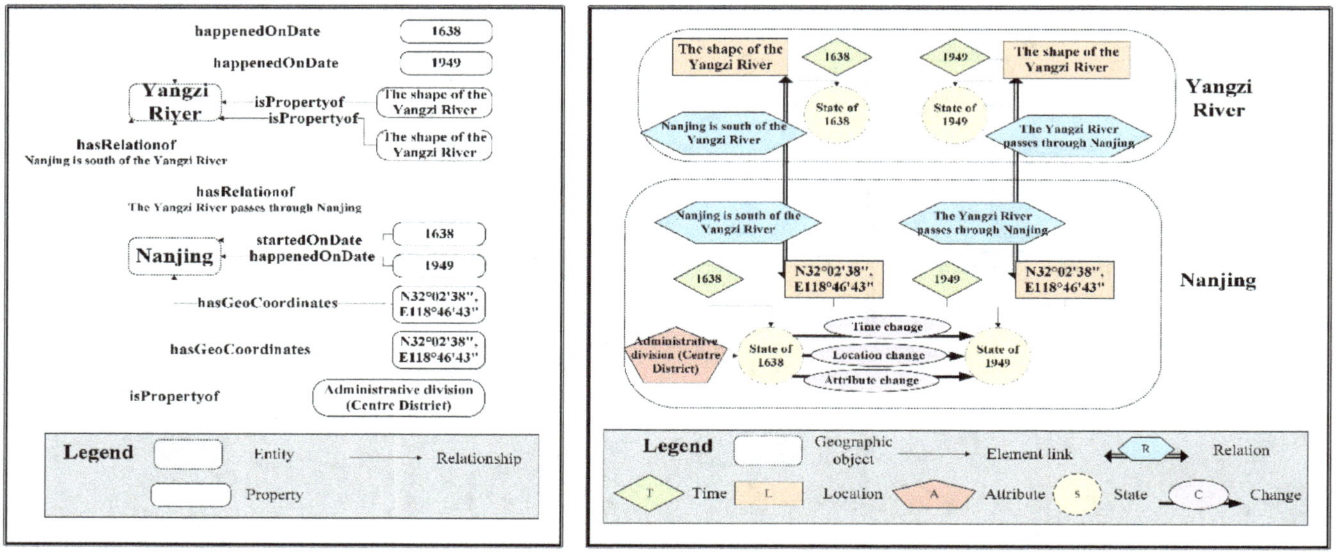

(a) the example in YAGO structure **(b) the example in GeoKG structure**

Figure 8. The examples with structures of the YAGO model and the GeoKG model. (**a**) the entities, properties and relationships in YAGO structure; (**b**) the elements in GeoKG structure.

In Figure 8a, there are only three kinds of elements: entity, property, and relationship. Each property links to a related entity by a relationship with a predicate. For example, "Nanjing" and "1638" have a relationship named "startedOnDate". Note that the YAGO structure does not contain the relationships between the properties. Thus, there are no semantic relationships between properties. In other words, the massive descriptive properties of an entity link to the entity independently. For example, two relationships happened on Nanjing and the Yangzi River: "Nanjing is south of the Yangzi River" and "The Yangzi River passes through Nanjing". It is difficult to understand this knowledge with no links between properties, whereas the GeoKG in Figure 8b sets six core elements and links these elements. With more integrated elements, the relationship of "Nanjing is south of the Yangzi River" can illustrate more clearly because this relationship links two locations in two different states of the two geographic objects. The different states providing this relationship happened on 1638 and the linked locations provide this relationship relate to different location descriptions. This knowledge cannot be provided without these links between the properties.

5.1.2. Construction

Both the GeoKG and YAGO were constructed manually by using the information about the case study of the administrative division evolution of Nanjing. The case study organized by the YAGO model was the classic SPO triple sets which has an open source ontology template. Additionally, the case study organized by the GeoKG model also stored by SPO triple sets that contain more predicates. The main supplement predicates include "isStateof", "isTimeof", "isLocationof", "isAttributeof", "isChangeof", "isRelationof", "isChangeto", and "isRelateto". All these predicates were applied to complete the semantic structure of the GeoKG model. From this perspective, the underlying storage mechanisms of GeoKG and YAGO are the same.

5.2. The Comparison of Knowledge Representation Ability between the GeoKG and the YAGO

5.2.1. Questions

Time, space, and attribute are three indispensable aspects on geoscience. These three kinds of questions can be defined as standard questions to evaluate whether the stored geographic knowledge is good. According to the differences between factual knowledge and inferential knowledge, each question was a set of two parts. To this case study, the questions are shown in Table 2.

Table 2. Questions to the GeoKG model and The YAGO model.

Question Types	Factual Question	Inferential Question
Time	When was Nanjing named?	When does Jiangning belong to Nanjing?
Space	Where is Nanjing?	What is the spatial relationship between Nanjing and Yangzi River?
Attribute	Which city does Gaochun belong to?	What administrative divisions belong to Nanjing?

5.2.2. Queries

Questions cannot be directly queried from the GeoKG and YAGO database. Thus, they need to be translated into SPARQL queries, because of either GeoKG or YAGO stored as triples in RDFs. For example, the factual question of time can be translated into SPARQL queries, as shown in Table 3.

Table 3. The SPARQL query of "When was Nanjing named?"

Steps	SPARQL Query	Semantic Meaning
1	PREFIX rdfs: <http://www.w3.org/2000/01/rdf-schema#>.	protocol
2	SELECT ?sTime WHERE {	Query content "?sTime" (start time)
3	?s rdfs:type :City.	Type is "City"
4	?s :cityName 'Nanjing'.	Get "Nanjing" geographic object
5	?s :hasName ?o.	Get time when named 'Nanjing'
6	?o :startedOnDate ?sTime.	Get started time
7	?o :usedName ?uName.	Constraint condition
8	FILTER regex(?uName, "^Nanjing")	Constraint condition setting
	}	

5.2.3. Comparison and Analysis

The collected items of YAGO and GeoKG on six questions are listed in Table 4. The comparisons will be conducted in terms of accuracy, completeness, and repetition.

a. Accuracy

In general, the results of the GeoKG are slightly better than the YAGO. Both of the two models can respond with accurate results to #Q1, #Q2, #Q3, #Q4, and #Q6. In #Q5, the result of the YAGO model returned two items and the results of the GeoKG model returned four items. Actually, "Zhenjiang" and "Nanjing" from the YAGO model are the misleading answers to the question of "Which city does Gaochun belong to?" Though the results from the GeoKG model: "Zhenjiang(Gaochun, state of 1949)", "Zhenjiang(Zhenjiang, state of 1949)", "Nanjing(Gaochun, state of 2018)" and "Nanjing(Nanjing, state of 2018)" are similar to the front, these results contain the geographic object and relevant state information which is a benefit for the users to understand the results. From this perspective, these state information from GeoKG provided more accurate information than the YAGO model.

Table 4. The results of YAGO and GeoKG on SPARQL queries.

Question Types	Questions	Results	
		YAGO	GeoKG
Time	#Q1: When was Nanjing named?	➢ 1368	➢ 1368 (Nanjing, state of 1368)
	#Q2: When does Jiangning belong to Nanjing?	➢ 1949 ➢ 2018	➢ 1949 (Jiangning, state of 1949) ➢ 2018 (Jiangning, state of 2018) ➢ 1949 (Nanjing, state of 2018) ➢ 2018 (Nanjing, state of 2018)
Space	#Q3: Where is Nanjing?	➢ N32°02′38″, E118°46′43″ ➢ N32°02′38″, E118°46′43″ ➢ N32°02′38″, E118°46′43″	➢ N32°02′38″, E118°46′43″ (Nanjing, state of 1368) ➢ N32°02′38″, E118°46′43″ (Nanjing, state of 1949) ➢ N32°02′38″, E118°46′43″ (Nanjing, state of 2018)
	#Q4: What is the spatial relationship between Nanjing and Yangzi River?	➢ Nanjing is south of the Yangzi River ➢ The Yangzi River passes through Nanjing ➢ The Yangzi River passes through Nanjing	➢ Nanjing is south of the Yangzi River (Nanjing, state of 1368) ➢ Nanjing is south of the Yangzi River (Yangzi River, state of 1368) ➢ The Yangzi River passes through Nanjing(Nanjing, state of 1949) ➢ The Yangzi River passes through Nanjing(Yangzi River, state of 1949) ➢ The Yangzi River passes through Nanjing(Nanjing, state of 2018) ➢ The Yangzi River passes through Nanjing(Yangzi River, state of 2018)
Attribute	#Q5: Which city does Gaochun belong to?	➢ Zhenjiang ➢ Nanjing	➢ Zhenjiang(Gaochun, state of 1949) ➢ Zhenjiang(Zhenjiang, state of 1949) ➢ Nanjing(Gaochun, state of 2018) ➢ Nanjing(Nanjing, state of 2018)
	#Q6: What administrative divisions belong to Nanjing?	➢ Centre District ➢ Jiangning ➢ Jurong ➢ Dangtu ➢ Luhe ➢ Pukou ➢ Hexian ➢ Lishui ➢ Gaochun ➢ Qixia	➢ Centre District(Nanjing, state of 1368) ➢ Centre District(Nanjing, state of 1949) ➢ Centre District(Nanjing, state of 2018) ➢ Jiangning(Nanjing, state of 1949) ➢ Jiangning(Nanjing, state of 2018) ➢ Jurong(Nanjing, state of 1949) ➢ Dangtu(Nanjing, state of 1949) ➢ Luhe(Nanjing, state of 1949) ➢ Luhe(Nanjing, state of 2018) ➢ Pukou(Nanjing, state of 1949) ➢ Pukou(Nanjing, state of 2018) ➢ Hexian(Nanjing, state of 1949) ➢ Lishui(Nanjing, state of 2018) ➢ Gaochun(Nanjing, state of 2018) ➢ Qixia(Nanjing, state of 2018) ➢ more items

b. Completeness

Although both of these two models can return the complete results, the results of the GeoKG contains more semantic integrity. In #Q6, YAGO returned 10 items: Centre District, Jiangning, Jurong, Dangtu, Luhe, Pukou, Hexian, Lishui, Gaochun, and Qixia. Among these divisions, Centre District belonged to Nanjing since 1368. Jiangning, Luhe and Pukou belonged to Nanjing since 1949. Jurong, Dangtu and Hexian belonged to Nanjing in 1949. Lishui, Gaochun, and Qixia belonged to Nanjing in 2018. As the question does not have an explicit time constraint condition, YAGO returned all the items, whereas GeoKG returned 30 items and each item recorded the target object and its relevant geographic object and state. It contains the item of "Centre District (Nanjing, state of 1368)" and the item of "Centre District (Centre District, state of 1368)", because of the relation existed oppositely.

c. Repetition

The results of the GeoKG has more repeat items than the results from the YAGO. The results from the YAGO have repeat items in #Q3 and #Q4, because of the records are repeat. However, the GeoKG model is different. In #Q2, #Q4, #Q5, and #Q6, the results of the GeoKG have many repeat items; for example, the items of "1949 (Jiangning, state of 1949)" and "1949 (Nanjing, state of 2018)" in #Q2. The query target object "1949" is the same. In spite of these two items sourced from different geographic objects (Jiangning and Nanjing), these two items are still quite similar, which pushed more redundant information to the users.

In summary, the results of the GeoKG model are more accurate and complete than the YAGO model with the enhancing state information. It can decrease the influence from the fuzziness questions and obtain answers with more semantic meaning (e.g., geographic object and its relevant state). Meanwhile, the GeoKG model could generate more pairs results (e.g., "Nanjing is south of the Yangzi River (Nanjing, state of 1368)" vs. "Nanjing is south of the Yangzi River (Yangzi River, state of 1368)"), because the relation is stored oppositely in a different geographic object.

5.2.4. User Evaluation

An online questionnaire survey is also given in order to verify the results of comparative analyses. The questionnaire is divided into eight parts. The first part is the basic information survey that asks individuals four aspects of information (gender, familiarity to the research area, background, and education level). The statistics of these basic information are shown in Figure 9. The 2nd–7th parts correspond to the questions #Q1–#Q6 and ask the questions about the best answer, accuracy, completeness, and repetition. The 8th part are summary questions including the overall evaluation, scores on YAGO and scores on GeoKG on different aspects. The scores are set as 1–5 corresponding to very bad, bad, normal, good, and very good, and each score group includes an overall score, accuracy score, completeness score, and repetition score. There are 106 valid feedbacks we finally received.

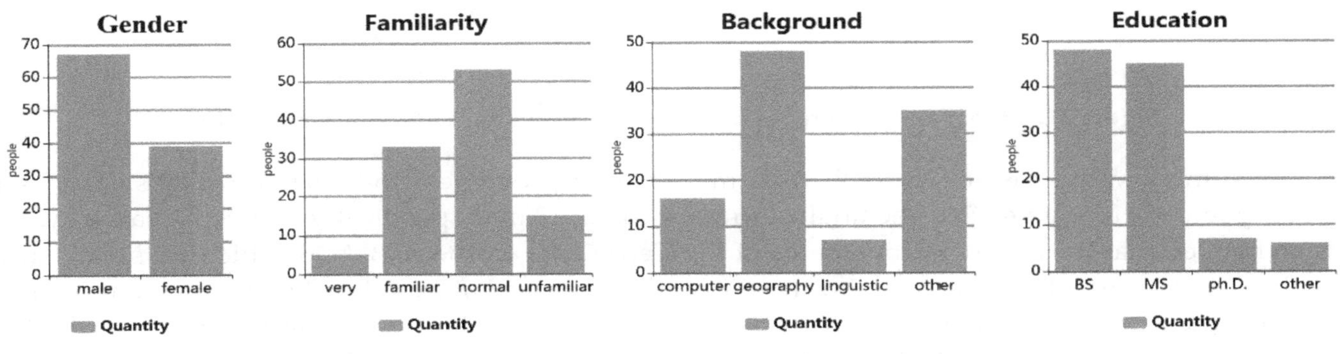

Figure 9. The statistics of the four main types of the basic information about the survey.

Figure 10 shows the best answers on #Q1–#Q6 and the overall scores of the YAGO and the GeoKG. In the best answer histogram, the overall results show 54.72% individuals support the GeoKG, which is 23.59% higher than the YAGO at 31.13%. Specifically, the quantities of #Q1 and #Q2 are quite close but the quantities of #Q3–#Q6 are not. The quantities of the GeoKG are much higher than the YAGO among the last four questions, especially in #Q5. The line charts of overall scores on YAGO and GeoKG also show that the evaluation of GeoKG is better than the YAGO. A 7.8% improvement on the average score from the YAGO (3.15) to the GeoKG (3.49) is obtained.

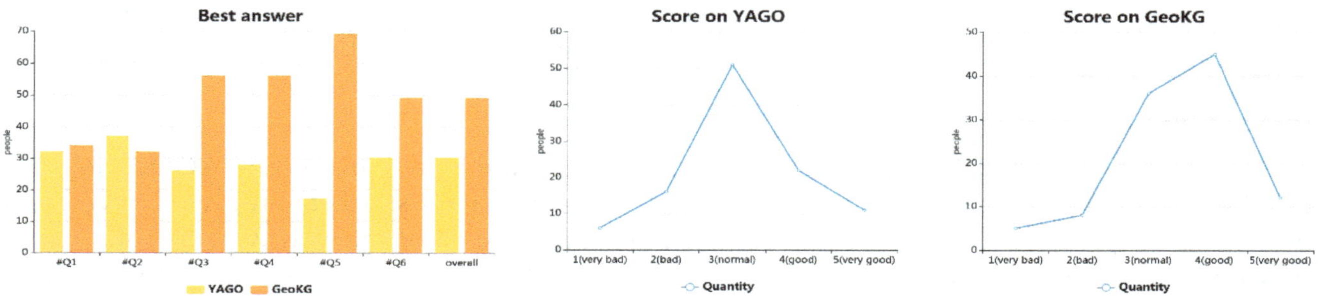

Figure 10. The best answer on #Q1–#Q6 and the overall scores of the YAGO and the GeoKG.

From the sub-aspects (accuracy, completeness, and repetition) of the point of view, different quantities can immediately show the scores from the YAGO and the GeoKG. Different quantities show the ability from the model (details in Figure 11). Nearly all three aspects of the YAGO obtained the score 3, whereas the GeoKG was different: a score of 4 on accuracy, a score of 4–5 on completeness, and a score of 3–4 on repetition. Comparing these scores, it can be seen that there is little promotion on the accuracy from the 3.11 average score in YAGO to the 3.78 average score in GeoKG. An overwhelming improvement shows on the completeness of the answers from a 2.99 average score in YAGO to a 3.87 average score in GeoKG. Additionally, the GeoKG also obtains a higher repetition from a 3.01 average score in YAGO to a 3.42 average score in GeoKG.

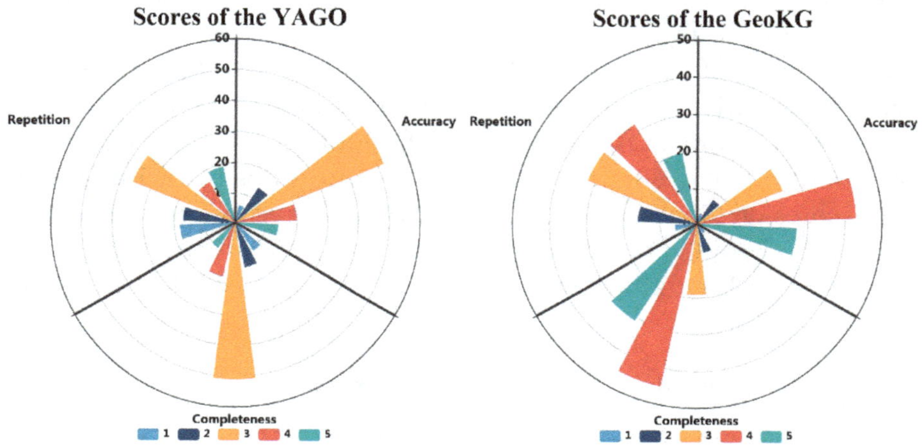

Figure 11. Rose maps of scores of different aspects on the YAGO and the GeoKG.

In summary, the answers from GeoKG makes an improvement to those of YAGO's. The user evaluation objectively verified the analyses in Section 5.2.3 and specifically showed clear answers. It can be seen that the main improvements of the GeoKG are on the #Q3–#Q6, which are spatial and attribute questions. These answers to these questions require more related state information and temporal information, which need the links between the elements (Figure 8). This is the reason why the GeoKG is better than the YAGO. In addition, the GeoKG contains more redundancy information than the YAGO because of the bi-directionality of the relation element. This could be a focus of continuous further research on the index and applications in the future.

6. Conclusions

Given that much attention has been paid to the representation of geographic knowledge, this paper is focused on the development of current geographic knowledge representations. We analyzed the problems of current geographic knowledge representation and found that two issues must be improved: the elements of geographic knowledge representation and the supplement of the construction operators of DL.

Following the basic idea of the six core geographical questions, we designed a conceptualized model called GeoKG based on the six elements around the geographical questions, then supplemented the construction operators of DL and finally provided the formalizations of the model with these operators. Additionally, an evolution case of administrative divisions of Nanjing was formalized and illustrated. Then, the knowledge graphs were constructed by both the GeoKG model and the YAGO model by using the case study. After setting a group of standard geographic questions, the query results were finally compared. The results showed that the results of GeoKG are more accurate and complete than the YAGO results, which are verified by the following user evaluation. This comparison indicates the GeoKG model displays its ability to organize geographic knowledge in computers and is a promising and powerful model for geographic knowledge representation.

Author Contributions: Conceptualization: S.W., X.Z., and P.Y.; data curation: S.W.; formal analysis: S.W.; funding acquisition: X.Z.; investigation: S.W., M.D., Y.L., and H.X.; methodology: S.W., X.Z., and M.D.; supervision: X.Z.; validation: P.Y., M.D., and Y.L.; visualization: M.D.; writing—original draft: S.W.; writing—review and editing: S.W. and P.Y.

Acknowledgments: The authors thank Mingguang Wu, Junzhi Liu and Jie Zhu for their critical reviews and constructive comments. This research is supported by the National Natural Science Foundation of China grants no. 41631177 and no. 41671393 and the National Key Research and Development Program of China, no. 2017YFB0503602.

References

1. Golledge, R.G. The Nature of Geographic Knowledge. *Ann. Assoc. Am. Geogr.* **2015**, *92*, 1–14. [CrossRef]
2. Haubrich, H. International Charter on Geographical Education. *J. Geogr.* **1997**, *96*, 33–39.
3. Davis, R. What Is a Knowledge Representation? *AI Mag.* **1993**, *14*, 17–33.
4. Zhang, Y.; Gao, Y.; Xue, L.L.; Shen, S.; Chen, K. A common sense geographic knowledge base for GIR. *Sci. Technol. Sci.* **2008**, *51*, 26–37. [CrossRef]
5. Kuhn, W. Modeling Vs Encoding for the Semantic Web. *Semant. Web* **2010**, *1*, 11–15.
6. Baader, F.; Sattler, U. An Overview of Tableau Algorithms for Description Logics. *Stud. Log.* **2001**, *69*, 5–40. [CrossRef]
7. Hoffart, J.; Suchanek, F.M.; Berberich, K.; Weikum, G. YAGO2: A spatially and temporally enhanced knowledge base from Wikipedia. *Artif. Intell.* **2013**, *194*, 28–61. [CrossRef]
8. Guarino, N.; Oberle, D.; Staab, S. What Is an Ontology? *Handb. Ontol.* **2009**, 1–17. [CrossRef]
9. Ding, Y.; Foo, S. Ontology research and development. Part 1: A review of ontology generation. *J. Inf. Sci.* **2002**, *28*, 123–136.
10. Couclelis, H. Ontologies of geographic information. *Int. J. Geogr. Inf. Sci.* **2010**, *24*, 1785–1809. [CrossRef]
11. Siricharoen, W.V.; Pakdeetrakulwong, U. A Survey on Ontology-Driven Geographic Information Systems. In Proceedings of the Fourth International Conference on Digital Information and Communication Technology and It's Applications, Bangkok, Thailand, 6–8 May 2014.
12. Gruber, T.R. Toward principles for the design of ontologies used for knowledge sharing? *Int. J. Hum.-Comput. Stud.* **1995**, *43*, 907–928. [CrossRef]
13. Fonseca, F.T.; Egenhofer, M.J. Ontology-driven geographic information systems. In Proceedings of the 7th ACM International Symposium on Advances in Geographic Information Systems, Kansas City, MO, USA, 2–6 November 1999; Volume 71, pp. 14–19.
14. Jun, X.U.; Tao, P.; Yao, Y. Conceptual Framework and Representation of Geographic Knowledge Map: Conceptual Framework and Representation of Geographic Knowledge Map. *J. Geo-Inf. Sci.* **2010**, *12*. [CrossRef]
15. Chen, J.; Deng, S.; Chen, H. Crowdgeokg: Crowdsourced Geo-Knowledge Graph. In Proceedings of the China Conference on Knowledge Graph and Semantic Computing, Chengdu, China, 26–29 August 2017.
16. Arvor, D.; Durieux, L.; Andrés, S.; Laporte, M.-A. Advances in Geographic Object-Based Image Analysis with ontologies: A review of main contributions and limitations from a remote sensing perspective. *ISPRS J. Photogramm. Remote. Sens.* **2013**, *82*, 125–137. [CrossRef]
17. Brown, S.H. *Knowledge Representation and the Logical Basis of Ontology*; Springer: London, UK, 2012; pp. 11–50.

18. Pittet, P.; Cruz, C.; Nicolle, C. Modeling Changes for Shoin(D) Ontologies: An Exhaustive Structural Model. In Proceedings of the IEEE Seventh International Conference on Semantic Computing, Irvine, CA, USA, 16–18 September 2013.
19. Sattler, U.; Horrocks, I. A description logic with transitive and inverse roles and role hierarchies. *J. Log. Comput.* **1999**, *9*, 385–410. [CrossRef]
20. Horrocks, I.; Sattler, U.; Tobies, S. Practical Reasoning for Expressive Description Logics. In Proceedings of the International Conference on Logic for Programming and Automated Reasoning, Tbilisi, Georgia, 6–10 September 1999.
21. Horrocks, I.; Sattler, U.; Tobies, S. Practical reasoning for very expressive description logics. *Log. J. IGPL* **2000**, *8*, 239–263. [CrossRef]
22. Aachen, R.; Informatik, L.T.; Horrocks, I.; Sattler, U.; Tobies, S. Pspace-Algorithm for Deciding Alcnir+-Satisfiability. In *LTCS-Report 98-08*; ACM Digital Library: Aachen, Germany, 1998.
23. Mei, J. From Alc to Shoq(D):A Survey of Tableau Algorithms for Description Logics. *Comput. Sci.* **2005**, *32*, 1–11. [CrossRef]
24. Singhal, A. *Official Google Blog: Introducing the Knowledge Graph: Things, Not Strings*; Northwestern University: Evanston, IL, USA, 2012.
25. Suchanek, F.M.; Kasneci, G.; Weikum, G. Yago: A Core of Semantic Knowledge. In Proceedings of the 16th International Conference on World Wide Web (WWW), Banff, AB, Canada, 8–12 May 2007; Volume 272, pp. 697–706.
26. Bollacker, K.; Cook, R.; Tufts, P. Freebase: A Shared Database of Structured General Human Knowledge. In Proceedings of the AAAI Conference on Artificial Intelligence, Vancouver, BC, Canada, 22–26 July 2007.
27. Wu, W.; Li, H.; Wang, H.; Zhu, K.Q. Probase: A Probabilistic Taxonomy for Text Understanding. In Proceedings of the 2012 ACM SIGMOD International Conference on Management of Data (SIGMOD'12), Scottsdale, AZ, USA, 20–24 May 2012.
28. Lehmann, J. Dbpedia: A Large-Scale, Multilingual Knowledge Base Extracted from Wikipedia. *Semant. Web* **2015**, *6*, 167–195.
29. Li, J.; Liu, R.; Xiong, R. A Chinese Geographic Knowledge Base for Gir. In Proceedings of the IEEE International Conference on Computational Science and Engineering, Guangzhou, China, 21–24 July 2017.
30. Kauppinen, T.; Espindola, G.M. Ontology-Based Modeling of Land Change Trajectories in the Brazilian Amazon. In Proceedings of the Geoinformatik, Münster, Germany, 15–17 June 2013.
31. Zhu, Y.; Zhou, W.; Xu, Y.; Liu, J.; Tan, Y. Intelligent Learning for Knowledge Graph towards Geological Data. *Sci. Program.* **2017**, *2017*, 1–13. [CrossRef]
32. William, M. (Ed.) *The American Heritage Dictionary of the English Language*; New College Edition; Houghton Mifflin Company: Boston, MA, USA, 1980.

7

Parallel Cellular Automata Markov Model for Land use Change Prediction over MapReduce Framework

Junfeng Kang [1,2], Lei Fang [3,*], Shuang Li [1] and Xiangrong Wang [3]

1 School of Earth Sciences, Zhejiang University, Hangzhou 310027, China; junfeng.kang@zju.edu.cn (J.K.); namespacezhang@mail.jxust.edu.cn (S.L.)

2 School of Architectural and Surverying & Mapping Engineering, Jiangxi University of Science and Technology, Ganzhou 341000, China

3 Department of Environmental Science and Engineering, Fudan University, Shanghai 200438, China; xrxrwang@fudan.edu.cn

* Correspondence: fanglei@fudan.edu.cn

Abstract: The Cellular Automata Markov model combines the cellular automata (CA) model's ability to simulate the spatial variation of complex systems and the long-term prediction of the Markov model. In this research, we designed a parallel CA-Markov model based on the MapReduce framework. The model was divided into two main parts: A parallel Markov model based on MapReduce (Cloud-Markov), and comprehensive evaluation method of land-use changes based on cellular automata and MapReduce (Cloud-CELUC). Choosing Hangzhou as the study area and using Landsat remote-sensing images from 2006 and 2013 as the experiment data, we conducted three experiments to evaluate the parallel CA-Markov model on the Hadoop environment. Efficiency evaluations were conducted to compare Cloud-Markov and Cloud-CELUC with different numbers of data. The results showed that the accelerated ratios of Cloud-Markov and Cloud-CELUC were 3.43 and 1.86, respectively, compared with their serial algorithms. The validity test of the prediction algorithm was performed using the parallel CA-Markov model to simulate land-use changes in Hangzhou in 2013 and to analyze the relationship between the simulation results and the interpretation results of the remote-sensing images. The Kappa coefficients of construction land, natural-reserve land, and agricultural land were 0.86, 0.68, and 0.66, respectively, which demonstrates the validity of the parallel model. Hangzhou land-use changes in 2020 were predicted and analyzed. The results show that the central area of construction land is rapidly increasing due to a developed transportation system and is mainly transferred from agricultural land.

Keywords: CA Markov; land-use change prediction; Hadoop; MapReduce; cloud computing

1. Introduction

Studying land use/land cover changes in different times and places and predicting land-use structures and spatial layouts can provide scientific support for the utilization of regional land resources, the protection of regional ecological environments, and sustainable social and economic development [1,2].

Many researchers have proposed their own land-use-change simulation and prediction models, such as CLUE [3], CLUE-S [4], cellular automata (CA) [2,5,6], Markov chain [7,8], SLEUTH [9,10], and the spatial logistic model [11]. Since the early 1980s when Wolfram first proposed the CA model [12], many studies have been conducted to use the CA model to simulate urban land-use changes [2,5,6,13] and researchers have integrated other methods or models, such as neural networks [14], support vector machines (SVM) [15], ant-colony optimization [16], and the Markov chain [17], into the CA model to simulate and monitor land-use change.

The CA-Markov model was one of the most widely used extended CA models and was used in the prediction and simulation of land-use changes in many countries, such as the United States [18], Brazil [19], Portugal [20], Egypt [21], Ethiopia [22], Bangladesh [23], Malaysia [24], and China [25]. It has also been applied to the research of the evolution of urban-settlement patterns [26], the process of spatial dynamic vegetation changes [27], and land transfer across metropolitan areas [28].

As land-use change simulation and prediction involves tremendous numbers of data and calculations, in recent years, some studies have designed parallel CA algorithms on Central Processing Unit (CPU) parallel computing [29], Message Passing Interface (MPI) [30], Graphics Processing Unit (GPU) parallel [31], and GPU/CPU hybrid parallel [32] to simulate urban growth. However, the parallel CA method cannot deal with the connection between partitions after a research area is divided into several pieces, resulting in different final-prediction results, whereas the traditional Markov method is able to maintain the integrity of the entire study area but results in a lack of spatial relationship of the land cell. On the other hand, with the development of Big Data technologies and applications, MapReduce is a promising method to improve traditional-serial-algorithm running efficiency and has been applied and proven effective in many cases. Rathore et al. [33] proposed a real-time remote-sensing image processing and analysis application. Raojun et al. [34] proposed a parallel-link prediction algorithm based on MapReduce. Wiley K. et al. [35] analyzed astronomical graphics based on MapReduce, while Almeer [36] used Hadoop to analyze remote-sensing images, improving batch-reading and -writing efficiency. For the CA Markov model, the MapReduce framework is not only capable of efficient parallel processing, but can also be a coupling to the CA-Markov model: The "Map" corresponds to the CA process to realize the parallelism of land-use-unit-change prediction; "Reduce" refers to the Markov process to achieve overall prediction of land-use changes. However, because the key problem of segmentation and connection remains unresolved, there is little research on the parallel CA-Markov model for land-use-change prediction over the MapReduce framework.

Based on in-depth analysis of the parallelism of the CA Markov model, this paper first proposes a parallel solution that uses the MapReduce framework to improve the CA Markov model for land-use-change prediction. The parallel CA-Markov model can not only solve the contradiction that the traditional CA-Markov model cannot simultaneously realize the integrity and segmentation for land-use change simulation and prediction, but can also ensure both efficiency and accuracy and realize land-use change prediction in a cloud-computing environment.

2. Materials and Background Technologies

2.1. Study Area and Data

Hangzhou is located in the southeast coast of China, which is the political, economic, cultural, and financial center of Zhejiang Province. Hangzhou has a complex topography: The west is a hilly area with the main mountain range, including Tianmu Mountain, and the east is a plain area with low-lying terrain and dense river networks.

The 2006 Landsat TM and the 2013 Landsat8 remote-sensing image with 30 m resolution of the study area were downloaded from http://www.gscloud.cn/. Other experiment datasets included DEM with a 30 m resolution, road-network data, traffic-site data, and location-address data.

Nowadays, many researches develop their own auto image-identification method to classify the high-resolution image, especially unmanned aerial vehicle (UAV) images [37,38]. Because Landsat images were the medium-resolution image, we chose to use a semi-manual method to preprocessed and interpreted to obtain land-use data using ENVI 5.3 and ArcGIS 10.2. [39]. As shown in Figure 1, the workflow of Landsat images classification included four main steps, namely definition of the classification inputs, preprocessing, region of interest, and classification. The processing steps mainly included geocorrection, geometric rectification, or image registration, radiometric calibration and atmospheric correction, and topographic correction [40]. A support-vector-machine (SVM) classifier was selected for land-use classification [41].

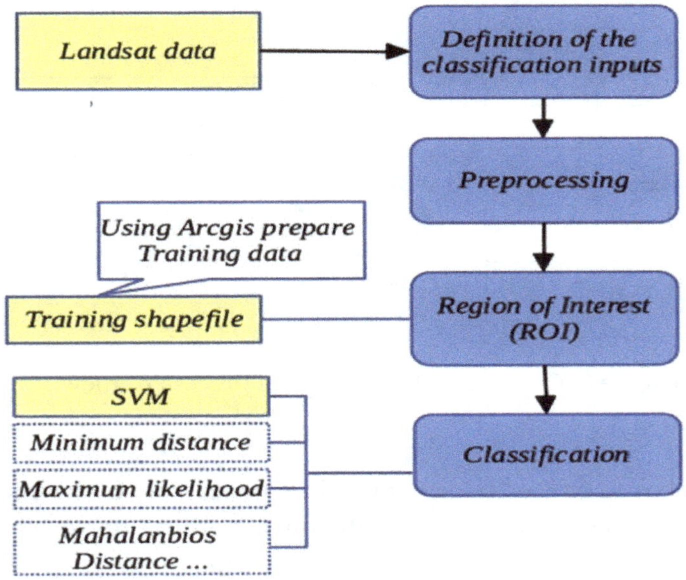

Figure 1. Landsat-image classification flow.

In general, land-use types include cultivated land, forest, grassland, waters, construction land, gardens, transportation land, unused land, and swampland. In our experiment, four land-use types were defined in the training shapefile after reclassifying. The land-use-type reclassification method of construction land (B), agricultural land (A), nature reserve (N), and water area (W) are defined as shown in Table 1.

Table 1. Land-use reclassifications.

Level 1.	Level 2	Definition
Construction land (B)	Land for construction (B1), land for transportation (B2)	Land for buildings and structures.
Agricultural land (A)	Cultivated land (A1), garden (A2)	Land for agricultural production.
Water area (W)	Waters (W1), swampland (W2)	River surface, lake surface, swamp.
Nature reserve (N)	Forest (N1), grassland (N2), unused land (N3)	Land with little or no human activity that did not include agricultural land, construction land, and waters.

2.2. *MapReduce*

The MapReduce [42] program consists of two functions, Map and Reduce. Both of these functions take key/value (key-value pair) as input and output. The Map function receives a user-entered key-value pair (k_1, v_1) and processes it to generate a key-value pair (k_2, v_2) as an intermediate result. Then, the corresponding values of all the same intermediate keys (k_2) are aggregated to generate a list of values for the k_2 key list (v_2), which is used as an input to the Reduce function and processed by the Reduce function to obtain the final result list (k_3, v_3). The process can be expressed by the following formula:

$$\text{Map} : (k_1, v_1) - \text{list}(k_2, v_2) \tag{1}$$

$$\text{Reduce} : (k_2, \text{list}(v_2)) - \text{list}(k_3, v_3) \tag{2}$$

The MapReduce framework is shown in Figure 2:

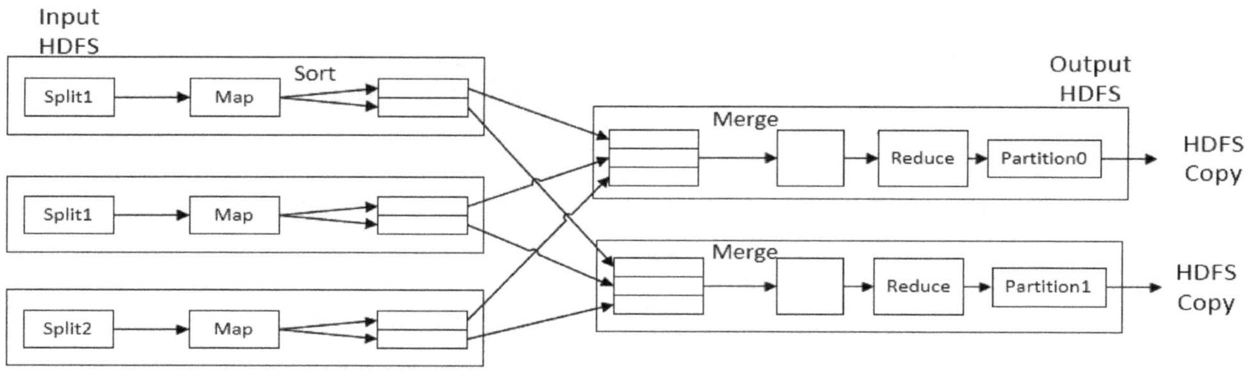

Figure 2. Overview of MapReduce process framework.

2.3. CA Markov Model

The Markov model is based on the theory of random processes. In this model, given the initial state and state-transition probabilities, simulation results have nothing to do with the historical condition before the current condition, which can be used to describe land-use changes from one period to another. We can also use this as a basis to predict future changes. Change was found by creating a land-use-change transition-probability matrix from periods t to $t + 1$, which is the basis to predict future land-use changes [43].

$$S(t+1) = P_{ij} \times S(t) \tag{3}$$

$S(t + 1)$ denotes the state of the land-use system at times $t + 1$ and t, respectively. P_{ij} is the state-transition matrix.

CA has four basic components: Cell and cell space, cell state, neighborhood, and transition rules. The CA model can be expressed as follows [44]:

$$S(t+1) = f(S(t), N) \tag{4}$$

In the formula, S is a state set of a finite and discrete state, t and $t + 1$ are different moments, N is the neighborhood of the cell, and f is the cell-transition rule of the local space.

Usually, in order to make cellular automata better simulate a real environment, space constraint variable β needs to be introduced to express the topographic terrain, as well as the adaptive constraints and restrictive constraints of the spatial-influence factors on the cells. The formula becomes

$$S(t+1) = f(S(t), N, \beta). \tag{5}$$

The separate Markov model lacks spatial knowledge and does not consider the spatial distribution of geographic factors and land-use types, while the CA-Markov model adds spatial features to the Markov model, uses a cellular-automata filter to create weight factors with spatial character, and changes the state of the cells according to the state of adjacent cells and the transition rules [21].

3. Methods

3.1. Parallel CA-Markov Model Overview

3.1.1. Parallel CA-Markov Structure

Figure 3 is the structure of parallel CA-Markov over MapReduce framework. The structure contains four layers from bottom to top: The data layer, parallel CA-Markov model layer, land-use transition-direction determination layer, and land-use prediction layer.

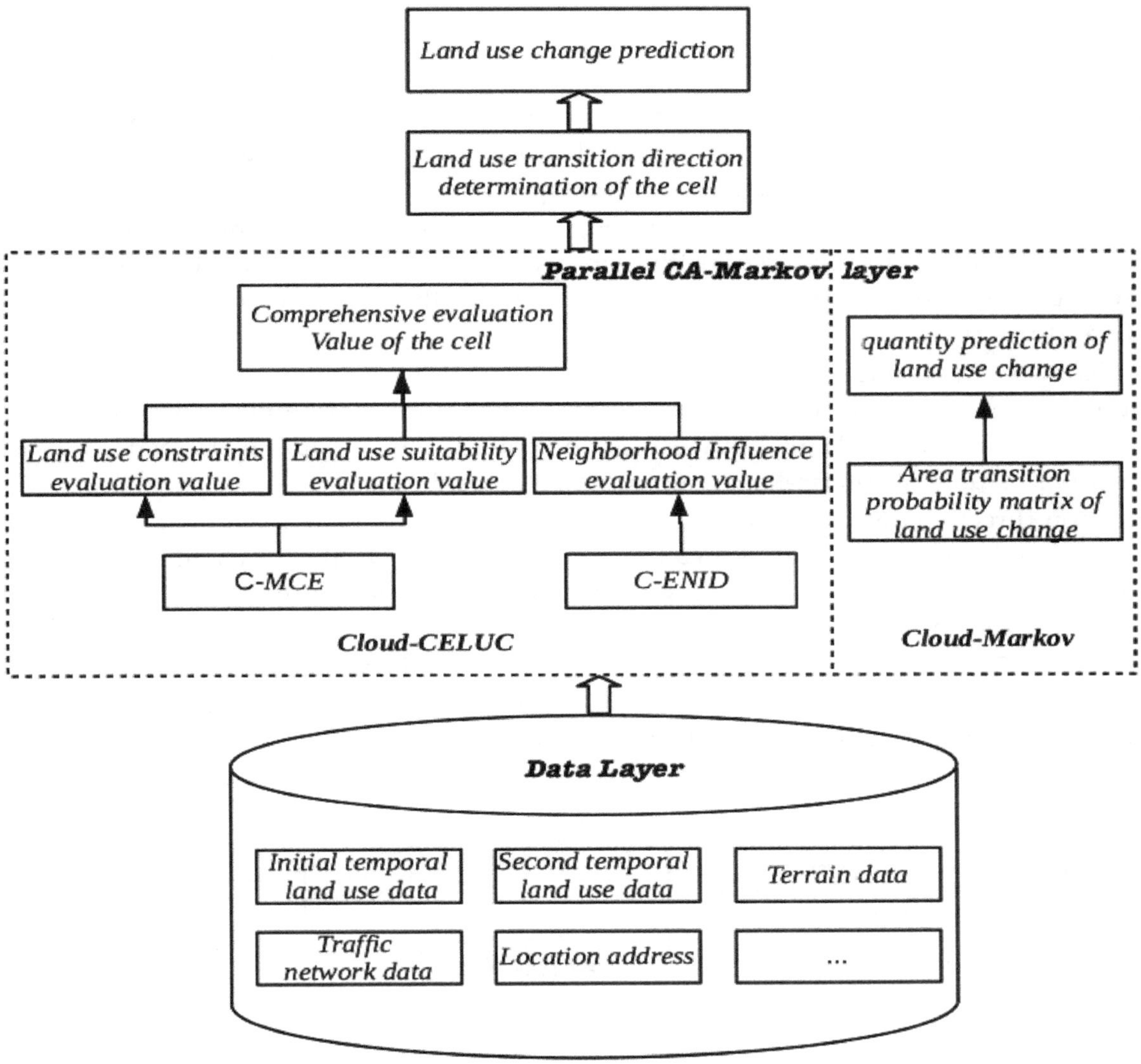

Figure 3. Structure of parallel cellular-automata Markov over MapReduce.

The experiment data include two-phase remote-sensing images, terrain data, traffic-network data, and location-address data.

The parallel CA-Markov model can mainly be divided into two parts: A parallel Markov algorithm based on MapReduce (Cloud-Markov), and a comprehensive evaluation method of land-use changes based on MapReduce (Cloud-CELUC).

Cloud-Markov was used to calculate the area transition-probability matrix of land-use changes. The algorithm of Cloud-Markov is discussed in Section 3.2.

Cloud-CELUC includes three main parts: Evaluation of the influence of neighborhoods under a cloud-computing environment (C-ENID), multicriteria evaluation under a cloud-computing environment (C-MCE), and comprehensive evaluation of land use. Among them, C-ENID is a parallel CA model over MapReduce, and C-MCE is a MapReduce algorithm to calculate constrained-evaluation and suitability-evaluation values. The cell's neighborhood influences and C-ENID's calculation method is discussed in Section 3.3.1. C-MCE design factors are discussed in Section 3.3.2, and the Cloud-CELUC algorithm is discussed in detail in Section 3.3.3.

After using Cloud-Markov to calculate the transition-probability matrix of land-use changes and using Cloud-CELUC to calculate each cell's comprehensive evaluation value, each cell's land-use transition direction was determined, which is discussed in Section 3.4. Finally, land-use-change predictions were obtained, and the land-use prediction experiment is discussed in Section 4.

3.1.2. Parallel CA-Markov Workflow

As shown in Figure 4, the flow of the parallel CA-Markov model has five major steps, as follows:

(1) Data-processing: Preprocessing and interpreting remote-sensing images to land-use maps, designing multicriteria-evaluation factors, and storing images, land-use data, and multicriteria-evaluation factors into the Hadoop HDFS.

(2) Parallel Markov: Using the overlay method, analyzing two-phase images and land-use data to obtain each cell's land-use-type transition probability, and calculating the total number of cells in each land-use-type transition direction, and counting the area transition-probability matrix of each land-use type.

(3) Parallel CA: In Cloud-CELUC, C-ENID was used to calculate the cell's neighborhood influence value. C-MCE was designed to calculate multicriteria-evaluation values, including constraint-evaluation and suitability-evaluation values. These values were then used to calculate the statistical table of comprehensive evaluation of land use.

(4) Transition-direction determination stage: Loop reading each cell's transition probability from the statistical table of comprehensive-evaluation values in the parallel CA stage and combining the area transfer-probability matrix of each land-use type in the parallel Markov stage to decide a cell's land-use-type transition direction.

(5) Land-use-change prediction: In our experiment, we used data from 2006 to predict 2013 land-use changes and then evaluate the precision of the parallel CA-Markov model with a Kappa coefficient. Land-use change prediction for 2020 was then obtained.

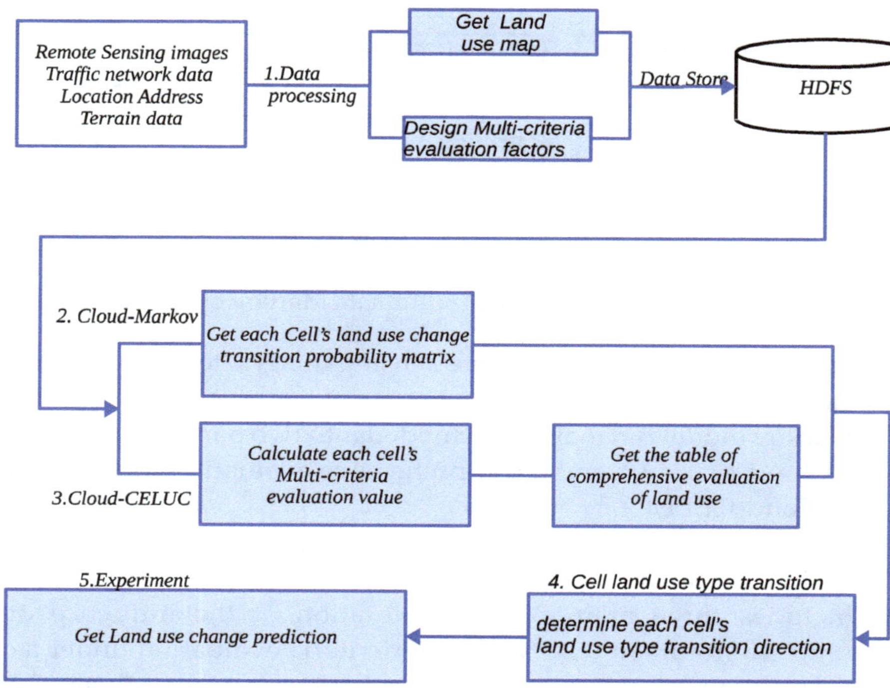

Figure 4. Parallel Cellular Automata-Markov (CA-Markov) flow.

3.2. Markov Model Parallel Processing (Cloud-Markov)

Using an overlay method to analyze two-phase raster images and land-use data with the same spatial position, we obtained the transition direction of each cell and calculated the number of cells in each transition direction, then obtained the area transfer matrix of each land-use type and calculated the area probability matrix of land use. Each year's area transfer matrix was then obtained using the probability matrix divided by the intervals of these two raster images. Then, MapReduce functions of the Markov model given in the study area are as follows:

$$\text{Map} : (N, (T_1, T_2)) \rightarrow List\left(C_{mk,}i\right) \tag{6}$$

$$\text{Combiner} : M\left(C_{mk,}list(i)\right) \rightarrow List\left(C_{mk,}s\right) \tag{7}$$

$$\text{Reduce} : L\left(C_{mk,}list(i)\right) \rightarrow List\left(C_{mk,}s\right) \tag{8}$$

where N is the row offset of the input row, T_1 represents all cells' land-use type of the earlier raster image, T_2 represents all cells' land-use type of the later raster image, C_{mk} represents the cell conversion from land-use type m to land-use type k, i indicates the cell transfer number of C_{mk}, M is the number of land-use types in a MapReduce node, s is the total cell number of C_{mk} in a MapReduce node, L is the total number of land-use types, and q is the combined value of the total number of cells and transition probability for C_{mk} conversion in the entire study area. For example, "100-3.12%" means that there were 100 cells with C_{mk} conversion, and transition probability was 3.12%. Figure 5 is the flow of Cloud-Markov algorithm. After summing the number of same land-use type transition direction in each node, the area transition matrix was calculated at the Reduce stage.

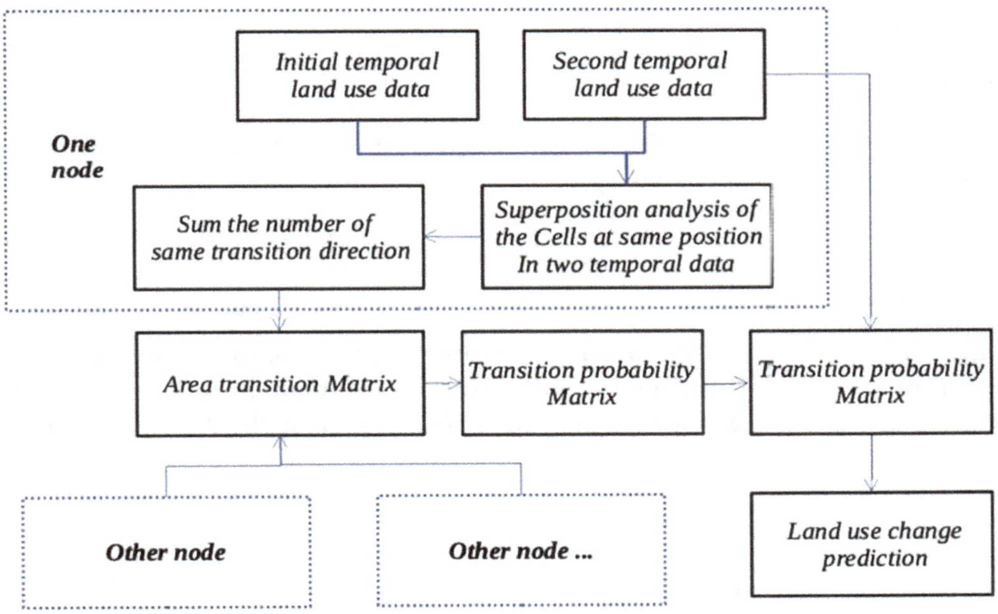

Figure 5. Cloud-Markov algorithm flow.

The steps are as follows:

(1) Map stage:

① Input <Key,Value>

② Raster cell's land-use type conversion analysis

In this step, by comparing the cells in the same position between these two raster images, we obtained a list of C_{mk}. If the value of C_{mk} is 'B-A', it means that the land-use-type of the cell with the same position in different raster images is converted from B into A.

③ Output <Key,Value>

where key is conversion direction C_{mk}, and Value is an integer equal to 1.

(2) Combiner stage:

① Enter <Key,Value>
② Calculate the number of each land-use-type conversion direction in each node

<Key,Value> is a key-value pair (C_{mk},s), where C_{mk} is the conversion direction and s is the total number of cells in a MapReduce node where the C_{mk} land-use-type conversion direction occurs.

③ Output <Key,Value>

Output key-value pairs (C_{mk},s).

(3) Reduce stage:

① Enter <Key,Value>
② Calculate transition probability

The transition-probability-matrix calculation formula is defined as follows:

$$P_{mk} = \frac{V_{mk}}{S_m} \tag{9}$$

where V_{mk} is the value that summed C_{mk} from all MapReduce nodes and S_m is the number of the initial raster image's cells whose land-use type is m.

③ Output <Key,Value>

<Key,Value> is a key-value pair (V_{mk},P_{mk}), where V_{mk} is the land-use-type area-conversion matrix and P_{mk} is the transition-probability matrix.

3.3. Cloud-CELUC

3.3.1. Cell Neighborhood Processing

Obtaining a cell's neighborhood-influence value requires reading the state of the neighbor cells. The general neighborhood-influence evaluation methods include Von Neumann and Moore. Figure 6 shows how to read cellular-neighborhood, where we chose the 3 × 3 Moore method to design our algorithm.

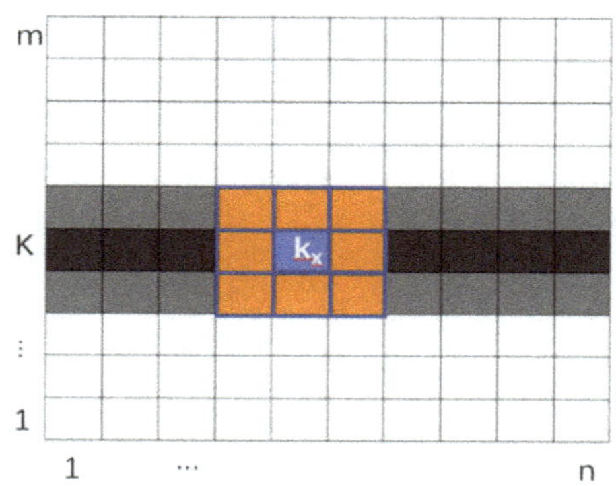

Figure 6. Cellular-neighborhood reading.

Figure 7 shows the process of cellular-dimension reduction, where a list structure was used to store all cells, so we could read each cell's neighborhood cells through the cell's row index and column index. The two-dimensional raster image was reduced into a one-dimensional array that could reduce the data exchange between each node during MapReduce processes. For example, the neighborhood cells of cell K_x were from line K − 1, K, and K + 1, were recorded as $K-1_{x-1}$, $K-1_x$, $K-1_{x+1}$, K_{x-1}, K_{x+1}, $K+1_{x-1}$, $K+1_x$, and $K+1_{x+1}$, and then stored as an array structure into the HDFS.

Figure 7. Cellular-dimension reduction.

3.3.2. Multicriteria Evaluation Factors

This research used the multicriteria evaluation (MCE) method to calculate constraint-evaluation and suitability-evaluation values. The purpose of MCE is to select the optimal decision solution in limited (infinite) solutions between which there are conflicts and coexistence [45].

The evaluation criteria of MCE are divided into two parts: Suitable factors and constraint factors. Suitable factors are used to normalize the influencing factor to the continuous measurable values, and constraint factors are utilized to categorize spatial features by their spatial characteristics. The value of the factors is a Boolean type with a value of 0 or 1.

Suitability factors are defined and standardized in Table 2, where eight factors are defined to distinguish the distance from a cell to different typical destination, and according to different land-use type, the weighted value of each factor is defined in Table 3. Then, the analytic hierarchy process (AHP) method and the expert scoring method were used to calculate the weights of the suitability factors [46], which are shown in Table 2. The waters and gradient factors were defined as the constraint factors.

Table 2. Suitability-factor classification.

Factor Name	Definition	Classification
FreLev TownLev	Distance from cell to highway. Distance from cell to town center.	Cell distance from main road or town center: 0–250, 250–500, 500–750, 750–1000, and 1000–1250 m.
SubLev BusLev MainLev	Distance from cell to subway station. Distance from cell to bus stop. Distance from cell to other roads.	Cell distance to subway or bus station, other roads: 0–100, 100–200, 200–300, 300–400, and 400–500 m.
TraLev StaLev	Distance from cell to train station. Distance from cell to bus station.	Cell distance to train or bus station: 0–200, 200–400, 400–600, 600–800, and 800–1000 m.
CityLev	Distance from cell to county center.	Cell distance to main road or county center: 500–1000, 1000–1500, 1500–2000, and 2000–2500 m.

Table 3. Weight parameters of suitability factors.

Factor Name	Agricultural Land	Construction Land	Nature Reserve
FreLev	0.0485	0.0461	0.0781
TownLev	0.1239	0.1332	0.1010
SubLev	0.0621	0.1320	0.0133
BusLev	0.0721	0.1110	0.0513
MainLev	0.0921	0.1102	0.0749
TraLev	0.0423	0.1333	0.0201
StaLev	0.0623	0.1321	0.0203
StaLev	0.0923	0.2021	0.0103

The value of the constraint factors is Boolean, which is determined by land-use type. This research defined the constraint factors of hills with gradients greater than 25 degrees, and waters and ecological reserves equal to 0, because these land-use types would rarely change.

3.3.3. Parallel Cloud-CELUC Algorithm

Cloud-CELUC only needs the Map function to calculate factors and obtain comprehensive-evaluation values. The Reduce function of Cloud-CELUC was only used to output the results. The Map function was defined as follows:

$$\text{Map} : (N, (i, H_1, H_2, H_3, H_4, \ldots H_m)) \rightarrow ((i, j), (L_1, L_2, L_3, \ldots L_m)) \quad (10)$$

where N is the line offset of the input line, i is the line index of the raster image, j is the column index of the raster image, H_1 is the cell-state value to be calculated, and H_2 and H_3 indicate the uplink and downlink cell-state values of H_1. $H_4, \ldots H_m$ are various constraint factors and suitability factors corresponding to the cell, and L_m is the value combined by the cell's composite-evaluation value of the transition direction and the cell's corresponding transition direction. For example, If L_1 is 'ba-1.2234', the evaluation value of the cell (i, j) from the initial land-use type 'b' to the final land-use type 'a' is '1.2234'.

The flow of Cloud-CELUC algorithm is shown in Figure 8, where the table of comprehensive evaluation value was obtained after each node by calculating the neighborhood impact value, suitability evaluation value, and constrained evaluation value. The steps are as follows:

① Enter <Key,Value>

② Calculate neighborhood-influence evaluation value (*NID*)

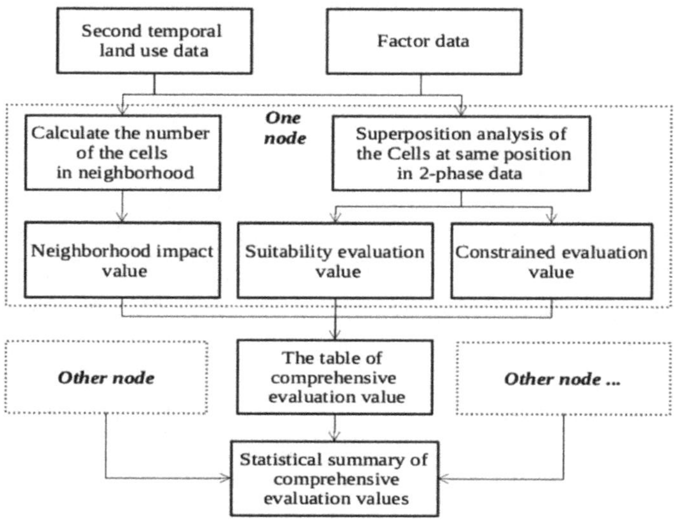

Figure 8. Cloud-comprehensive-evaluation value (CELUC) algorithm.

According to H_1 and H_3, the neighborhood-influence degree of each cell in H_2 was calculated. The calculation formula for the evaluation value of the neighborhood-influence degree of the cell (i,j) corresponding to a class at a certain time is as follows:

$$NID_a = \frac{1}{h-1} \sum Yes\left(S_{ij} == a\right) \quad (11)$$

where h is the number of neighborhood cells, and $Yes\left(S_{ij} == a\right)$ is used to judge whether the neighbor cell's land-use type of cell (i, j) is a or not. If a is 1, then $Yes\left(S_{ij} == a\right)$ returns 1, otherwise, it returns 0.

③ Calculate Suitability-Evaluation Value (*SEV*)

The calculation formula of the suitability-evaluation value is as follows:

$$SEV_a = \sum V_{a\delta} \times DIS_{ij\delta} \quad (12)$$

where $V_{a\delta}$ is the weight of factor δ corresponding to land-use type 'a', $DIS_{ij\delta}$ are the suitability factors of cell (i,j) that are defined in Table 1.

④ Calculate constraint-evaluation value (*CEV*)

The constraint-evaluation formula is as follows:

$$CEV_a = \prod Yes\big(K_{ij}\big) \tag{13}$$

where $Yes\big(K_{ij}\big)$ represents the constraint-evaluation value of cell (i, j) corresponding to constraint factor 'k'. If the cell is constrained, $Yes\big(K_{ij}\big)$ returns 0, otherwise, it returns 1.

⑤ Calculate comprehensive-evaluation value (*CELUC*)

Based on the three evaluation values above, *NID, SEV,* and *CEV,* the comprehensive-evaluation formula was defined as follows:

$$CELUC_a = NID_a \times SEV_a \times CEV_a \tag{14}$$

where a is a land-use type.

⑥ Output <Key,Value>

<Key,Value> is a key-value pair ((i, j), *CELUC*) where i is the cell's row index, j is the cell's column index, and *CELUC* is the comprehensive-evaluation value of the cell.

3.4. Cell Land-Use-Type Conversion

The multi-objective land-use competition method was used to achieve cells' land-use-type conversion, which was to solve the problem when the cell had confliction in land-use-type conversion [47]. For example, if there were N types of land-use types, cell (i, j) may have had N kinds of conversion possibilities. According to constraint factors, suitability factors, and neighborhood conditions, each cell's conversion possibility should be given a comprehensive evolution value of land use. If the evaluation value of the cell's transition direction is determined by the biggest conversion possibility, the value may lead to a proliferation of dominant land-use types and cause oversimulation of dominant land-use types and inadequate simulation of weak land-use types. Hence, both land-use comprehensive evaluation values and the area transfer matrix of land-use changes were used to decide the cell's conversion direction. Figure 9 shows the flow of a cell's land-use-type conversion process.

The steps are as follows:

① Calculating the maximum evaluation value from the table of statistical summary of comprehensive evaluation that was obtained from Cloud-CELUC.

② Loop reading each row of the table. Each row was a key-value pair ((i, j), CELUCs), where (i, j) is the position of the cell, CELUCs means cell (i, j) has N kinds of land-use conversion possibilities (CELUC), and CELUCi means the ith CELUC of the cell.

③ Determining whether the area of the converted land-use type reached the upper area limit of this land-use-type conversion or not.

The upper area limit of the land-use-type conversion was obtained from the area transition-probability matrix of each land-use type, defined as a key-value pair (V_{mk}, P_{mk}) where V_{mk} is the land-use-type area conversion matrix and P_{mk} is the land-use transition-probability matrix at the CLOUD-Markov stage.

④ If reaching the upper area limit, the CELUCi of the cell should be marked as 0, meaning that one of the cell (i, j)'s CELUCi was deleted to make sure the CELUCi would not be used in the subsequent steps. Then, it returns to the first step.

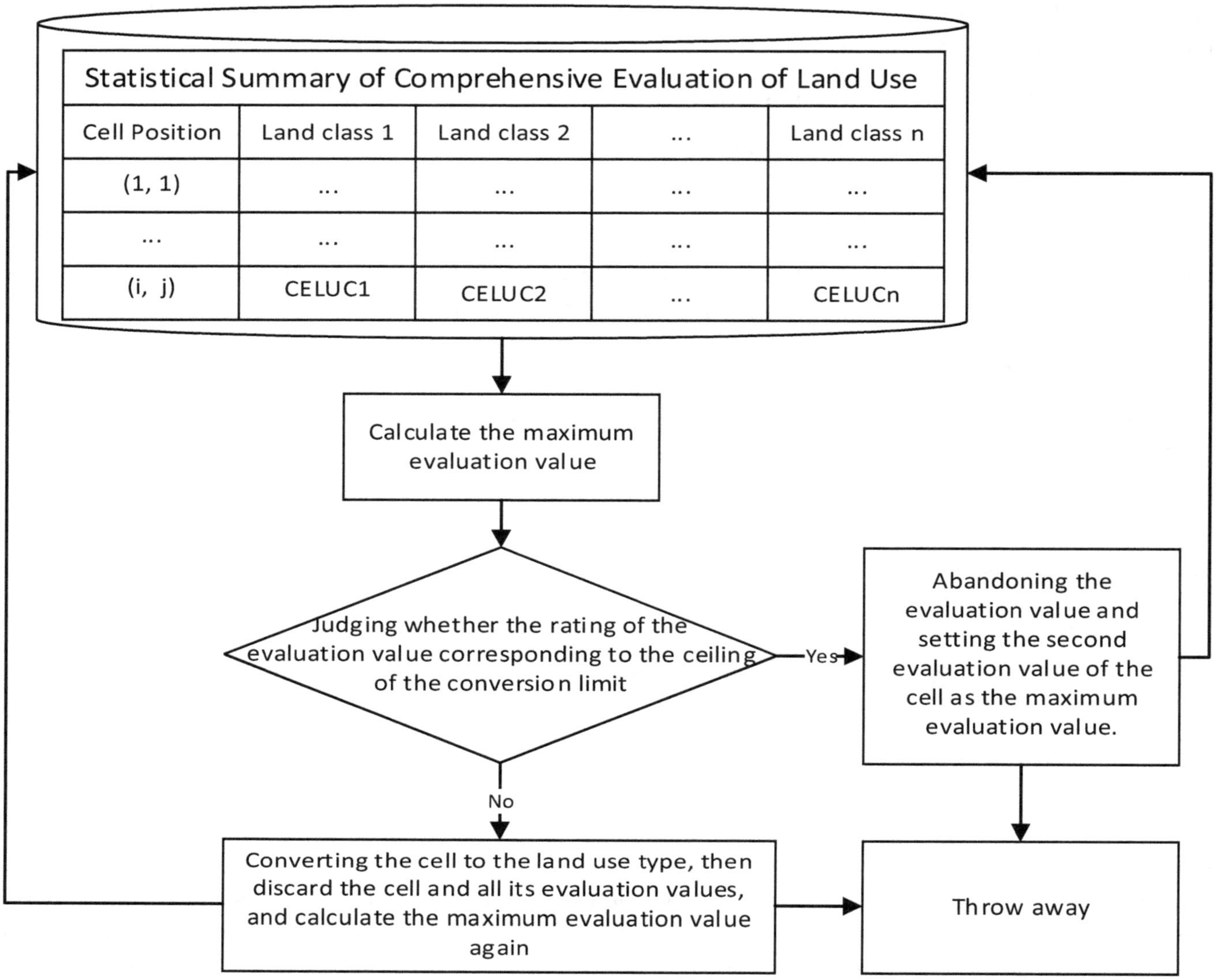

Figure 9. Cell's land-use-type conversion flow.

If the upper area limit is not reached, the land-use type of the cell would be converted into a new land-use type and stored as a key-value pair ((*i*, *j*), CELUC*i*) into an array, discarding the other CELUCs to make sure the cell is not used in the subsequent steps. Then, it returns to the first step.

⑤ Repeating the above steps until all cells completed conversion, and finally obtaining the prediction of the whole land-use-change distribution, which was stored as an array. Each item of the array was a key-value pair ((*i*, *j*), CELUC).

4. Results and Discussion

4.1. Model-Efficiency Analysis

One machine was used as the master node for the work of NameNode and JobTracker, and four other machines were used as the slave nodes for the work of DataNode and TaskTracker. The operating environment of the machines was the CentOS 7.1.1503 system with Java version 1.8.0_112 and Hadoop version 2.7.3. The experiment environment configuration is shown in Table 4. The hardware environment of the serial algorithm was the same as the hardware of the Hadoop node.

Table 4. Hadoop experiment environment.

IP Address	Node Role	CPU	RAM
192.168.128.1	Master/Namenode/Jobtracker	Four-core 2.4 Ghz	4 G
192.168.128.2	Slaves/Datanode/Tasktracker	Four-core 2.4 Ghz	4 G
192.168.128.3	Slaves/Datanode/Tasktracker	Four-core 2.4 Ghz	4 G
192.168.128.4	Slaves/Datanode/Tasktracker	Four-core 2.4 Ghz	4 G
192.168.128.5	Slaves/Datanode/Tasktracker	Four-core 2.4 Ghz	4 G

The running efficiency and acceleration ratio results of the serial-Markov algorithm relative to Cloud-Markov algorithm are shown in Figure 10. The running efficiency and acceleration ratio results of the serial-CELUC algorithm relative to Cloud-CELUC algorithm are shown in Figure 11. As shown in Figures 10a and 11a, the abscissa axis indicates the number of the cells (1 n is approximately 9,000,000) and the ordinate axis indicates running time.

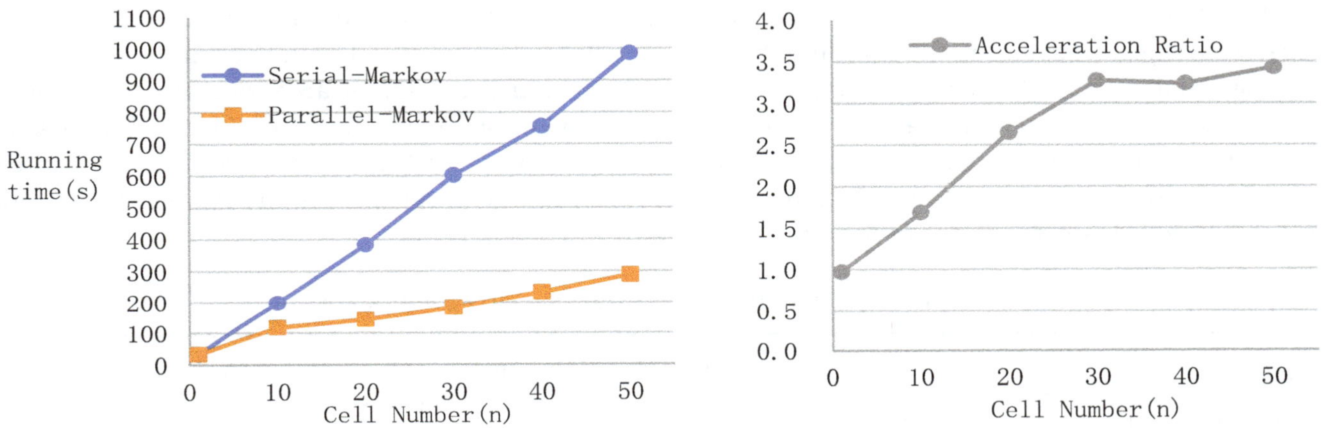

(**a**) (**b**)

Figure 10. Running efficiency and acceleration ratio of Cloud-Markov relative to serial-Markov. (**a**) Running efficiency comparison, (**b**) acceleration ratio.

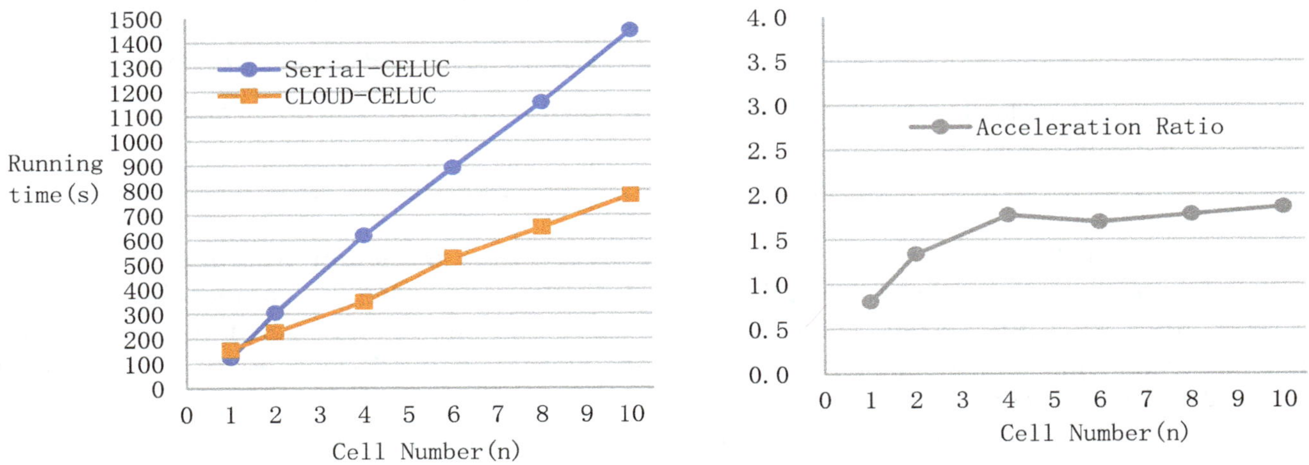

(**a**) (**b**)

Figure 11. Running efficiency and acceleration ratio of serial-CELUC relative to Cloud-CELUC. (**a**) Running efficiency comparison, (**b**) acceleration ratio.

The results showed that the execution time of Cloud-Markov was less than that of the serial Markov algorithm, and with the increase of input data, the acceleration ratio increased and tended to smooth. The acceleration ratio of the Cloud-Markov algorithm to the serial Markov algorithm tended to be steady at 3.27, and the acceleration ratio of Cloud-CELUC to serial CELUC tended to be steady at 1.77. The highest acceleration ratio of Cloud-Markov could reach 3.43, and the highest acceleration ratio of Cloud- CELUC could reach 1.86.

The efficiency of the parallel Markov model based on a cloud environment was remarkable because the MapReduce system effectively distributed the workload of the two phases of cell matching and quantitative statistics. The efficiency of Cloud-CELUC was also improved because the Map phase we defined ran very fast. However, when the output of the Map phase was input into the Reduce phase, it occupied a relatively large part of the long running time, which reduced the efficiency of the operation.

4.2. Precision Evaluation and Result Analysis

4.2.1. Precision Evaluation

The 2006 remote-sensing images and other data were defined as the initial data, and the 2013 land-use changes were simulated by suing the parallel CA-Markov model. In our experiment, the water area was fixed with no change. Based on the 2006 and 2013 land-use data, the area transition-probability matrix could be calculated. Table 5 is the 2006–2013 area transition matrix, in which each cell represents the total area of one land-use type transferring to another land-use type from 2006 to 2016. Table 6 is the 2006–2013 transition-probability matrix, in which each cell represents the probability of one land-use type transferring to another land-use type from 2006 to 2016.

Table 5. The 2006–2013 area transition matrix (unit: km^2).

2006 \ 2013	Agricultural Land	Construction Land	Nature Reserve	Total
Agricultural land	1282.95	409.71	95.67	1788.33
Construction land	210.94	1381.69	12.52	1605.15
Nature-reserve land	93.39	50.79	4409.60	4553.78
Total	1587.28	1842.19	4517.79	7947.26

Table 6. The 2006–2013 transition-probability matrix (unit: %).

2006 \ 2013	Agricultural Land	Construction Land	Nature Reserve
Agricultural land	71.74	22.91	5.35
Construction land	13.14	86.08	0.78
Nature-reserve land	2.05	1.12	96.83

As shown in Tables 5 and 6, the biggest land-use-type transition was agricultural land transferring to construction land; its ratio reached 22.91%. In order to evaluate the simulated precision, the 2013 land-use data were classified using real 2013 remote-sensing images at the data-processing stage. The simulated land-use data and the classified 2013 land-use data are shown in Figures 12 and 13, respectively.

After simulation, the precision evaluation experiments were done to correct all kinds of weight parameters defined in C-MCE. After a great number of repetitive experiments and weight-parameter corrections, the weight parameters of the suitability factors were obtained, which are listed in Table 2. At present, commonly used precision evaluation methods include visual comparison, dimensionality tests, pixel contrasts, and the Kappa coefficient test.

With visual comparison, it was found that the simulation of the nature reserve in the western and southern regions was the most accurate. It can be concluded that the suitability factor and the constrained factor were in line with the change trend of the nature reserve in the study area. The construction land of the central urban area had better simulation accuracy. However, the construction land of east and north, including the towns of Yuanpu and Linpu, was scattered and intertwined with agricultural land. Therefore, simulation error was relatively large.

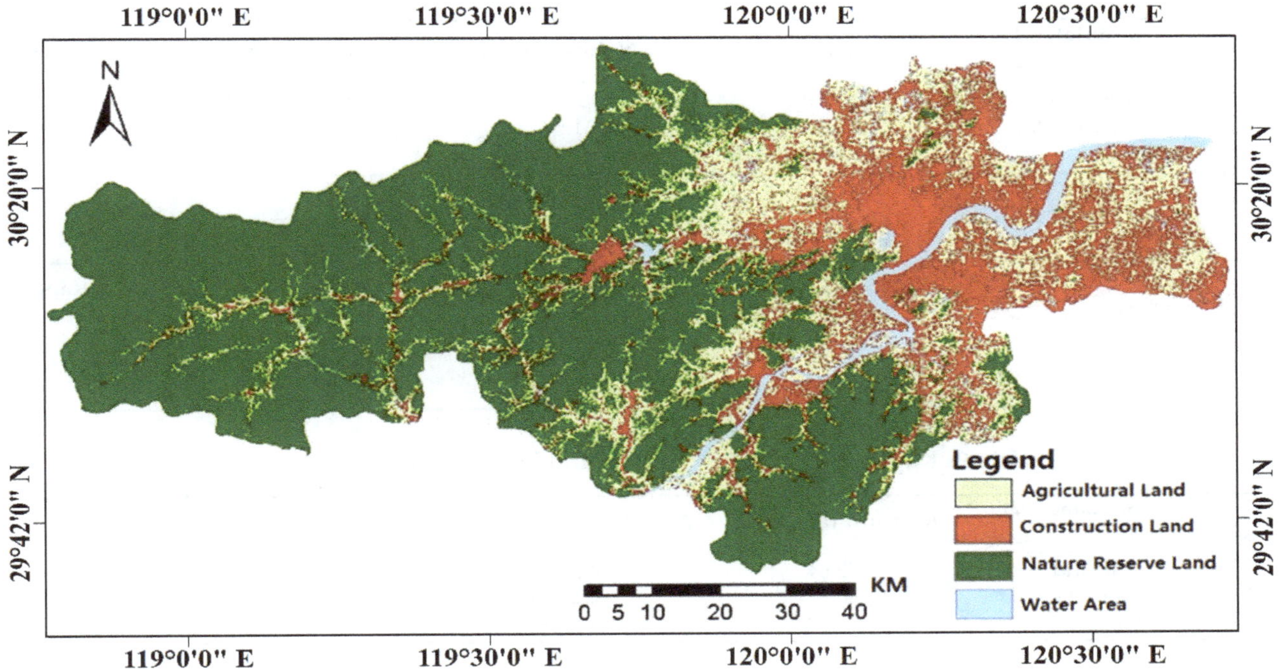

Figure 12. Map of land-use simulation in 2013.

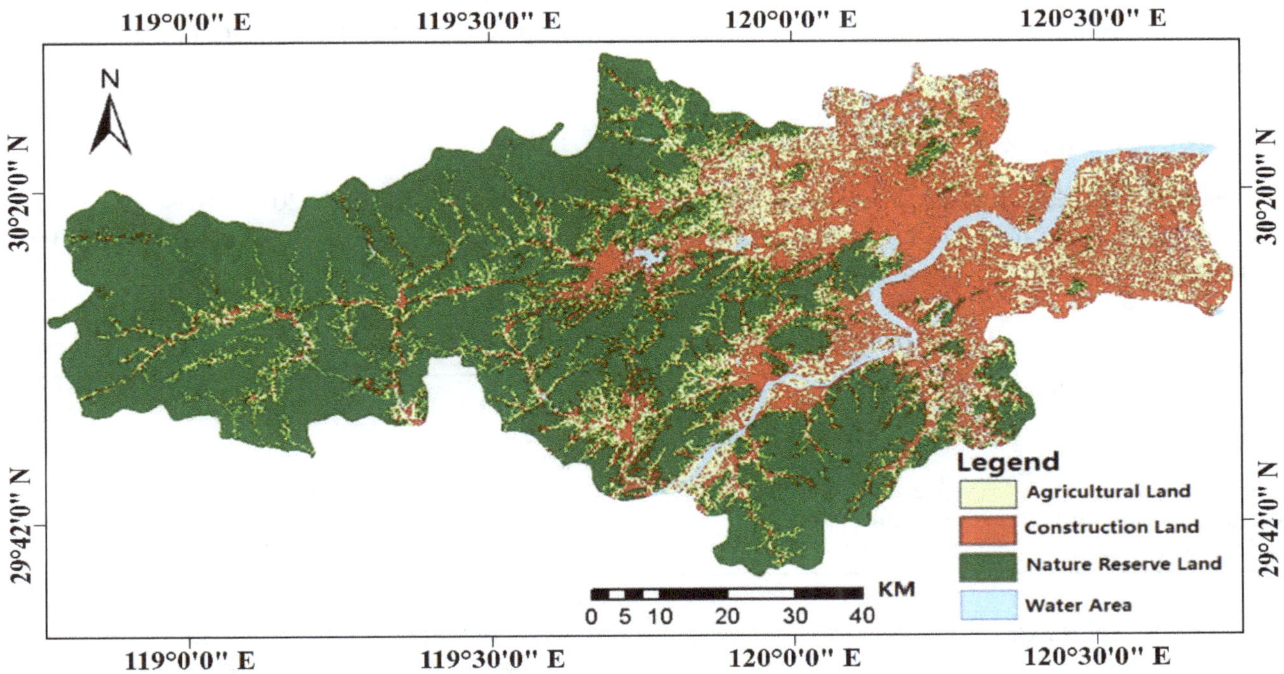

Figure 13. Classification map of remote-sensing images in 2013.

The Kappa coefficient test is the most commonly used quantitative test method [48]. When the Kappa coefficient is used to compare the consistency of data, commonly used criteria are as follows: If the two land-use maps are identical, then Kappa = 1; when Kappa > 0.8, consistency is almost perfect; when 0.6 < Kappa ≤ 0.8, consistency is substantial; when 0.4 < Kappa ≤ 0.6, consistency is moderate; when 0.2 < Kappa ≤ 0.4, consistency is slight; when 0 < Kappa ≤ 0.2, consistency is poor [49,50].

Simulated land-use data and actual classified land-use data of 2013 were compared, and the results of the Kappa coefficient are shown in Tables 7–9, respectively.

Table 7. Kappa coefficient test table of nature reserve land in 2013 (unit: km^2).

Classified Data \ Simulated Data	Nature Reserve	Non-Nature Reserve	Total	Accuracy	Kappa
Nature-reserve land	4221.35	288.47	4509.82	93.60%	
Non-nature-reserve land	296.44	3431.50	3727.94	92.05%	0.86
Total	4517.79	3717.97			

Table 8. Kappa coefficient of construction land in 2013 (unit: km^2).

Classified Data \ Simulated Data	Construction Land	Non-Construction Land	Total	Accuracy	Kappa
Construction land	1391.41	452.77	1844.18	75.45%	
Non-construction land	450.78	5942.80	6393.58	92.95%	0.68
Total	1842.19	6395.57			

Table 9. Kappa coefficient of agricultural land in 2013 (unit: km^2).

Classified Data \ Simulated Data	Agricultural Land	Non-Agricultural Land	Total	Accuracy	Kappa
Agricultural land	1152.64	427.17	1579.81	72.96%	
Non-agricultural land	434.65	6223.30	6657.95	93.47%	0. 66
Total	1587.29	6650.47			

Results showed that the Kappa coefficients for nature reserve, construction land, and agricultural land were 0.85, 0.6, and 0.65, respectively. This meant that the 2013 results of the simulation were quite accurate [38,51–53], and that using the parallel CA-Markov model to predict future land use would be highly reliable.

4.2.2. Land-Use-Change Prediction

Based on the classified 2013 land-use data and other experiment data, 2020 land-use changes were predicted using the parallel CA-Markov model. The results of the 2013–2020 area transition matrix are shown in Table 10, and the 2020 land-use prediction map is shown in Figure 14.

Table 10. The 2013–2020 area transition matrix (unit: km^2).

2013 \ 2020	Agricultural Land	Construction Land	Nature Reserve	Total
Agricultural land	1133.35	361.94	84.52	1579.81
Construction land	242.35	1587.45	14.38	1844.18
Nature-reserve land	92.49	50.30	4367.03	4509.82
Total	1468.19	1999.69	4465.93	7933.81

As can be seen from the 2020 land-use prediction map, construction land in the study area was on the rise as a whole, and this increase mainly came from the conversion of agricultural land. Agricultural land showed a downward trend, and natural-reserve land changed little in proportion to its vast size.

The overall growth of construction land was relatively large. In particular, due to expanding road and public-transport systems, construction land grew faster in the urban center of each county. The growth of construction land in the districts of Xihu, Gongshu, Xiacheng, Binjiang, and Xiaoshan was prominent. Due to the terrain and water-body restrictions, other counties and urban areas maintained stable acreage of construction land.

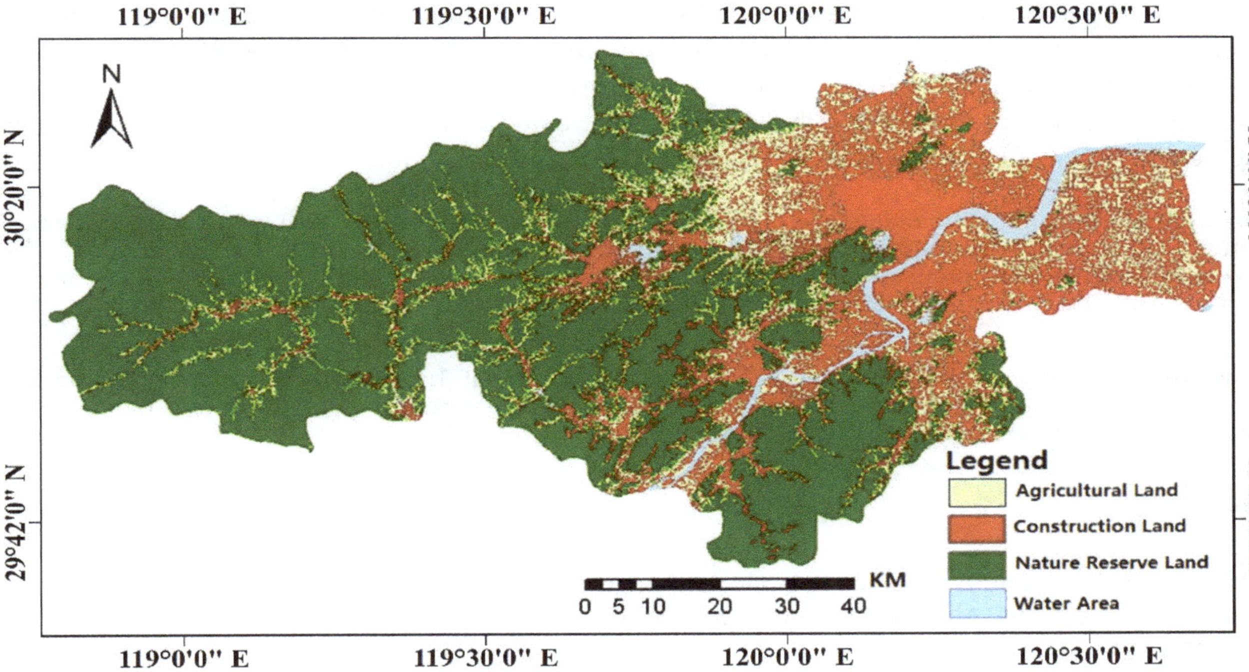

Figure 14. The 2020 land-use prediction map.

Agricultural land and construction land intertwined in large areas in the northwest and north of the study area. However, the evaluation value of construction land in this area was not high because it was far from the city center and the county center and the public transportation system was less developed. The acreage of agricultural land in the above area was basically stable. The agricultural land in the rest of the study area was affected by urban expansion, the road, and public transportation systems, and large acreage was converted to construction land.

Nature reserves were mostly found in the western and southern hilly areas, which were subject to various restraints for development, such as sloping terrain and ecological-protection restrictions, and an inconvenient transportation system, and therefore remained basically stable in acreage.

5. Conclusions

Experiments showed that the results of land-use simulation based on the CA-Markov model under a cloud environment were reliable, reflecting that the method proposed in this study was reliable and applicable. Meanwhile, MapReduce was effective in parallelizing the CA-Markov model to improve the processing speed of land-use-change prediction based on the CA-Markov model. This method parallelized the CA-Markov model in two parts: The parallel Markov model based on a cloud environment (Cloud-Markov), and the comprehensive evaluation method of land-use changes based on MapReduce (Cloud-CELUC). By selecting Hangzhou as the study area and setting up a Hadoop experiment environment, the experiments were designed to verify the reliability, precision, and operating efficiency of the method. Land-use changes in Hangzhou in 2020 were simulated and the results were analyzed. The experimental results showed that the method which simultaneously realized the integrity and segmentation for land use change simulation and prediction is also practical and effective.

This research has successfully applied the MapReduce framework to improve land-use-change prediction efficiency. However, there are still some important issues worth further investigation. First, land-use changes were restrained not only by natural conditions, but also by political, economic, demographic, and other complex factors. Due to limited data sources, this study built its model mainly on traffic, terrain, and location factors. If more data are available, the prediction module of the spatial pattern of land-use, social, economic, policy, demographic, and other factors should be taken into account in future research. We will also consider combining the auto image-identification method [37] into our current work to reduce manual preprocessing work. When Cloud-CELUC acquired the output result of the Map stage in the Reduce stage, input/output (IO) became a bottleneck in system performance. Therefore, more research efforts should be dedicated to testing increasing IO efficiency on the performance of cloud computing based on the CA-Markov model.

Author Contributions: Conceptualization, J.K. and L.F.; Methodology, J.K. and S.L.; Software, J.K. and S.L.; Validation, L.F.; Formal analysis, J.K. and S.L.; Investigation, J.K.; Resources, L.F. and X.W.; Data curation, S.L. and J.K.; Writing—original draft preparation, J.K.; Writing—review and editing, L.F., J.K., and S.L.; Visualization, J.K.; Supervision, L.F. and X.W.; Project administration, J.K. and X.W.; Funding acquisition, L.F. and X.W.

Acknowledgments: We would like to thank Zhejiang University GIS Lab for providing the data, and Kaibin Zhang for helping setup the experimental environment and evaluating our algorithms, and the anonymous reviewers for their valuable suggestions.

References

1. Veldkamp, A.; Lambin, E. Predicting land-use change. *Agric. Ecosyst. Environ.* **2001**, *85*, 1–6. [CrossRef]
2. Li, X. A review of the international researches on land use/land cover change. *ACTA Geogr. Sin. Ed.* **1996**, *51*, 558–565.
3. Wijesekara, G.N.; Farjad, B.; Gupta, A.; Qiao, Y.; Delaney, P.; Marceau, D.J. A comprehensive land-use/hydrological modeling system for scenario simulations in the Elbow River watershed, Alberta, Canada. *Environ. Manag.* **2014**, *53*, 357–381. [CrossRef] [PubMed]
4. Verburg, P.H.; Soepboer, W.; Veldkamp, A.; Limpiada, R.; Espaldon, V.; Mastura, S.S.A. Modeling the spatial dynamics of regional land use: The CLUE-S model. *Environ. Manag.* **2002**, *30*, 391–405. [CrossRef] [PubMed]
5. White, R.; Engelen, G. Cellular Automata and Fractal Urban Form: A Cellular Modelling Approach to the Evolution of Urban Land-Use Patterns. *Environ. Plan. A Econ. Spec.* **1993**, *25*, 1175–1199. [CrossRef]
6. Clarke, K.C.; Hoppen, S.; Gaydos, L. A self-modifying cellular automaton model of historical urbanization in the San Francisco Bay area. *Environ. Plan. B Plan. Des.* **1997**, *24*, 247–261. [CrossRef]
7. Zhang, C.; Li, W. Markov chain modeling of multinomial land-cover classes. *GISci. Remote Sens.* **2005**, *42*, 1–18. [CrossRef]
8. Fu, X.; Wang, X.; Yang, Y.J. Deriving suitability factors for CA-Markov land use simulation model based on local historical data. *J. Environ. Manag.* **2018**, *206*, 10–19.
9. Jantz, C.A.; Goetz, S.J.; Shelley, M.K. Using the SLEUTH urban growth model to simulate the impacts of future policy scenarios on urban land use in the Baltimore-Washington metropolitan area. *Environ. Plan. B Plan. Des.* **2004**, *31*, 251–271. [CrossRef]
10. Dietzel, C.; Clarke, K.C. Toward optimal calibration of the SLEUTH land use change model. *Trans. GIS* **2007**, *11*, 29–45. [CrossRef]
11. Serneels, S.; Lambin, E.F. Proximate causes of land-use change in Narok district, Kenya: A spatial statistical model. *Agric. Ecosyst. Environ.* **2001**, *85*, 65–81. [CrossRef]
12. Wolfram, S. Cellular automata as models of complexity. *Nature* **1984**, *311*, 419–424. [CrossRef]
13. Batty, M.; Xie, Y.; Sun, Z. Modeling urban dynamics through GIS-based cellular automata. *Comput. Environ. Urban Syst.* **1999**, *23*, 205–233. [CrossRef]
14. Li, X.; Yeh, A.G.-O. Neural-network-based cellular automata for simulating multiple land use changes using GIS. *Int. J. Geogr. Inf. Sci.* **2002**, *16*, 323–343. [CrossRef]
15. Yang, Q.; Li, X.; Shi, X. Cellular automata for simulating land use changes based on support vector machines. *Comput. Geosci.* **2008**, *34*, 592–602. [CrossRef]
16. Yang, X.; Zheng, X.-Q.; Lv, L.-N. A spatiotemporal model of land use change based on ant colony optimization, Markov chain and cellular automata. *Ecol. Model.* **2012**, *233*, 11–19. [CrossRef]
17. Rimal, B.; Zhang, L.; Keshtkar, H.; Wang, N.; Lin, Y. Monitoring and Modeling of Spatiotemporal Urban Expansion and Land-Use/Land-Cover Change Using Integrated Markov Chain Cellular Automata Model. *ISPRS Int. J. Geo. Inf.* **2017**, *6*, 288. [CrossRef]
18. Subedi, P.; Subedi, K.; Thapa, B. Application of a Hybrid Cellular Automaton—Markov (CA-Markov)

Model in Land-Use Change Prediction: A Case Study of Saddle Creek Drainage Basin, Florida. *Appl. Ecol. Environ. Sci.* **2013**, *1*, 126–132.
19. Bacani, V.M.; Sakamoto, A.Y.; Quénol, H.; Vannier, C.; Corgne, S. Markov chains-cellular automata modeling and multicriteria analysis of land cover change in the Lower Nhecolândia subregion of the Brazilian Pantanal wetland. *J. Appl. Remote Sens.* **2016**, *10*, 016004. [CrossRef]
20. Araya, Y.H.; Cabral, P. Analysis and modeling of urban land cover change in Setúbal and Sesimbra, Portugal. *Remote Sens.* **2010**, *2*, 1549–1563. [CrossRef]
21. Halmy, M.W.A.; Gessler, P.E.; Hicke, J.A.; Salem, B.B. Land use/land cover change detection and prediction in the north-western coastal desert of Egypt using Markov-CA. *Appl. Geogr.* **2015**, *63*, 101–112. [CrossRef]
22. Hishe, S.; Bewket, W.; Nyssen, J.; Lyimo, J. Analysing past land use land cover change and CA-Markov-based future modelling in the Middle Suluh Valley, Northern Ethiopia. *Geocarto Int.* **2019**, 1–31. [CrossRef]
23. Rahman, M.T.U.; Tabassum, F.; Rasheduzzaman, M.; Saba, H.; Sarkar, L.; Ferdous, J.; Uddin, S.Z.; Islam, A.Z.M.Z. Temporal dynamics of land use/land cover change and its prediction using CA-ANN model for southwestern coastal Bangladesh. *Environ. Monit. Assess.* **2017**, *189*, 565. [CrossRef] [PubMed]
24. Memarian, H.; Kumar Balasundram, S.; Bin Talib, J.; Teh Boon Sung, C.; Mohd Sood, A.; Abbaspour, K. Validation of CA-Markov for Simulation of Land Use and Cover Change in the Langat Basin, Malaysia. *J. Geogr. Inf. Syst.* **2012**, *4*, 542–554. [CrossRef]
25. Liang, J.; Zhong, M.; Zeng, G.; Chen, G.; Hua, S.; Li, X.; Yuan, Y.; Wu, H.; Gao, X. Risk management for optimal land use planning integrating ecosystem services values: A case study in Changsha, Middle China. *Sci. Total Environ.* **2017**, *579*, 1675–1682. [CrossRef] [PubMed]
26. Liu, X.H.; Andersson, C. Assessing the impact of temporal dynamics on land-use change modeling. *Comput. Environ. Urban Syst.* **2004**, *28*, 107–124. [CrossRef]
27. Mobaied, S.; Riera, B.; Lalanne, A.; Baguette, M.; Machon, N. The use of diachronic spatial approaches and predictive modelling to study the vegetation dynamics of a managed heathland. *Biodivers. Conserv.* **2011**, *20*, 73–88. [CrossRef]
28. Alsharif, A.A.A.; Pradhan, B. Urban Sprawl Analysis of Tripoli Metropolitan City (Libya) Using Remote Sensing Data and Multivariate Logistic Regression Model. *J. Indian Soc. Remote Sens.* **2014**, *42*, 149–163. [CrossRef]
29. Guan, Q.; Clarke, K.C. A general-purpose parallel raster processing programming library test application using a geographic cellular automata model. *Int. J. Geogr. Inf. Sci.* **2010**, *24*, 695–722. [CrossRef]
30. Li, X.; Zhang, X.; Yeh, A.; Liu, X. Parallel cellular automata for large-scale urban simulation using load-balancing techniques. *Int. J. Geogr. Inf. Sci.* **2010**, *24*, 803–820. [CrossRef]
31. Li, D.; Li, X.; Liu, X.P.; Chen, Y.M.; Li, S.Y.; Liu, K.; Qiao, J.G.; Zheng, Y.Z.; Zhang, Y.H.; Lao, C.H. GPU-CA model for large-scale land-use change simulation. *Chin. Sci. Bull.* **2012**, *57*, 2442–2452. [CrossRef]
32. Guan, Q.; Shi, X.; Huang, M.; Lai, C. A hybrid parallel cellular automata model for urban growth simulation over GPU/CPU heterogeneous architectures. *Int. J. Geogr. Inf. Sci.* **2016**, *30*, 494–514. [CrossRef]
33. Rathore, M.M.U.; Paul, A.; Ahmad, A.; Chen, B.; Huang, B.; Ji, W. Real-Time Big Data Analytical Architecture for Remote Sensing Application. *IEEE J. Sel. Top. Appl. Earth Obs. Remote Sens.* **2015**, *8*, 4610–4621. [CrossRef]
34. Rao, J.; Wu, B.; Dong, Y.-X. Parallel Link Prediction in Complex Network Using MapReduce. *J. Softw.* **2014**, *23*, 3175–3186. [CrossRef]
35. Wiley, K.; Connolly, A. Astronomical image processing with hadoop. *Astron. Data Anal. Softw. Syst.* **2010**, *442*, 93–96.
36. Almeer, M.H. Cloud Hadoop Map Reduce For Remote Sensing Image Analysis. *J. Emerg. Trends Comput. Inf. Sci.* **2012**, *3*, 637–644.
37. Dang, L.M.; Ibrahim Hassan, S.; Suhyeon, I.; Sangaiah, A.K.; Mehmood, I.; Rho, S.; Seo, S.; Moon, H. UAV based wilt detection system via convolutional neural networks. *Sustain. Comput. Inform. Syst.* **2018**, in press. [CrossRef]
38. Sankey, T.T.; McVay, J.; Swetnam, T.L.; McClaran, M.P.; Heilman, P.; Nichols, M. UAV hyperspectral and lidar data and their fusion for arid and semi-arid land vegetation monitoring. *Remote Sens. Ecol. Conserv.* **2018**, *4*, 20–33. [CrossRef]
39. Lanorte, A.; De Santis, F.; Nolè, G.; Blanco, I.; Loisi, R.V.; Schettini, E.; Vox, G. Agricultural plastic waste spatial estimation by Landsat 8 satellite images. *Comput. Electron. Agric.* **2017**, *141*, 35–45. [CrossRef]

40. Du, Z.; Li, W.; Zhou, D.; Tian, L.; Ling, F.; Wang, H.; Gui, Y.; Sun, B. Analysis of Landsat-8 OLI imagery for land surface water mapping. *Remote Sens. Lett.* **2014**, *5*, 672–681. [CrossRef]
41. Jia, K.; Wei, X.; Gu, X.; Yao, Y.; Xie, X.; Li, B. Land cover classification using Landsat 8 Operational Land Imager data in Beijing, China. *Geocarto Int.* **2014**, *29*, 941–951. [CrossRef]
42. Dean, J.; Ghemawat, S. MapReduce. *Commun. ACM* **2008**, *51*, 107. [CrossRef]
43. Guan, D.; Gao, W.; Watari, K.; Fukahori, H. Land use change of Kitakyushu based on landscape ecology and Markov model. *J. Geogr. Sci.* **2008**, *18*, 455–468. [CrossRef]
44. White, R.; Engelen, G.; Uljee, I. The use of constrained cellular automata for high-resolution modelling of urban land-use dynamics. *Environ. Plan. B Plan. Des.* **1997**, *24*, 323–343. [CrossRef]
45. CARVER, S.J. Integrating multi-criteria evaluation with geographical information systems. *Int. J. Geogr. Inf. Syst.* **1991**, *5*, 321–339. [CrossRef]
46. Saaty, T.L.; Vargas, L.G. *Models, Methods, Concepts & Applications of the Analytic Hierarchy Process*; Springer Science & Business Media: New York, NY, USA, 2001; ISBN 978-1-4613-5667-7.
47. Qiu, B.; Chen, C. Land use change simulation model based on MCDM and CA and its application. *Acta Geogr. Sin. Ed.* **2008**, *63*, 165–174.
48. Foody, G.M. Map comparison in GIS. *Prog. Phys. Geogr.* **2007**, *31*, 439–445. [CrossRef]
49. Landis, J.R.; Koch, G.G. The Measurement of Observer Agreement for Categorical Data. *Biometrics* **1977**, *33*, 159. [CrossRef]
50. Viera, A.J.; Garrett, J.M. Understanding interobserver agreement: The kappa statistic. *Fam. Med.* **2005**, *37*, 360–363.
51. van Vliet, J.; Bregt, A.K.; Hagen-Zanker, A. Revisiting Kappa to account for change in the accuracy assessment of land-use change models. *Ecol. Model.* **2011**, *222*, 1367–1375. [CrossRef]
52. Bu, R.C.; Chang, Y.; Hu, Y.M.; Li, X.Z.; He, H.S. Measuring spatial information changes using Kappa coefficients: A case study of the city groups in central Liaoning province. *Acta Ecol. Sin.* **2005**, *205*, 4.
53. Mora, A.D.; Vieira, P.M.; Manivannan, A.; Fonseca, J.M. Automated drusen detection in retinal images using analytical modelling algorithms. *Biomed. Eng. Online* **2011**, *10*, 59. [CrossRef] [PubMed]

8

Advanced Cyberinfrastructure to Enable Search of Big Climate Datasets in THREDDS

Juozas Gaigalas, Liping Di * and Ziheng Sun

Center for Spatial Information Science and Systems, George Mason University, Fairfax, VA 22030, USA; juozasgaigalas@gmail.com (J.G.); zsun@gmu.edu (Z.S.)

* Correspondence: ldi@gmu.edu

Abstract: Understanding the past, present, and changing behavior of the climate requires close collaboration of a large number of researchers from many scientific domains. At present, the necessary interdisciplinary collaboration is greatly limited by the difficulties in discovering, sharing, and integrating climatic data due to the tremendously increasing data size. This paper discusses the methods and techniques for solving the inter-related problems encountered when transmitting, processing, and serving metadata for heterogeneous Earth System Observation and Modeling (ESOM) data. A cyberinfrastructure-based solution is proposed to enable effective cataloging and two-step search on big climatic datasets by leveraging state-of-the-art web service technologies and crawling the existing data centers. To validate its feasibility, the big dataset served by UCAR THREDDS Data Server (TDS), which provides Petabyte-level ESOM data and updates hundreds of terabytes of data every day, is used as the case study dataset. A complete workflow is designed to analyze the metadata structure in TDS and create an index for data parameters. A simplified registration model which defines constant information, delimits secondary information, and exploits spatial and temporal coherence in metadata is constructed. The model derives a sampling strategy for a high-performance concurrent web crawler bot which is used to mirror the essential metadata of the big data archive without overwhelming network and computing resources. The metadata model, crawler, and standard-compliant catalog service form an incremental search cyberinfrastructure, allowing scientists to search the big climatic datasets in near real-time. The proposed approach has been tested on UCAR TDS and the results prove that it achieves its design goal by at least boosting the crawling speed by 10 times and reducing the redundant metadata from 1.85 gigabytes to 2.2 megabytes, which is a significant breakthrough for making the current most non-searchable climate data servers searchable.

Keywords: climate science; metadata; web cataloging service; big geospatial data; geospatial cyberinfrastructure

1. Introduction

Cyberinfrastructure plays an important role in today's climate research activities [1–6]. Climate scientists search, browse, visualize, and retrieve spatial data using web systems on a daily basis, especially as data volumes from observation and model simulation grow to large amounts that personal devices cannot hold entirely [7,8]. The big data challenges of volume, velocity, variety, veracity, and value (5Vs), have pushed geoscientific research into a more collaborative endeavor that involves many observational data providers, cyberinfrastructure developers, modelers, and information stakeholders [9]. Climate science has developed for decades and produced tens of petabytes of data products, including stationary observations, hindcast, and reanalysis, which are stored in distributed data centers in different countries around the globe [10]. Individuals or small groups of scientists face big challenges when they attempt to efficiently discover the data they require. Currently, most scientists acquire their knowledge about datasets via conferences, colleague recommendations, textbooks, and search engines. They become very familiar with the datasets they use, and every time they want to retrieve the data, they go directly to the dataset website to download the data falling within the requested time and spatial windows. However, these routines are less sustainable as the sensors/datasets become more varied, models evolve more frequently, and new data pertaining to their research is available somewhere else [9].

In most scenarios, metadata is the first information that researchers see, before they access and use the actual Earth observation and modeling data the metadata describes [11,12]. Based on the metadata, they decide whether or not the actual data will be useful in their research. For big spatial data, metadata is the key component backing up all kinds of users' daily operations, such as searching, filtering, browsing, downloading, displaying, etc. Currently, two of the fundamental problems in accessing and using big spatial data are the volume of metadata and the velocity of processing metadata [13,14]. Through manual investigation of Unidata THREDDS data repository (a metadata source we take as an example of typical geodata storage patterns) [15,16], it reveals that most of the metadata are highly redundant. The vast majority of metadata records contain identical information and only key fields representing spatial and temporal characteristics are regularly updated. However, there exists a regular pattern to how the redundant information is structured and how new information is added to the repository—but, the pattern varies according to data organization hierarchy and changes with the type of data being delivered (for example: Radar station vs. satellite observation vs. regular forecast model output).

To overcome these big data search challenges, we must confront practical problems in the information model, information quality, and technical implementation of information systems. Our study follows the connection between fundamental scientific challenges and existing implementations of geoscience information systems. This study aims to build a cataloging model capable of fully describing real-time heterogeneous metadata whilst simultaneously reducing data volume and enabling search within big Earth data repositories. This model can be used to efficiently represent redundant data in the original metadata repository and to perform lossless compression of information for lightweight efficient storing and searching. The model can shrink the huge amount of metadata (without sacrificing information complexity or variety available in the original repositories) and reduce the computational burden on searching among them. The model defines two types of objects: Collections and granules. It also defines their lifecycle and relationship to the upstream THREDDS repository data. Collection contains content metadata (title, description, authorship, variable/band information, etc.). Each collection contains one or more granules. Each granule contains only the spatiotemporal extent metadata. We prototyped the model as an online catalog system within EarthCube CyberConnector [17–20]. We have made the final system available online at: http://cube.csiss.gmu.edu/CyberConnector/web/covali. The system provides a near real-time replica of the source catalog (e.g., THREDDS), optimizes the metadata storage, and enables searching capability which was not available before. The system is like a clearinghouse with its own metadata database. Currently, the system is mainly used for searching operational time-series observations/simulations collected/derived from field sensors. Other datasets, like remote sensing datasets and airborne datasets, can be foreseen to be supported in the near future. The novelty of this research is that it turns the legacy data center repositories into lightweight flexible catalog services, which are more manageable by providing searching capabilities for petabytes of datasets. The work provides important references to people operating the operation of big climate data centers and advises on further improvements in those operational climate data centers to better serve the climate science community. This paper is organized as follows. Section 2 describes the background knowledge and history. Section 3 introduces related work. Section 4 introduces the proposed model. Section 5 shows the implementation of the model and the required cyberinfrastructure. Section 6 demonstrates the experiment results. Section 7 discusses the results of our approach. Section 8 concludes the paper.

The study described in this paper is an attempt to contribute to the global scientific endeavor on understanding and predicting the impacts of climate change. Understanding climate change and its impacts requires understanding Earth as a complex system of systems with behaviors that emerge from the interaction and feedback loops that occur on a range of temporal and spatial scales. However, new advances in these studies are obstructed by the challenges of interdisciplinary collaboration and the difficulty of data and information collaboration [21–27]. The difficulties of information collaboration can be understood in terms of long-standing big data problems of variety (complexity) and volume/velocity.

2. Background

Metadata is a powerful tool for dealing with big data challenges. We discuss the background work on metadata and interoperability of metadata catalogs as critical components of advanced cyberinfrastructure that we envision.

2.1. Metadata

The topic of metadata has been approached by two distinct scholarly traditions. Understanding them helps us clarify our approach to metadata in cyberinfrastructure. Library information scientists have described the metadata bibliographic control approach. Bibliographic principles allow information users to describe, locate, and retrieve information-bearing entities. The basic metadata unit is the "information surrogate" that derives its usefulness from being locatable (by author, title, and subject), accurately describing the information object (the data of the metadata) and identifying how to locate the object. The second (complementary) view of metadata originates in the computer science discipline and is called the data management approach. Complex and heterogeneous data (textual, graphical, relational, etc.) is not separated into information units, but is instead described by data models and architectures that represent "additional information that is necessary for data to be useful" [27]. The key difference is the bibliographic approach works with distinct information entities of limited types, while the data management approach works with models of data/information structures and their relationships.

This distinction between bibliographic and data management approaches is important in the context of ongoing efforts of metadata standardization [28–31]. The second approach is not conducive for standardization because the data management models are as complex and heterogeneous as the structures of the data being modeled. Consequently, in accordance with existing standards, the currently available metadata for large climate datasets follows the first approach, which provides bibliographic information and does not describe the data structures in a way that may permit new capacities of advanced cyberinfrastructure. Our paper describes the work to supplement and transform the existing bibliographical metadata with a custom metadata management model resulting in new applications for the existing data. Metadata standardization is a prerequisite for interoperability, which is a prerequisite for building distributed information systems capable of handling complex Earth system data [32].

2.2. Interoperability, Data Catalogs, Geoinformation Systems, and THREDDS

Data and information collaboration across disciplines is critical for advanced Earth science. Unfortunately, there is no strongly unified practice for data recording, storage, transmission, and processing that the entire scientific community follows [33–37]. Disparate fields and traditions have their own preferred data formats, software tools, and procedures for data management. However, Earth system studies generally work with data that follow a geospatial–temporal format [38–42]. All of the data can be meaningfully stored on a 4D (3 spatial and one temporal) dimension grid. This basic commonality has inspired standardization efforts with the goal of enabling wider interoperability and collaboration.

Following organic outgrowth from the community, the standardization efforts are now headed by Open Geospatial Consortium (OGC) and the International Organization for Standardization Technical Committee 211 (ISO TC 211) and have yielded successful standards in two areas relevant to us [43–47]. First is the definition of NetCDF as one of the standard data formats for storing geospatial data. The second is the metadata standardization. Those efforts are extremely relevant to our research and are further discussed in the Related Works section. For background, it is important to mention that the standard geospatial metadata models developed by OGC are still evolving capabilities for describing the heterogeneous, high-volume, or high-velocity big data we are studying. The commonly used OGC/ISO 19* series metadata standards have relatively limited relational features (aggregation only) and, in the repository we studied, each XML encoded metadata record contains mostly redundant information (for example, two metadata objects that represent two images from a single sensor mostly contain duplicated information that describes sensor characteristics). However, there are multiple lines of work that ISO TC 211 is pursuing that addresses these issues and suggests a trend for expanding the applicability of standardized metadata models and the integration of a greater variety of information.

The standard geographical metadata model was developed in conjunction with a standard distributed catalog registration model titled Catalog Services for the Web (CSW) [48,49]. The CSW standard is widely known and many Earth system data providers offer some information about their data holdings via the CSW interface. However, the CSW standard is also poorly suited to support big data collaborative studies for the Earth system. CSW follows the basic OGC metadata model in a way that makes it challenging to capture valuable structure and semantics of existing data holdings without storing extremely redundant information—which exhausts computing resources without taking advantage of the true value of large and complex Earth big data. However, OGC metadata stored in CSW is the existing standard that governs not only data distribution practices, but also how researchers think about data collaboration.

The next item this study works with is the UCAR Unidata THREDDS Data Server (TDS). The University Corporation for Atmospheric Research (UCAR) Unidata is a geoscience data collaboration community of diverse research and educational institutions. It provides the real-time heterogeneous Earth system data that this study targets. THREDDS is Unidata's Thematic Real-Time Environmental Distributed Data Services. TDS is a web server that provides metadata and data access for scientific datasets to climate researchers. TDS provides its own rudimentary hierarchical catalog service that is not searchable and does not support the CSW standard. However, it does support the OGC geospatial metadata standard—although not consistently or comprehensively. In order to make data hosted by TDS searchable, the TDS metadata must be copied to another server and a searchable catalog must be created for the metadata. This task is performed by a customized web crawler developed by this study.

This study attempts to build upon the existing infrastructure with its available resources and limitations to provide new capabilities. The limitations of the existing systems are two-fold. First is the limits of the CSW metadata registration model (it does not naturally support registering information about metadata lifecycle or sufficiently detailed aggregation information), and second is the incompleteness of information within metadata provided by THREDDS. This study attempts to erase the limits by first interpolating information to improve the quality of the existing metadata model and then by extending the model to provide advanced capabilities. It demonstrates how to integrate TDS metadata with CSW software and proposes several practical solutions that work around the limitations of the CSW metadata registration model. We do this to show that improvements in metadata and catalog capabilities can also reduce the challenges of big data in variability, volume, and velocity.

3. Related Work

This paper brings together several existing lines of work to confront the problems of integrating and searching vast and diverse climate science datasets. Existing research in areas of metadata modeling, geospatial information interoperability, geospatial cataloging, web information crawling, and search indexing provides the building blocks for our work to demonstrate and evaluate advanced climate data cyberinfrastructure capabilities.

3.1. Metadata Models

There are many studies exploring the fundamental relationship between metadata models and information capabilities. There exists diverse work in other areas that deal with the same basic issues and demonstrates that the creation of novel metadata models can be used as a method for solving information challenges. For example, Spéry et al. [50] have developed a metadata model for describing the lineage of changes of geographical objects over time. They used a direct acyclic graph and a set of elementary operations to construct their model. The model supports new application of querying historical cadastral data and minimizes the size of geographical metadata information. Spatiotemporal metadata modeling can be generalized as a description of objects in space and time, and relationships between objects conceived as flows of information, energy, and material to model interdependent evolution of objects in a system [51]. Provenance ("derivation history of data product starting from its original sources" [52]) modeling is an important part of metadata study. Existing metadata models and information systems have been experimentally extended with provenance modeling capabilities to enable visualization of data history and analysis of workflows that derive data products used by scientists [53,54]. An experiment to re-conceptualize metadata as a practice "knowledge management" yielded a metadata model that can support the needs of spatial decision-making by identifying issues of entity relationships, integrity, and presentation [55]. The proposed metadata model allows communicating more complex information about spatial data. This metadata model makes it possible to build an original geographic information application, named Florida Marine Resource Identification System, that extends the use of the existing environment and civil data to empower users with higher-level knowledge for analysis and planning. Looking outside the geospatial domains, we still observe that the introduction of specialized metadata approaches and models permits the development of new capabilities.

3.2. Geospatial Metadata Standardization, Interoperability, and Cataloging

The diversity of metadata models and formats developed by research has enabled new powerful geoinformation systems, but has also introduced a new set of problems of data reuse and interoperability. Public and private research, administrative, and business organizations have accumulated growing stores of geoinformation and data, but this data has not become easier to discover and access for users outside limited organization jurisdictions. This has led to significant resource wastage and duplication of effort for data producers and consumers. Cataloging has grown increasingly challenging because of this heterogeneity. In response, new spatial data infrastructures have been developed. They have attempted to integrate and standardize multiple metadata models and develop shared semantic vocabulary models to enable discovery by employing the "digital library" models of metadata. In this process, syntactic and semantic interoperability challenges have been identified. Syntactic operability refers to information portability—the ability of systems to exchange information. Semantic interoperability refers to domain knowledge that permits information services to understand how to meaningfully use the data from other systems [56].

Various techniques for achieving metadata interoperability have been explored [57]. Two related families of techniques can be identified. One approach attempts to create standard and universal models, the other creates mappings between several metadata representations of the same data. Transformation between several metadata models requires that syntactic, structural, and semantic heterogeneities can be reconciled. The reconciliation is accomplished with techniques called metadata crosswalks. A crosswalk is "a mapping of the elements, semantics, and syntax from one metadata schema to another". Once mappings are developed, they can be used to apply multiple metadata schemas to existing data [58].

The possibilities for interoperability have been advanced by the efforts led by the International Organization for Standardization Technical Committee 211 (ISO TC 211) to standardize metadata representation. It introduced the ISO 19* series of geospatial metadata standards for describing geographic information by the means of metadata [54,59–61]. The standards define mandatory and optional metadata elements and associations among elements. For example, spatiotemporal extent, authorship, and general description of datasets are required and recommended by the standard. Other kinds of information like sequencing of datasets in a collection, aggregation, and other relational data are optional in the standard. The ISO 19* series of standards also provides an XML schema for the representation of the metadata in XML [62].

Looking at existing metadata interoperability work, we see a recurrence of similar problems such as diversity of metadata representations and complexity of mapping between them. Several authors discuss the practical challenges of developing software and systems for translation [27,59,63]. There exists a proliferation of study efforts and results that advance the goals of interoperability by identifying key understanding of the challenges of interoperability and demonstrating systems, services, and models that address common challenges. Our work attempts to preserve existing interoperability advances while exploring the possibilities of expanding existing metadata models to support new possibilities use of existing data.

Standardized metadata is often stored and made available using catalog services. Catalogs allow users to find metadata using queries that describe the desired spatial, temporal, textual, and other information characteristics of the searched data [64]. The OGC Catalog Service for the Web (CSW) is one of the widely used catalog models in the geoscience domain to describe geographic information holdings [6,65].

3.3. Web Harvesting and Crawling

One critical capacity of metadata cyberinfrastructure is the ability to integrate metadata from remote web repositories. The process of finding and importing web linked data in a metadata repository is called "crawling" and is accomplished using a software system called "metadata web crawler". A web crawler is a computer program that browses the web in a "methodical, automatic manner or in an orderly fashion" [66]. A crawler is an internet bot, it is a program that autonomously and systematically retrieves data from the world wide web. It automatically discovers and collects different resources in an orderly fashion from the internet according to a set of built-in rules. Patil and Patil [66] summarize this general architecture of web crawlers and also provide a definition of several types of web crawlers. A focused crawler is a type that is designed to eliminate unnecessary downloading of web data by incorporating an algorithm for selecting which links to follow. An incremental crawler first checks for changes and updates to pages before downloading their full data. It necessarily involves an index table of page update dates and times. We follow these two strategies in the design of our crawler. The authors also outline common strategies for developing distributed and parallelized crawlers. Our crawler runs on a single machine, but we use a multithreaded process model with a shared queue mechanism—a common parallelization strategy identified by the authors [67].

A fairly recent review collected by Desai et al. [68] shows that web crawler research is an active area of work—however, most of this work is focused on the needs of general web search engine index construction. There exists an area of research called "vertical crawling" which contends with the problems of crawling non-traditional web data: News items, online shopping lists, images, audio, video. There does not appear any publications regarding efficient crawling of heterogeneous Earth system metadata.

There exists substantial previous work to show the feasibility of crawling this metadata. One recent paper summarizes the state of the art. Li et al. [69] present a heterogenous Earth system metadata crawling and search system named PolarHub—a web crawling tool capable of conducting large-scale search and crawling of distributed geospatial data. It uses existing textual web search engines (Google) to discover OGC standards-compliant geospatial data services. It presents an interactive interface that allows users to find a large variety and diversity of catalogs and related data services. It has a sophistical distributed multi-threaded software system architecture. PolarHub shows that it is possible to present data from many sources in a single place. However, it does not present datasets, only endpoints that users must further explore on their own. It does not download, summarize, or harmonize the metadata stored on the remote catalogs. It shows the feasibility of cyberinfrastructure that integrates a variety of data based on interoperable standards but does not discuss data volume and velocity challenges that arise when deeper and fuller crawling is done. PolarHub users can find a large number of catalogs and services that contain, for instance, "surface water temperature" data but they cannot use metadata crawler following this catalog hub strategy to discover datasets that hold "surface water temperature inside X spatial and temporal extent with Y spatial and temporal resolution".

A complementary strategy is discussed by Pallickara et al. [70], who present a metadata crawling system named GLEAN, which provides a new web catalog for atmospheric data based on the extraction of fine-grained metadata from existing large-scale atmospheric data collections. It solves the data volume problem by introducing a new metadata scheme based on custom synthetic datasets that represent collections (or subsets or intersections) of multiple existing datasets. This reduces metadata overhead greatly and permits high performance and precise discovery and access of specific datasets inside vast atmospheric data holdings. Unlike PolarHub, GLEAN avoids the data variety challenge by limiting its processing to one type of data format used in atmospheric science. They also do not contend with the interrelated velocity and near real-time access problems—in GLEAN crawling, the discovery of updated datasets is initiated by manual user request. They do not use the OGC catalog or metadata standards to support interoperability.

BCube project (part of EarthCube initiative) attacks similar problems with another approach [71]. EarthCube is a National Science Foundation initiative to create open community-based cyberinfrastructure for all researchers and educators across the geosciences. EarthCube cyberinfrastructure must integrate heterogeneous data resources to allow forecasting the behavior of the complex Earth system. EarthCube is composed of many building blocks. Our work is part of the EarthCube Cyberway building block. BCube (The Brokering Building Block) offers a different approach for heterogeneous geodata interoperability. BCube adopts a brokering framework to enhance cross-disciplinary data discovery and access. A broker is a third party online data service that contains a suite of components, called accessors. Each accessor is designed to interface with a different type of geodata repository. A broker allows users to access multiple repositories with a single interface without requiring data providers to implement interoperability measures. BCube supports metadata brokering. It can search, access, and translate heterogeneous metadata from multiple sources. It demonstrates deeper interoperability than other approaches discussed here, but does not attempt to solve data volume or velocity problems [72]. The BCube approach is very relevant to us; however, BCube has very few documents available and the system is inaccessible. We were unable to compare some of the details of our different approaches.

Song and Di [73] studied the same problem with the same example repository: Unidata TDS. The authors determined the volume and velocity characteristics of the target repository metadata. Like our study, they propose modeling it with concepts of collection and granule. They implemented a crawler that is able to crawl some of the TDS archive. Their work is the previous progress in the same project as ours and is highly relevant to this study. However, their approach did not perform well using real-world TDS data, which led us to take it in a different direction. We rebuilt their work to demonstrate real-time search and the possibility of processing all of TDS by using a more sophisticated metadata model, and a more advanced integrated search client and indexing service that permits true real-time search.

Reviewing existing work reveals tremendous advances toward solving the challenges of creating interoperable Earth system cyberinfrastructures that can practically process a large volume and variety of observation and model data that are generated in high-velocity data production processes. Lines of work in metadata modeling, standardization, interoperability, repository crawling, and processing provide the basis for the materials for our study. Our contribution is to synthesize these approaches to explore how interoperability and performance could be achieved simultaneously.

4. Materials and Methods

To enable searching of big climate data, we propose a new big data cataloging solution, which includes the following steps. (1) Analyze the target geodata repository that provides a good example of data challenges for cross-disciplinary Earth system scientific collaboration. (2) Analyze the qualities and characteristics of the data in the selected repository. (3) Construct a model of the repository. (4) Use the repository model to construct an efficient metadata resource model. (5) Develop a crawler system that uses repository and metadata resource models to optimize its crawling algorithm and metadata representation. (6) Demonstrate advanced interoperable big geodata search and access capabilities that our approach permits. The completed cyberinfrastructure model and system architecture (derived from our metadata model) is shown in Figure 1.

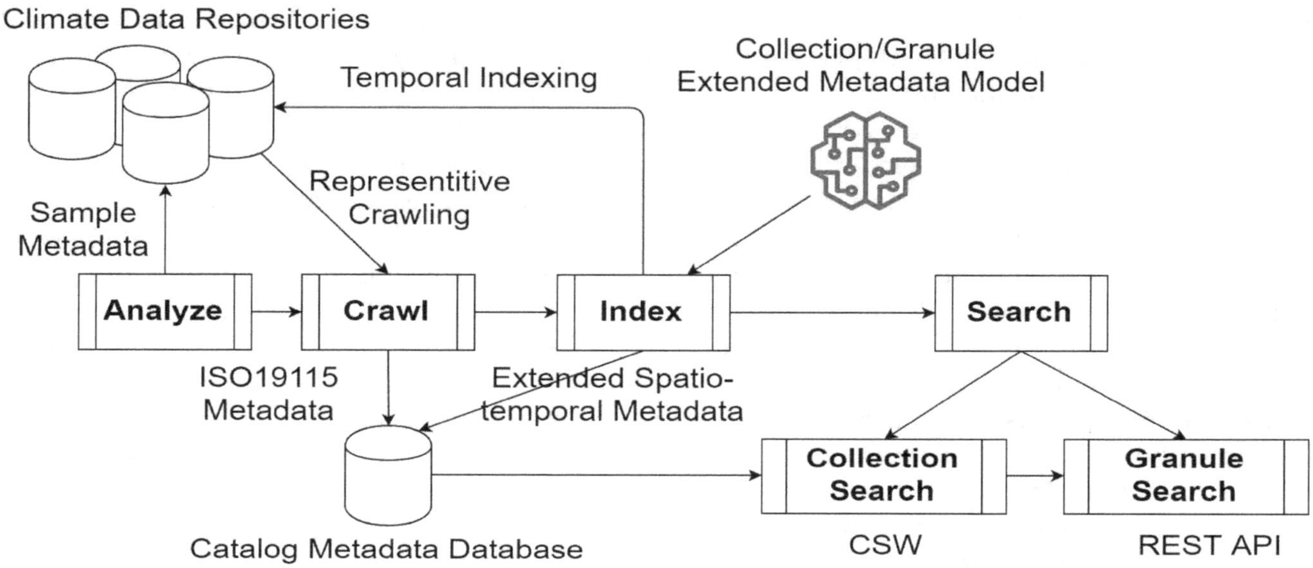

Figure 1. The proposed big climate data cataloging solution. Abbreviations: CSW, Catalog Services for the Web; REST API, Representational State Transfer Application Programming Interface.

4.1. *Metadata Repository Selection*

We took Unidata THREDDS Data Server (TDS) as our example target geodata repository platform. TDS was chosen because it is widely used by atmospheric and other related Earth science fields. It supports a good variety of open metadata and data standards and there exist many data centers that use TDS. It supports basic catalog features but lacks advanced search capabilities. It gives users and administrators large latitude of how the data is organized and updated inside the TDS catalog. The geodata stored across many TDS installations meets our broad criteria for real-world data variety, volume, and velocity.

A single TDS instance was selected as a target for our experiment. UCAR Unidata TDS (thredds.ucar.edu) repository was determined as a suitable target system and a good example of diverse uses of TDS. Unidata TDS contains a requisite variety of data. It has near real-time data that demonstrate the data velocity challenge. It contains a variety of data granularity and a good range in the size and complexity of datasets available. The volume of data and the volume of metadata is sufficiently challenging. The catalog structure is heterogeneous—different types of data are organized on different principles. On initial inspection, Unidata TDS was determined to be a great example of the challenges we wanted to explore.

Using manual inspection and basic statistical analysis via custom Python scripts, we started mapping out the characteristics of the Unidata TDS information system. We tried to answer the following questions: (a) What is the hierarchical structure of data organization in this repository?; (b) how frequently are new records are added and removed?; (c) which parts of the catalog exhibit regular patterns in the information structure that can be generalized and which parts contain unique information?; (d) what are the size and content of the metadata resources stored in the catalog?; (e) how is information in metadata resources related to metadata resources location within the hierarchy of catalog structure?; and (f) what are the data transmission qualities of the Unidata TDS network system—what portion of the TDS information can be transferred and copied to our system?

4.2. Repository Analysis

The following figures show some of the surface structure of the Unidata TDS catalog retrieved using a web browser from http://thredds.ucar.edu/thredds/catalog.html. Figure 2 shows the top level of the catalog hierarchy. Each listed item is a folder (a catalog). Most catalogs contain several levels of nested catalogs (Figure 3) in a tree-like hierarchy similar to a file system. At the bottom (leaf) tree level (Figure 4), the catalogs contain a list of data resources. Catalogs are presented in two formats. First is the HTML format, suitable for manual web browsing. Second is the XML format that contains additional metadata about the catalogs and the data resources. The XML representation follows THREDDS Client Catalog Specification. The specification extends the basic filesystem-like structure with temporal, spatial, and data variable description metadata annotations [74].

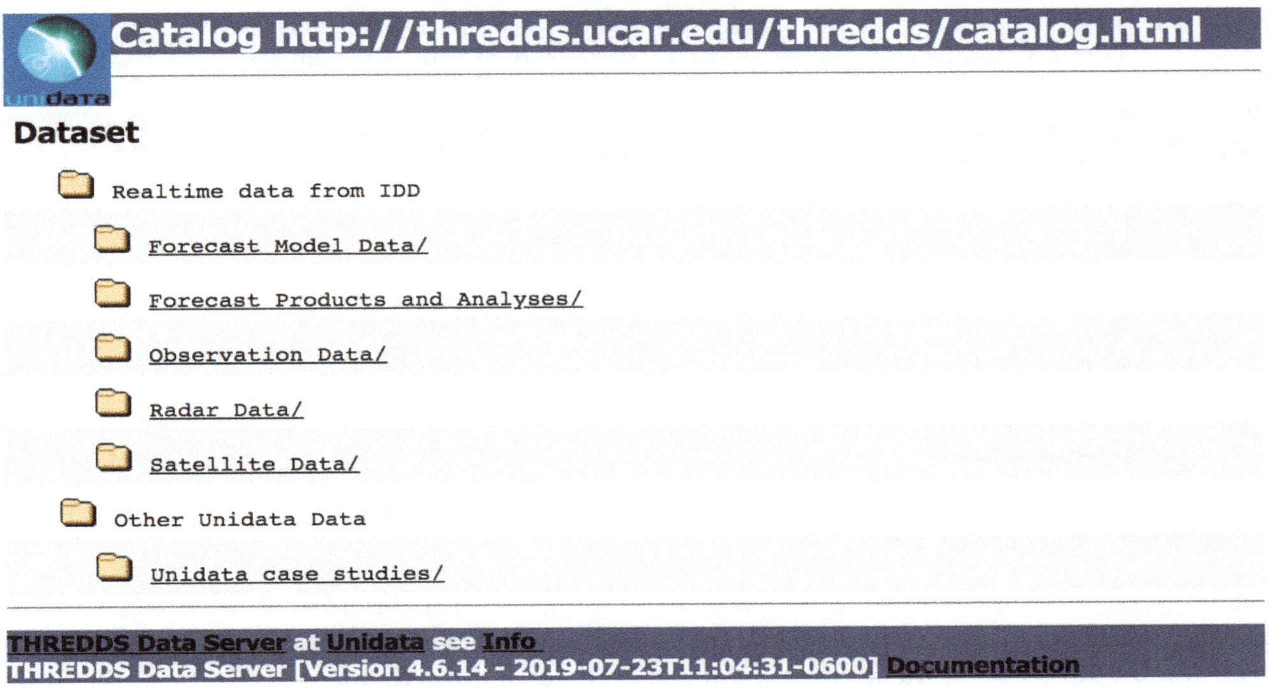

Figure 2. Top level Unidata THREDDS Data Server (TDS) catalog listing.

The TDS catalog provides a powerful general catalog hierarchy model. However, the practical use of this model by scientists who produce geodata is what determines the possibility of data collaboration and harmonization—as well as the specific shapes and possible solutions for big data problems. Email correspondence with Unidata explained that the data placed in different sub-catalogs is produced and organized by different teams of scientists. Although Unidata TDS acts as a unified repository for diverse Earth data, there are no mandatory overarching organizing principles to enable data harmonization [75].

Catalog http://thredds.ucar.edu/thredds/idd/forecastModels.html

Dataset

- NCEP Forecast Models
  - Global Ensemble Forecasting System (GEFS)
    - GEFS Members - Analysis/
    - GEFS Members - Forecasts/
    - GEFS Derived Forecast Products/
  - Global Forecast System (GFS)
    - GFS Quarter Degree - Global Coverage
      - GFS Quarter Degree Analysis/
      - GFS Quarter Degree Forecast/
    - GFS Half Degree - Global Coverage
      - GFS Half Degree Analysis/
      - GFS Half Degree Forecast/
    - GFS One Degree - Global Coverage
      - GFS One Degree Analysis/
      - GFS One Degree Forecast/
    - GFS Global 1.0 Degree (NOAAPORT)/
    - GFS Pacific 20km/

Figure 3. Nested catalogs. Only first 11 entries shown here. More entries omitted.

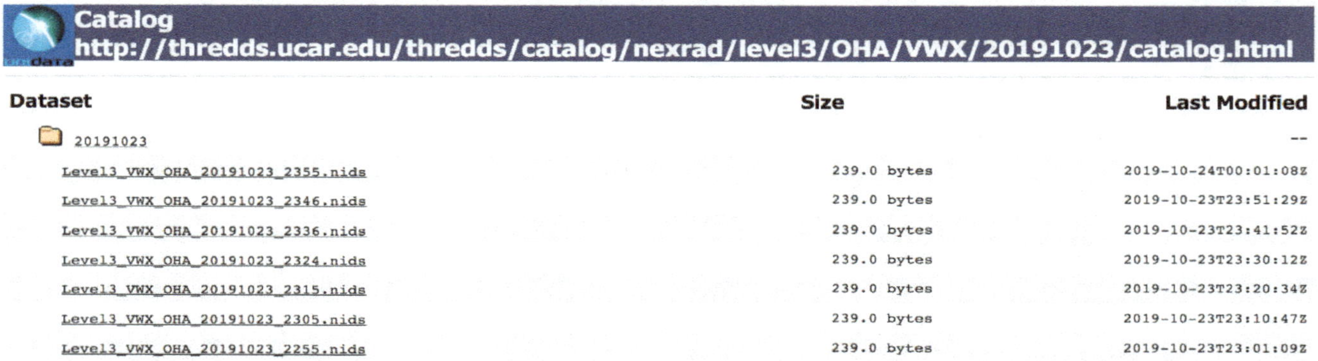
Catalog http://thredds.ucar.edu/thredds/catalog/nexrad/level3/OHA/VWX/20191023/catalog.html

Dataset	Size	Last Modified
20191023		--
Level3_VWX_OHA_20191023_2355.nids	239.0 bytes	2019-10-24T00:01:08Z
Level3_VWX_OHA_20191023_2346.nids	239.0 bytes	2019-10-23T23:51:29Z
Level3_VWX_OHA_20191023_2336.nids	239.0 bytes	2019-10-23T23:41:52Z
Level3_VWX_OHA_20191023_2324.nids	239.0 bytes	2019-10-23T23:30:12Z
Level3_VWX_OHA_20191023_2315.nids	239.0 bytes	2019-10-23T23:20:34Z
Level3_VWX_OHA_20191023_2305.nids	239.0 bytes	2019-10-23T23:10:47Z
Level3_VWX_OHA_20191023_2255.nids	239.0 bytes	2019-10-23T23:01:09Z

Figure 4. Data resource (dataset) listing at the bottom of the catalog hierarchy. Only first seven entries shown here. More entries omitted.

That being the case, the next step was to understand and describe different sub-structures organically adopted by different teams. After manual inspection and basic statistical analysis performed with custom Python scripts, the following information was compiled to broadly describe the different patterns of sub-catalog utilization (Table 1).

Table 1. Unidata TDS subcatalog big data characteristics.

Catalog	Sub-Catalogs/ Granules	Estimated Catalogs Size	New Granule Production	Granularity	Regularity
NCEP Forecast	5300/5300	500 MB	6 h	Coarse	Regular
Observations	8/186	20 MB	Irregular	Coarse	Irregular
Satellite	1500/71,000	150 MB	10 min	Fine	Regular
Radar	25,000/7 million	25 GB	5–10 min (irregular)	Very fine	Irregular

Four general types of data are simultaneously held in the Unidata TDS repository: (1) Forecast model output, (2) observations (time series from in-situ instruments), (3) satellite imagery, and (4) radar imagery from stationary radar network (NEXRAD, Next Generation Weather Radar) [76]. Each type contains much additional variety in its own hierarchy of sub-catalogs, but at this level, there are some clear and useful broad differences in data qualities that can guide our experiment.

In Table 1, the estimated catalog size is the total size of the metadata held in the catalog. Most of this metadata is completely redundant, but without knowing the deeper structure of this data, we would have to mirror all of this data in order to enable search and discovery capabilities that THREDDS does not support. We calculated maximum data transfer throughput of 4 MB/s or 5 min to load 1 GB of catalog data. It appears to be possible to mirror the entire Unidata TDS metadata catalog in several hours, but data throughputs we observed were not consistent, often slowing down by one order of magnitude. Furthermore, the speed of data processing (indexing and registering with a standard-compliant OGC CSW catalog) is also very time, and compute and storage resource, consuming. We do not have the capabilities to register and search millions of records mostly containing redundant information. Furthermore, the Unidata TDS data were added in near real-time according to specific patterns and structure in the sub-catalogs. If we attempted to copy and register all of that metadata, then we would not have been able to provide near real-time capabilities.

The last two columns in Table 1 show two critical qualities that determine what approach we needed to take to integrate that metadata into our systems.

If final datasets have "coarse" granularity, that means each dataset is a very file and the size of metadata is small in relation to the data size—for "coarse" datasets, we can copy, harvest, and index the metadata into our search system. "Fine" datasets stretch technical capabilities to transfer and process metadata. "Very fine" records are too numerous (the data files too small) for us to be able to effectively synchronize or process their metadata.

If datasets are produced in a regular way (predictable spatiotemporal attributes), then we can harvest minimal information and model the entire catalog. However, for NEXRAD radar metadata, there is no regular pattern to metadata production. A new record could be added every 5 min or every 15 min—and their regularity/irregularity also varies in time and depending on different radar sites (different sub-catalogs). This fine-grained irregular data is the most challenging, because it can neither be harvested wholesale nor modeled in an accurate way. It requires a targeted combination approach. Additional considerations arise when tracking what datasets have expired and been removed—ideally, this should be accomplished without performing an expensive full scan of the TDS repository.

Further examination of the sub-catalogs structure for irregular (and regular) highly granular data revealed additional useful structural information. Some catalogs are "dynamic" (or "live" or "streaming")—they are updated with new data resources with regular (or irregular) frequency. Other catalogs are archival—they can be assumed to never change (until they expire and are deleted entirely). Three distinct types of sub-catalogs can be identified:

- Pure archival directories: These folders only contain old collections and granules and will never be updated or deleted.
- Mixed archival directories: Some of the sub-folders contain archive material, some contain live, streaming near real-time data granules and collections.
- Daily archival directories: Folders that contain streaming data for a given day; when the day passes, this directory becomes an archive folder and does not need to be mirrored again. When daily archives expire, all the data resources for that day are deleted together.

4.3. *Crawling*

Big data catalogs normally need to complete a lot of crawling tasks to grab metadata files, and repeat scanning to capture metadata of the newly observed datasets on a regular basis. Crawling is the fundamental information source of metadata, and how to intelligently crawl is one of the largest challenges in big data searching due to the repeated computational burden and the complexity of the content. When designing our crawling strategy, we considered the observation update frequency, time window, observatory network organization, and made the crawler only touch the folders of those updated sensors at collection (sensor) level. Although a sensor has millions of metadata records, we only crawl the metadata at the sensor level. In other words, only one metadata is crawled for each sensor (or instrument). Using this strategy, we can save numerous hours in crawling and metadata transferring over network, especially when the network is unstable. After applying a parallel worker mechanism, we can have dozens of crawlers working on scanning and capturing new/updated metadata of petabytes of climate datasets.

Our crawler is different from most existing crawlers in the literature, because it is not a general-purpose search engine crawler. Typical crawlers download the entire web page, find links to follow, and add those links to the work queue. We cannot do a similar thing, because the web content we are crawling (TDS catalog) contains vastly redundant information that is not possible to download and process in its entirety without overloading available computing and network resources. There are various sensors in the climate monitoring networks and the sensors are dynamically changing, with new sensors added or old sensors removed. We had to crawl the THREDDS Data Server to make sure all the observations were fully synchronized in our catalog. Our crawler design must incorporate knowledge of metadata and metadata structure its processing and queueing algorithms in order to download only essential information.

4.4. Indexing

The third step is indexing, which extracts the spatiotemporal information from the crawled metadata and creates indexes for data granules of times series by each instrument. CSW provides the basic metadata registration and query model. However, the large granularity of metadata objects (and lack of aggregation/relational capabilities) makes CSW inefficient for storing and querying large numbers of datasets and that have only small variations in their metadata. A more efficient model is needed. This is a long explored and essentially solved problem in computer science and informatics. Theodoridis et al. [77] summarize the basic approach. For a time-evolving spatiotemporal object, a snapshot of its evolution can be represented by a triplet $\{o\_id, s_i, t_i\}$—object id, space-stamp, and time-stamp. This information allowed us to create a "repository production model" (Figure 5). We identified patterns in the catalog hierarchical structure that allowed us to identify which paths in the catalog folder hierarchy are "live" and which ones are "archival". In our crawler implementation (discussed in the next section), we used the structure path patterns to drive the crawler algorithm in two stages—"full sync" stage, which copies the archival data, and "update" stage, which monitors and refreshes the listing from "live" catalog paths.

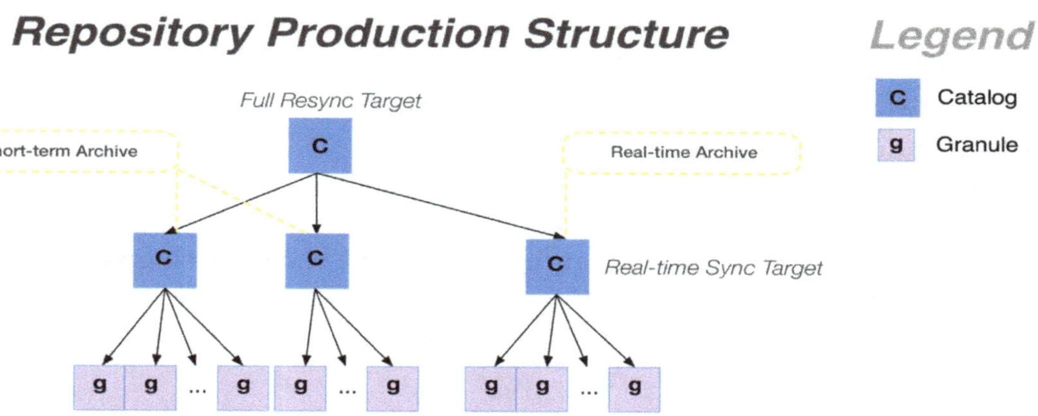

Figure 5. Repository metadata production model.

The repository production model allows targeted crawling—however, the number of metadata resources remains too large to harvest, process, and index in its entirety, even when done in two stages to avoid redundant harvesting. We needed a second model that encompasses the metadata information structure (Figure 6). There are two issues we needed to solve: First is that most of the metadata in the catalog is completely redundant; second is that metadata information scope is not consistent in the catalog. The two issues have the same source: Catalogs, and sub-catalogs and data granules all can have metadata attached.

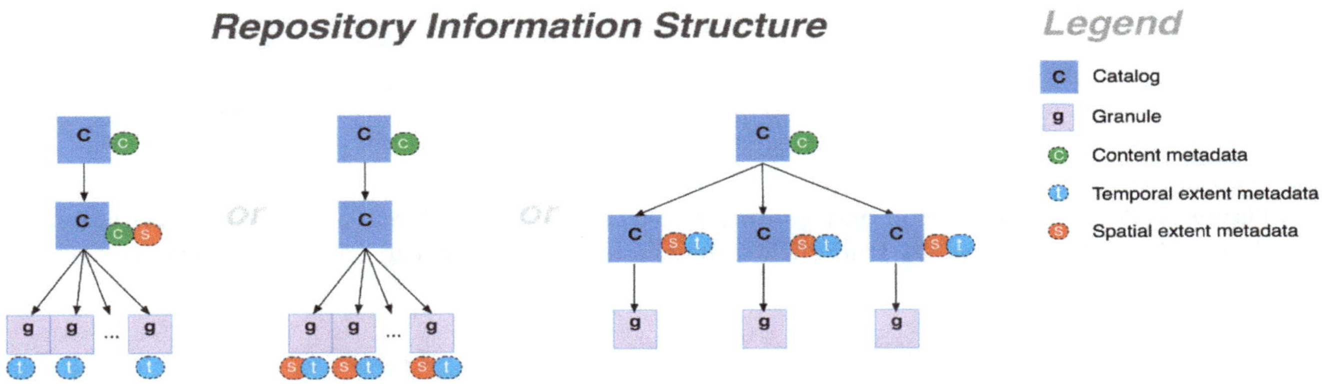

Figure 6. Repository information model.

In these examples from Unidata TDS, we see that metadata is attached to the hierarchical catalog structure in various ways. In the first example, a catalog contains some content metadata (for example: Authorship), the sub-catalog contains additional content metadata (ex: Variable names) and spatial metadata, while each granule contains temporal metadata. In the next two examples, the distribution of metadata between catalogs and granules is different. The last example is a case where each catalog only contains a single data record (granule). In some cases, the metadata is simply duplicated between several catalog levels, while in others, one specific layer contains all metadata. Another important detail is that the catalog hierarchy, the names of parent catalogs is also metadata for the data resources.

When combined, these two perspectives (information change model and information structure model) produce a model of the Unidata TDS repository that can be used to develop efficient (non-redundant) harvesting and representation of all contained metadata. By applying the production model to our crawler design, we were able to harvest only the information we know had changed. Knowing the structure of data changes also allowed us to perform targeted incremental harvesting for near real-time discovery capability. We defined two types of objects: Collections and granules. Collection contains content metadata (title, description, authorship, variable/band information, etc.). Each collection contains one or more granules. Each granule contains only the spatiotemporal extent metadata. The OGC CSW catalog standard does not support the composition of collections and granules, so we used CSW to represent collections only, while granules had to be stored externally. We used popular PyCSW software to hold collection metadata. We extended PyCSW with a PostgreSQL relational database to store relations between collections and granules and granule metadata (Figure 7).

4.5. Two-Step Search Process

When the metadata is harvested into PyCSW and temporal granule index is saved in PostgreSQL, the search clients can use these two data sources to retrieve final results for access. The search process takes place in two steps. Initially, the client searches the PyCSW store using standard search methods and queries. This returns a list of collection level results. To get a list of granules, the search client sends a second query to the crawler service. The crawler service queries the granule index, refreshes the index with the latest granules if needed, and returns a list of granules for the requested collection. The search client can then use the collection level CSW record and combine it with selected granule information to produce granule level CSW information. Figures 1 and 8–10 show these interactions from systems architecture and event sequence perspectives.

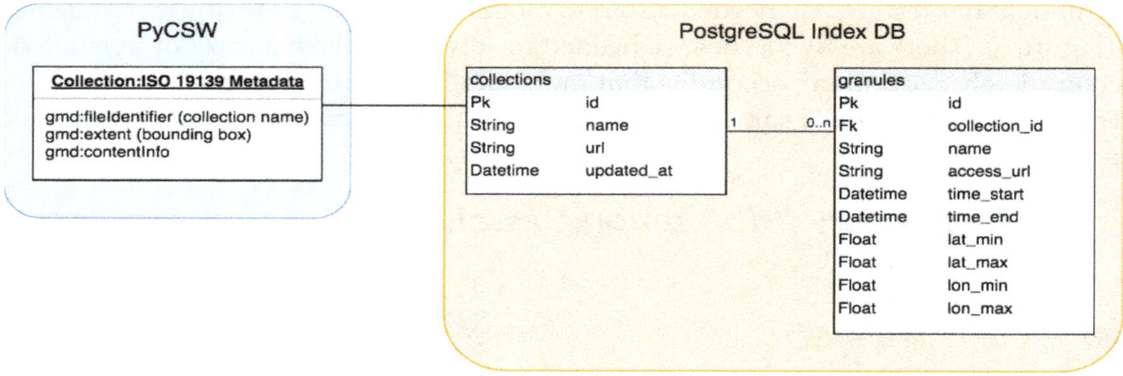

Figure 7. Metadata collection and granule resources stored in referentially linked PyCSW and PostgreSQL databases. PyCSW gmd: fileIdentifier corresponds as a key to collections table name field in the SQL database.

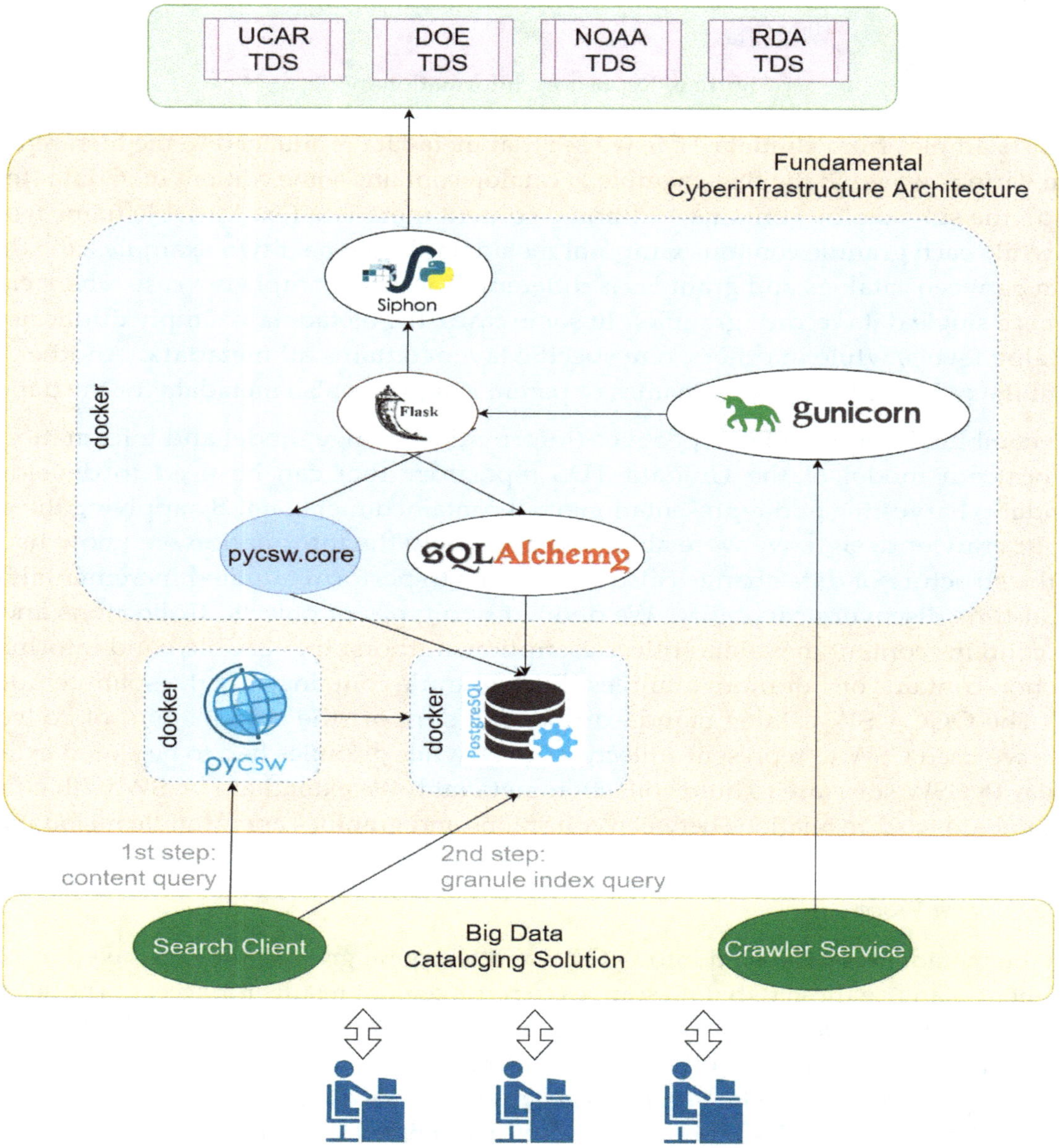

Figure 8. Implementation architecture for searching big data served via Unidata THREDDS Data Server (TDS). Although in our study only UCAR TDS is used, the system is designed to support any TDS repository as a data source.

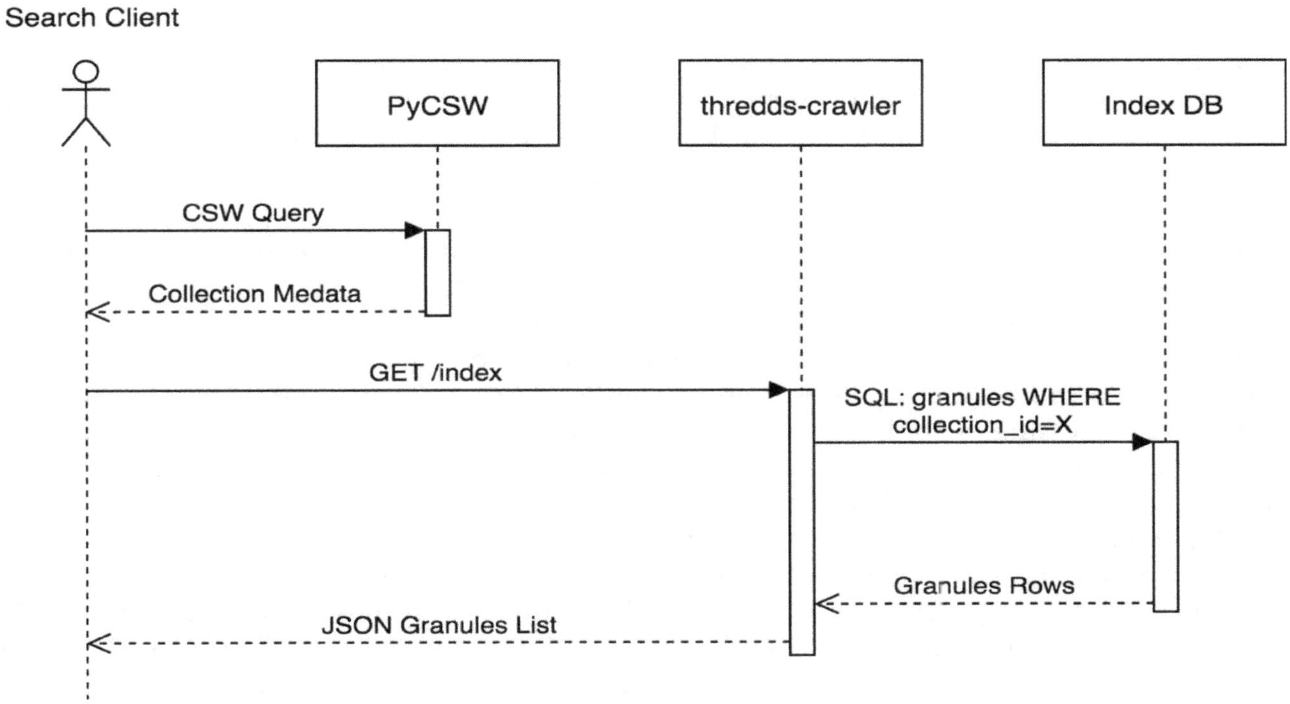

Figure 9. Simple granule index retrieval during search.

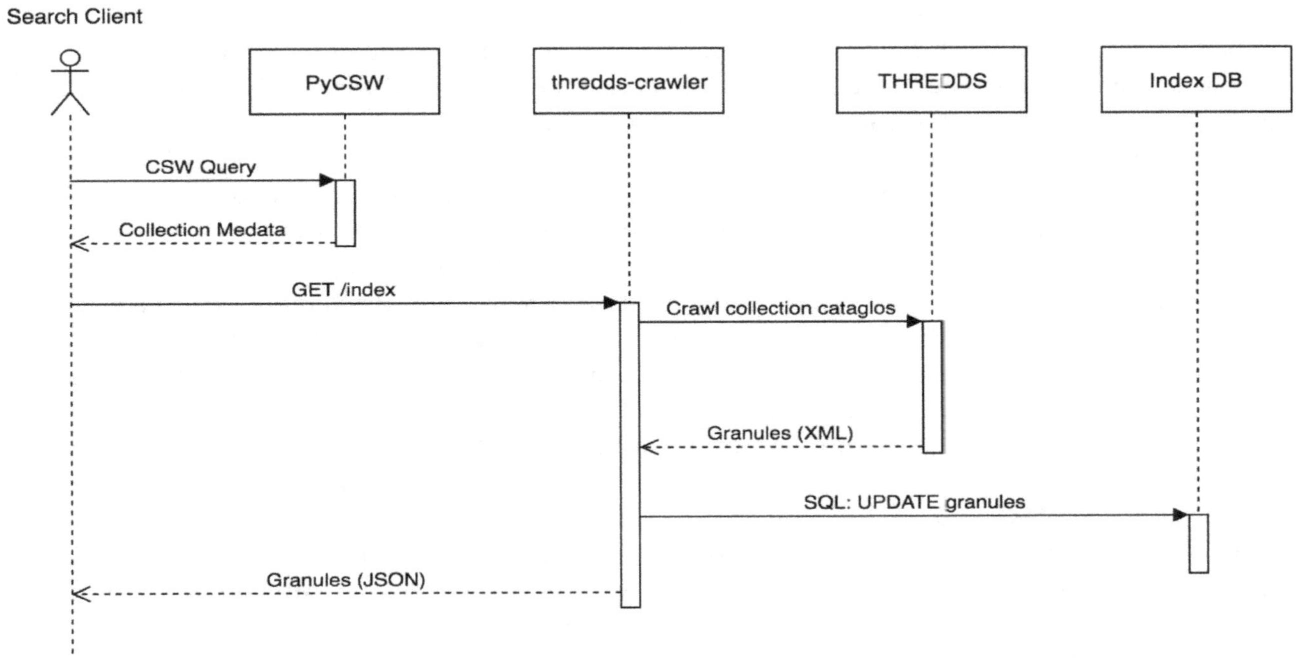

Figure 10. Granule index retrieval when granule temporal range is outside the range stored in the index. Search triggers an additional step that immediately crawls the TDS catalog and updates the real-time granule metadata. Abbreviations: DB, Database.

So far, we have analyzed the Unidata TDS repository structure, built a model of the repository that can inform an effective crawling strategy, and defined the model for the product output for the crawler. We have also described how a search client should function. To complete our experiment, we built a crawler that follows our metadata model and demonstrates a web search capability for the entire contents of Unidata TDS.

5. Implementation

We implemented a module system within EarthCube CyberConnector [17,78,79] to realize the proposed mode (Figure 8). The implementation included the searching server system and the client system. We will introduce the searching capabilities enabled by these systems.

5.1. Crawler Service Implementation

We built a web crawler that traverses Unidata TDS and extracts and stores essential metadata without using unnecessary resources. It is named 'thredds-crawler' and the source code is available via a public GitHub repository: https://github.com/CSISS/thredds-crawler.

The crawler is written in Python. It was built using common open source libraries for HTTP API interaction (Flask [https://www.fullstackpython.com/flask.html], Gunicorn [https://gunicorn.org/]), general XML processing (libxml [https://lxml.de/}]), and database abstraction (SQLAlchemy [https://www.sqlalchemy.org/]). It uses native Python threading libraries to support concurrency. For traversing the Unidata THREDDS catalog and retrieving metadata, it uses the Unidata provided Python Siphon library [https://github.com/Unidata/siphon].

To support our big-data experiment requirements, the crawler is tightly integrated with a catalog software PyCSW [https://pycsw.org/] and a PostgreSQL [https://www.postgresql.org/] database. The crawler, PyCSW, and the database each run in a separate Docker [https://www.docker.com/] container. For the sake of this demonstration, all three services run on the same machine and communicate over the local network. The Docker-compose tool is used to connect and orchestrate the three containers. This architecture allows simple scaling out to multiple machines using containers, which allows for potential substantial improvement in system performance.

The crawler docker container runs as a web service hosted by Gunicorn—a python HTTP server widely used for hosting web applications. It serves three HTTP API endpoints that perform the following functions: Harvest, create index, and read index.

The harvest function loads the Unidata Catalog XML from specified catalog_url using Siphon library. The catalog contains a list of datasets. TDS has a feature to translate its dataset metadata into the ISO/OGC-compatible XML format. For each dataset being harvested, the harvester constructs a query to TDS to retrieve ISO/OGC metadata for the dataset. The ISO/OGC metadata returned by TDS is, however, often incomplete, inaccurate, or inconsistent in some way. The crawler harvesting process then applies a chain of XML filters to the ISO/OGC metadata to rectify it with information from native TDS dataset metadata. Once the metadata is downloaded and processed, it is saved in the PyCSW database directly by using a PyCSW compatibility library.

Indexing is similar to harvesting, but involves a strategy for targeting datasets to be harvested and additional processing steps. During index creation, TDS catalogs and datasets are turned into collections and granules in our model. For each TDS dataset encountered, we determine the collection name. Crucially, the collection name is not the name of the catalog containing the dataset. We found that catalog names are inconsistent, but that TDS dataset ids contain consistent identification information. In the TDS, dataset ids are kept unique by including timestamps in the dataset id. For example, in a dataset with id "NWS/NEXRAD3/PTA/YUX/20190830/Level3_YUX_PTA_***20190830_1713***.nids", the portion "***20190830_1713***" is a timestamp. To turn TDS catalogs with datasets into collections with granules, we remove the temporal information to construct the collection id. Then, we download the dataset in the ISO/OGC XML format and transform its XML content with a filter function called "collection builder". This function updates the dataset metadata to turn it into a more general form that describes the collection. It changes identifiers stored in the metadata. It also adds standard-compliant additional fields that identify the metadata for describing "series" ("series" in ISO/OGC model, "collection" in our model). This process needs to be done only for the first dataset encountered for each collection. When processing additional datasets, the existing collection is reused. In TDS, the dataset spatiotemporal extent information is part of the catalog metadata, which means that we only need to download a single dataset metadata to build the collection metadata and we can index the remainder of granules from catalog metadata. This solves the redundancy issue that previously prevented TDS from being searchable. We also correct TDS identifiers to ensure that the namespace authority portion of the identifier is correctly set.

The following tables help illustrate the process of extract collections from a granule identifier for multiple types of data. Table 2 shows the catalog paths of TDS datasets. Table 3 shows how the collection identifier is generated, and Table 4 shows the final result collection name.

Table 2. Catalog paths of TDS datasets for three types of data. The catalog path hierarchy is marked in green. The dataset filename is marked in red.

Data Type	Example TDS Catalog Path
RADAR	Radar Data › NEXRAD Level III Radar › PTA › YUX › 20190830 › **Level3_YUX_PTA_20190830_1713.nids**
Model	Forecast Model Data › GEFS Members › Analysis › GEFS_Global_1p0deg_Ensemble_ana_20190731_0000.grib2 › **GEFS_Global_1p0deg_Ensemble_ana_20190731_0000.grib2**
Satellite	Satellite Data › GOES West Products › CloudAndMoistureImagery › Mesoscale-2 › Channel16 › 20190831 › **OR_ABI-L2-CMIPM2-M6C16_G17_s20192430003570 _e20192430003570_c20192430003570.nc**

Table 3. TDS dataset identifiers for three types of data. The portion of the identifiers that contain temporal information is highlighted.

Data Type	Example TDS Dataset Identifier
RADAR	NWS/NEXRAD3/PTA/YUX/ **20190830** /Level3_YUX_PTA_ **20190830_1713** .nids
Model	grib/NCEP/GEFS/Global_1p0deg_Ensemble/members-analysis/ GEFS_Global_1p0deg_Ensemble_ana_ **20190731_0000** .grib2
Satellite	goes-west-products/CloudAndMoistureImagery/Mesoscale-2/Channel16/ 20190831/OR_ABI-L2-CMIPM2-M6C16_G17 _ **s20192430003570_e20192430003570_c20192430003570** .nc

Table 4. Collection identifiers for the example dataset IDs. They are calculated by removing temporal information and prefixing an authority namespace field.

Data Type	Calculated Example Dataset Collection Identifier
RADAR	edu.ucar.unidata:NWS/NEXRAD3/PTA/YUX/Level3_YUX_PTA.nids
Model	edu.ucar.unidata:grib/NCEP/GEFS/Global_1p0deg_Ensemble/members-analysis/ GEFS_Global_1p0deg_Ensemble_ana.grib2
Satellite	edu.ucar.unidata:goes-west-products/CloudAndMoistureImagery/Mesoscale-2/ Channel16/OR_ABI-L2-CMIPM2-M6C16_G17.nc

When the index harvesting is complete, the collection information (OGC/ISO 19139 XML metadata format) is stored in PyCSW. The granule information is stored in a compact SQL index store (Figure 7). Once the index is created, it can be retrieved from the crawler web service using HTTP API (GET/index). These requests take a collection name and temporal extent as parameters. Although our data model includes granule spatial and temporal extent, at the time of publication, only temporal index queries were implemented. It checks the index data store to see if the latest available granules are newer than the requested time extent. If more recent granules are not required, the crawler returns a list of granules in compact JSON format (Figure 9). However, if the index does not contain recent enough granules, then the index service performs a partial "refresh" indexing of the TDS repository. It uses the TDS catalog link stored in our PyCSW collection and re-runs the index process described here. (Figure 10). However, as we discussed in the Experiment section, the TDS catalog is organized with some sub-catalogs storing archival information, while others contain near real-time "live data". The crawler index refresh process takes advantage of that structure. It ignores the old sub-catalogs and only indexes those that contain more recent and unknown data. This makes near real-time index retrieval fast and efficient.

Both harvesting and index creation use the same multi-threaded queue strategy to achieve higher performance. Normally, most of the time is spent waiting for data to be transmitted over the network. By using many threads, we can increase the saturation of both the network and local computer and memory resources, which allows the metadata to become available much faster.

5.2. Search System Implementation

The search system is implemented based on the previously developed EarthCuber CyberConnector infrastructure building block [17]. CyberConnector is a Java-based web application that supports discovery and visualization of data from CSW catalogs [17]. We extended CyberConnector to support accessing metadata harvested and indexed by the thredds-crawler described in the previous section. We modified the CyberConnector Search Client to perform a two-stage search. The web application user selects "Search" function (Figure 11). They select a time range, which is used by the thredds-crawler index service to determine if granule refresh is needed. The web browser sends an AJAX request to the CyberConnector web application with search parameters. CyberConnector queries thredds-crawler PyCSW service for collections that match query parameters. It returns a list of collections. To see the granules available in a collection, the user clicks the "List Granules" button (Figure 12). This issues another request to CyberConnector for a granules list in the specified temporal extent. CyberConnector web application proxies the granules list request to thredds-crawler indexing service, which returns a list of granules (Figure 9); or thredds-crawler harvests TDS to update the index and then returns a list of granules (Figure 10). The client receives a list of granules, which can then be downloaded or visualized (Figure 13).

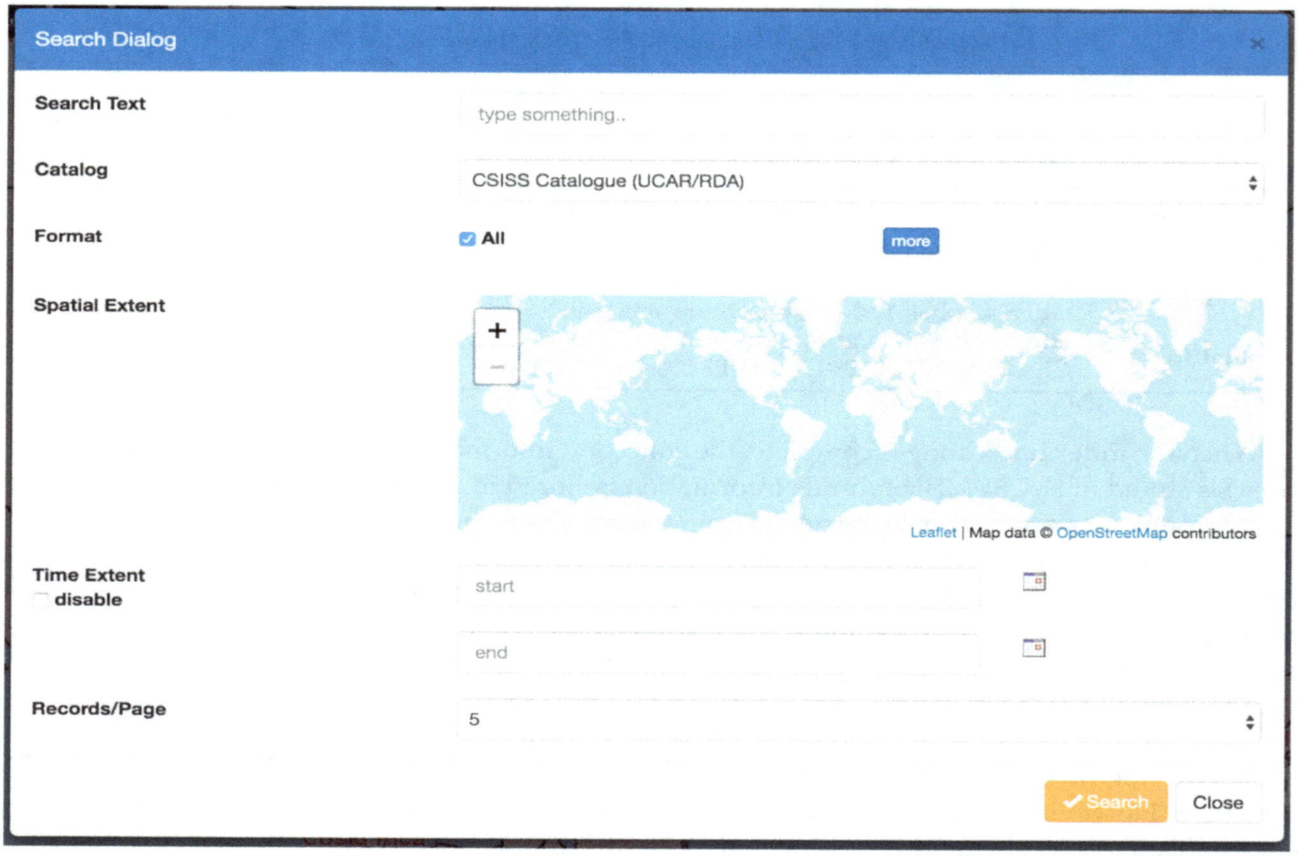

Figure 11. Search client web interface.

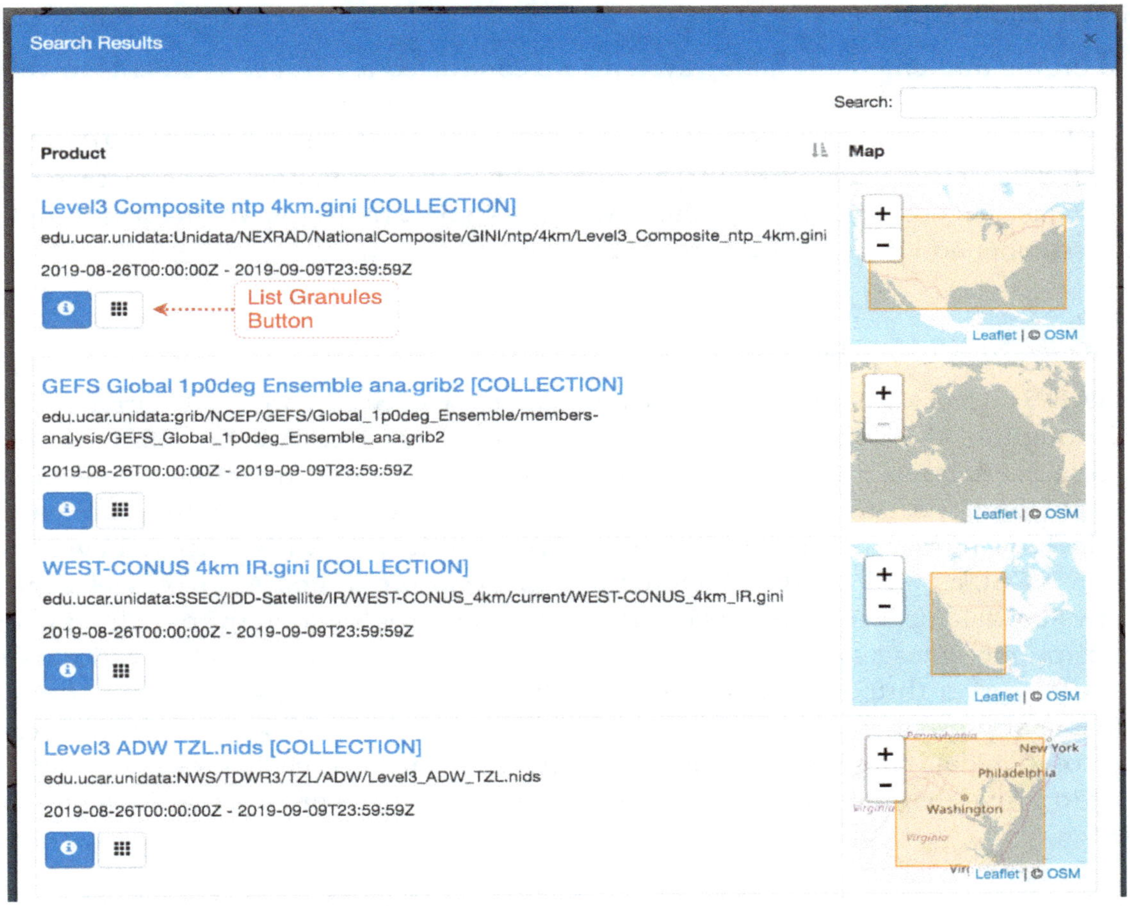

Figure 12. Search results with "List Granules" button.

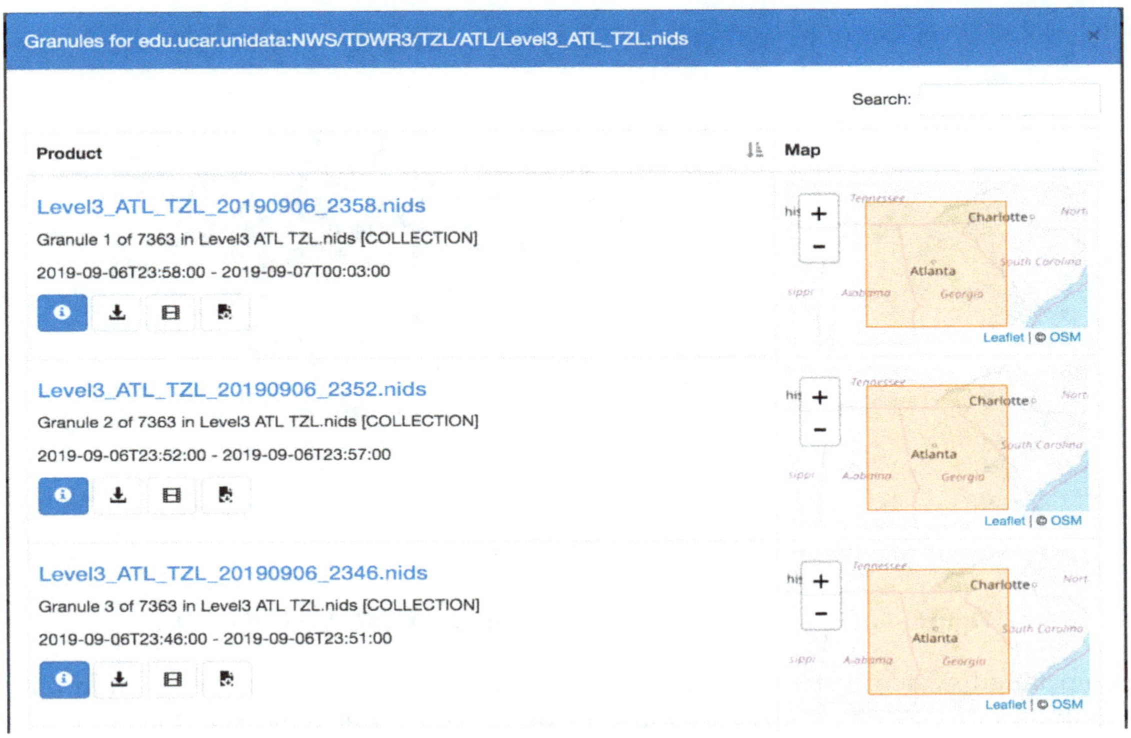

Figure 13. Granules list for a collection. The buttons allow users to view metadata, download the dataset, or visualize it.

6. Experiment and Results

Based on the implemented catalog system, we conducted several experiments to validate the feasibility of the proposed approach. The datasets for climate science are generally very large because of their long-term running and high temporal resolution. We took the UCAR NEXRAD dataset [80] and the RDA ASR dataset (53.09 terabytes) [81] as our demonstration examples. The searching capabilities on the two datasets were established in the EarthCube CyberConnector. We made a complete set of tests on the searcher and the results are introduced below.

6.1. Searching the NEXRAD Dataset

NEXRAD is a very important dataset for climate science research. It currently comprises 160 sites throughout the United States and selected overseas locations (as shown in Figure 14). The basic original datasets, including three meteorological base data quantities: Reflectivity, mean radial velocity, and spectrum width, are called Level II. The derived products are called Level III, which include numerous meteorological analysis products. All NEXRAD Level-II data are available via NCEI, as well as NOAA big data plan cloud providers, Amazon web service (http://thredds-aws.unidata.ucar.edu/thredds/catalog.html) and Google Cloud (https://cloud.google.com/storage/docs/public-datasets/nexrad). UCAR provides the near real-time observed data via their THREDDS data server (http://thredds.ucar.edu). Unfortunately, all these data repositories are still non-searchable at present, because it is a huge challenge for any catalog to index and search such big amount of metadata files for the frequently updated radar data records (every 6 min). We used this dataset to prove that the proposed cataloging approach can work well on frequently updated big datasets.

Figure 14. NOAA NCDC Radar Data Map (NEXRAD Level II and III).

The completed system consists of the harvester/indexer service and the search client that is available to the user as a web application. As a result, users are able to search diverse heterogeneous Earth system observation and modeling datasets simultaneously. Once the metadata is found, users can use the CyberConnector visualization system to simultaneously visualize near real-time NEXRAD radar, satellite observation, and forecast simulation model product data. The system performance characteristics of this approach are significantly improved over the existing naive method of harvesting all of the datasets' metadata.

6.2. Searching UCAR RDA (Research Data Archive) TDS Repository

NSF-funded NCAR CISL (Computational & Information System Lab) maintains Research Data Archive (RDA), which stores over 11,000 terabytes of climate datasets in its high-performance data storage system.

RDA hosts many climate datasets at present, and the Arctic System Reanalysis (ASR) is one of them. ASR is a demonstration regional reanalysis for the greater Arctic developed by Ohio State University. The ASR version 2 dataset (the latest version) is served via RDA with a total volume of 53.04 terabytes. The horizontal resolution is 15 km and the temporal coverage is from 2000 to 2016. It has 34 pressure levels (71 model levels), 31 surface (including 3 soil variables), and 11 upper air analysis variables, 71 surface (including 3 soil variables), and 17 upper air forecast variables.

RDA provides TDS for most of its archived datasets. We harvested the metadata of ASR from its TDS and made them publicly available in CyberConnector. As shown in Figure 15, scientists can search the ASR dataset by providing keywords, spatial extent, or temporal range. The ASR data is in NetCDF format, which is displayable in COVALI. We demonstrated searching ASR dataset in COVALI and visualized the temperature at 2 m above the surface within 12 h. COVALI and RDA were deployed in two remotely distributed facilities. The interactions between COVALI and RDA big data storage were conducted via the standard service interface and over the network. The experiment proves that the proposed solution works well for enabling search on remote big data.

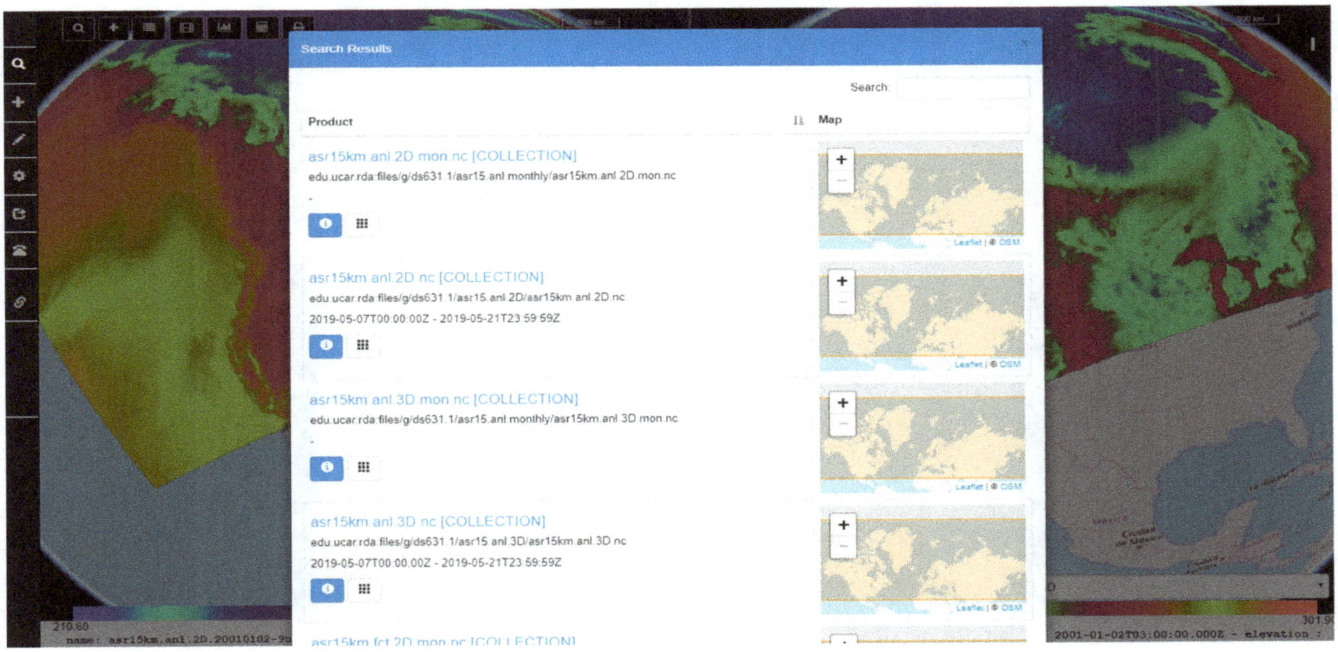

Figure 15. ASR (Arctic System Reanalysis) search results and visualization of the temperature at 2 m above the surface in the CyberConnector COVALI visualization system.

6.3. Performance Evaluation

The traditional approach for cataloging climate datasets is fully harvesting all the metadata files of every single data record. We implemented the searcher using the traditional method before, but the performance was very slow and sustained operation not possible for the practical scenario for big data cataloging. After we applied the new cataloging strategy, we tested it by crawling several hundreds and thousands of records from UCAR THREDDS Data Server. We tested using different sets of parallel workers: 40 workers, 20 workers, 10 workers, 5 workers, and a single worker, respectively, to measure the improvements of parallel crawling. Figure 16 displays the time cost of the test to compare the performance of the traditional approach and the proposed approach. The results demonstrate that the proposed approach outperforms the traditional approach at least ten times on the overall time cost (from ~10 to ~1 s) and has significant improvements on harvesting speed, storage use, and search speed based on the number of datasets being processed.

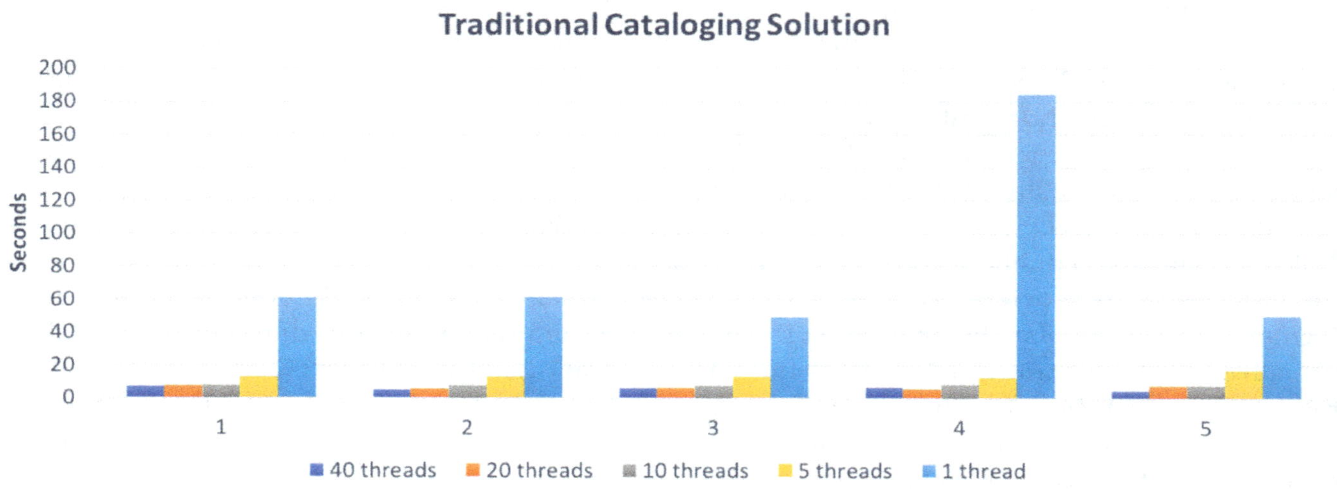

(**a**)

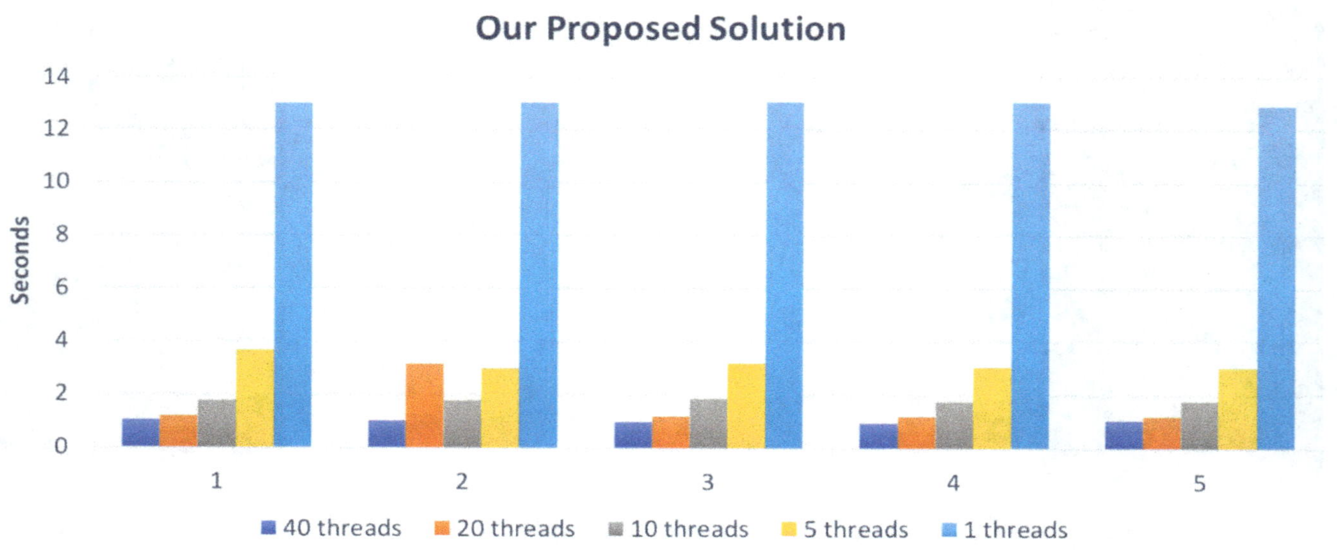

(**b**)

Figure 16. Performance comparison (time in seconds) of the traditional harvesting approach (**a**) and our approach (**b**), sampled 5 times for crawling 125 records.

Search time cost has two components. The time to search for collections in the catalog and the time to retrieve the granules list from the granule index. Figure 17 shows that search result retrieval is extremely fast in our system.

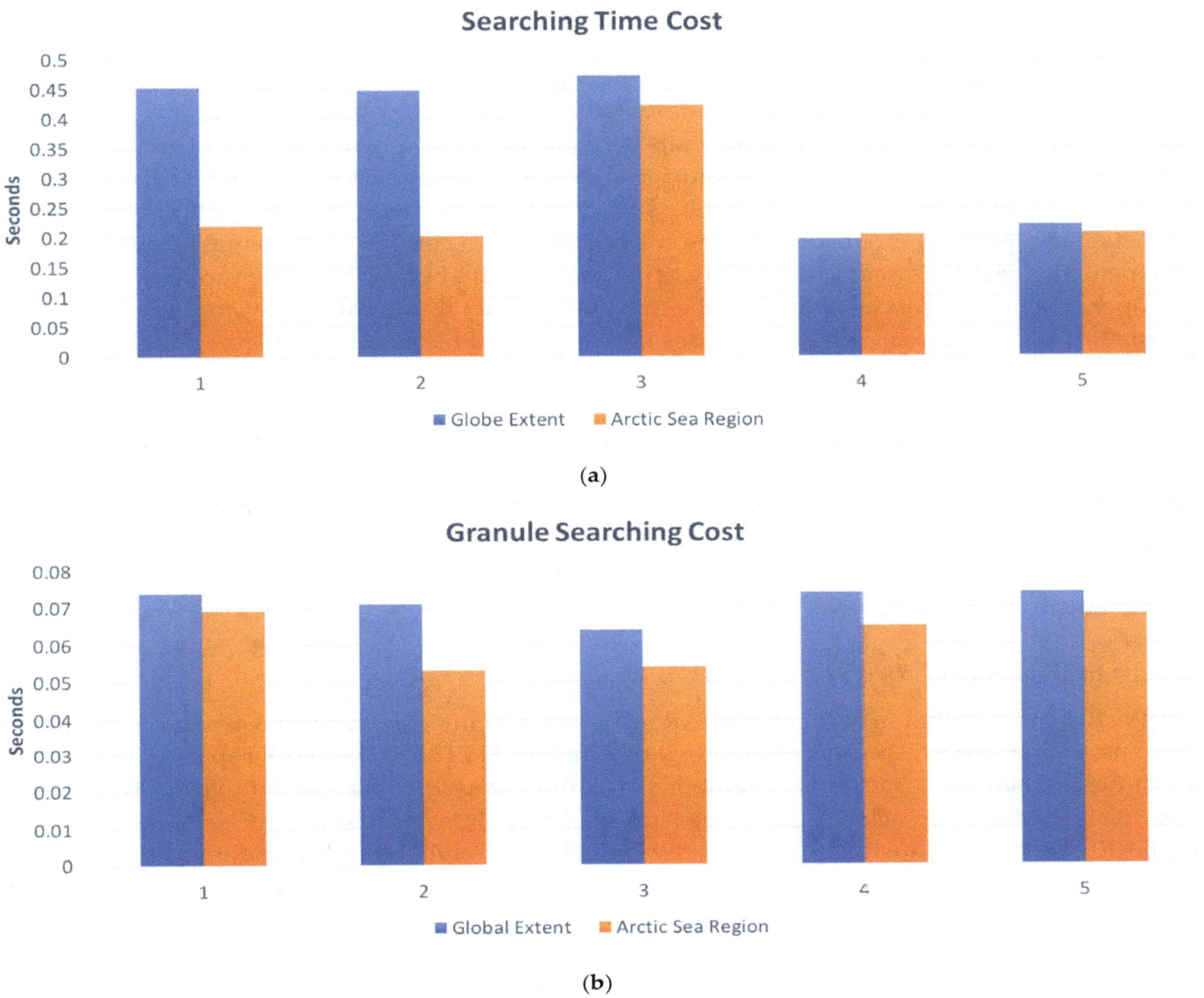

Figure 17. Search performance (time in seconds) with two different spatial extent parameters. (**a**) Time to query the catalog (collection searching, step 1); (**b**) time to query the granule index (granule searching, step 2).

The search currently supports filters, including keywords, data format, and spatiotemporal extents. All of them are fixed filters with less uncertainties. Therefore, the returned results stay the same as long as the metadata base does not add new records or delete existing records. The result completeness is 100% accurate because correct records match the filter conditions. Users can narrow down spatiotemporal extent based on their interest and provide one or more keywords which could match the data field names. The first page relevance of the search results depend on the relationships between the inputted keywords and the metadata field values. Based on our experiences with climate scientists, we find that they normally do not input any keywords and only use spatiotemporal filters to check out what can be searched in a catalog. Once they have a region of interest or a time window, they then have an impression about what possible data is available. They come to the catalog just to find the access URL to download or visualize the data files. Normally, results from our search client are numerous because of the loose filters the scientists give. The results on the first page are usually very well related to the scientists' needs. A more intelligent search, such as a semantics-based search, which could find more accurate first-page results with higher relevance, will be studied in the next stage of work.

7. Discussion

Our solution to the big data volume, variety, and velocity challenges discussed in the paper consists of a novel metadata model, and cyberinfrastructure architecture and implementation that is derived from the model. The metadata model combines the description of metadata content (the "information model") with the description of metadata repository structure and behavior. The cyberinfrastructure consists of a crawler service that takes advantage of the metadata model to optimize THREDDS crawling strategy to eliminate the transfer and processing of redundant metadata information. Additionally, the metadata repository model permits the crawler service to perform incremental metadata transfer, which enables real-time search capability. The demonstrated cyberinfrastructure also includes an interoperable catalog service that uses the metadata model to minimize the storage of redundant information. Finally, a search client that uses the catalog and the crawler services is implemented.

7.1. Can the Proposed Solution Address the Volume Challenge?

Metadata volume is ~25 GB for the UCAR RADAR dataset. The traditional method for harvesting metadata (as discussed in Section 6.3) is able to process approximately one record (with an approximate size of 100 KB) per second. To completely ingest all of THREDDS RADAR metadata at the observed harvesting rate, it would take 250,000 s or ~70 h. By using the proposed metadata model and cataloging system, we observe harvesting rates that are at least 10 times faster. This permits daily synchronization of all Unidata TDS metadata.

7.2. Can the Proposed Solution Address the Velocity Challenge?

We determined that new (live) RADAR metadata is being generated at 330 records per minute. Our maximum harvest capacity (constrained by Unidata THREDDS network capacity) is 60 records per minute. Using the traditional method, we cannot keep up with the data velocity. Using the indexing harvester approach, we can process up to 1400 records per minute. This exceeds the velocity of THREDDS data production. Additionally, by using incremental index update during the client search request exchange, we can target the indexing harvest process to the exact sub-catalog containing the updated information and thus provide real-time search capability for this high-velocity data.

7.3. Can the Proposed Solution Reduce Metadata Crawling Redundancy?

The solution demonstrated here is able to reduce redundancy in crawling and storage resource consumption. For example, using the traditional method with Forecast Models catalog, ~7000 records are downloaded. The total storage used is 1.85 GB. The same metadata can be processed using our approach by downloading only 45 sample metadata records (2.2 MB) that represent collection level information. This represents a 99% reduction in data transmission and storage costs.

7.4. What Are the Benefits and Drawbacks of the Proposed Solution Compared to Other Big Data Searching Strategies?

The solution demonstrates the expected benefits described at the beginning of this study. The main drawback of this solution is the model and software system complexity. Custom software has to be developed to intelligently process catalogs as they are being harvested. To get complete and accurate results, the ingested metadata must be cleaned and transformed to fill in missing pieces of information and to make it conform to our model. Although our approach is general enough to work with multiple TDS repositories, in practice, inconsistencies and additional varieties from each repository must be reconciled using custom code. Our work demonstrates that it is possible to build a unified and highly efficient searchable catalog system for large and heterogeneous Earth system data repositories that supports real-time queries; however, every solution has its limitations and costs. In this case, the costs are complexity in software and systems architecture, which means increased software development and maintenance costs.

8. Conclusions

This paper proposed and demonstrated a novel cyberinfrastructure-based cataloging solution to enable an efficient two-step search on big climatic datasets by leveraging the existing data centers and state-of-art web service technologies. We used the huge datasets served by UCAR THREDDS Data Server (TDS), which serves Petabyte-level ESOM data and updates hundreds of terabytes of data every day, as our study dataset to validate its feasibility. We analyzed the metadata structure in TDS and created an index for data parameters. A developed metadata registration model, which defines constant information, delimits variable information, and exploits spatial and temporal coherence in metadata, was constructed. The model derives a sampling strategy for a high-performance concurrent web crawler bot which is used to mirror the essential metadata of the big data archive without overwhelming network and computing resources. The metadata model, crawler, and standard-compliant catalog service form an incremental search cyberinfrastructure, allowing scientists to near real-time search in big climatic datasets. We experimented with the approach on both UCAR TDS and NCAR RDA TDS, and the results prove that the proposed approach achieves its design goal, which is a significant breakthrough for the current most non-searchable climate data servers. The solution identified redundant information and determined the sampling frequencies to keep unpredictable parts of the source catalog synchronized with our downstream mirror catalog. An automated hierarchical crawler–indexer and a complimentary search system using the pre-existing EarthCube CyberConnector were implemented. Metadata crawling and access performance validates our integrated approach as an effective method for dealing with big data challenges posed by heterogeneous, real-time Earth System Observation and Model data. However, although the proposed approach outperforms the traditional searching solution for big data, it is still time-consuming in both crawling and searching processes, and may be out of pace dealing with real-time streaming data. In the future, we will study to further reduce the time spent in crawling redundant metadata and to find a high-performance method for rapid and intelligent search.

Author Contributions: Conceptualization, Liping Di and Ziheng Sun; methodology, Ziheng Sun and Juozas Gaigalas; software, Juozas Gaigalas and Ziheng Sun; validation, Juozas Gaigalas, Ziheng Sun; formal analysis, Liping Di, Ziheng Sun, and Juozas Gaigalas; investigation, Juozas Gaigalas; resources, Liping Di; data curation, Juozas Gaigalas, Ziheng Sun, and Liping Di; writing—original draft preparation, Juozas Gaigalas; writing–review an Editing, Liping Di and Ziheng Sun; visualization, Juozas Gaigalas and Ziheng Sun; supervision, Liping Di and Ziheng Sun; project administration, Liping Di and Ziheng Sun; funding acquisition, Liping Di.

Acknowledgments: We sincerely thank UCAR, UCAR Unidata Support Team, and the authors of the software, libraries, tools, and datasets that we have used in this work.

References

1. Wright, D.J.; Wang, S. The emergence of spatial cyberinfrastructure. *Proc. Natl. Acad. Sci. USA* **2011**, *108*, 5488–5491. [CrossRef] [PubMed]
2. *CyberinfrastruCture Vision for 21st Century DisCoVery*; National Science Foundation Cyberinfrastructure Council: Arlington, VA, USA, 2007.
3. Yang, C.; Goodchild, M.; Gahegan, M. Geospatial Cyberinfrastructure: Past, present and future. *Comput. Environ. Urban Syst.* **2010**, *34*, 264–277. [CrossRef]
4. Yue, P.; Gong, J.; Di, L.; Yuan, J.; Sun, L.; Sun, Z.; Wang, Q. GeoPW: Laying Blocks for the Geospatial Processing Web. *Trans. GIS* **2010**, *14*, 755–772. [CrossRef]
5. Di, L. Geospatial Sensor Web and Self-adaptive Earth Predictive Systems (SEPS). In Proceedings of the Earth Science Technology Office (ESTO)/Advanced Information System Technology (AIST) Sensor Web Principal Investigator (PI) Meeting, San Diego, CA, USA, 13 February 2007.
6. Zhao, P.; Yu, G.; Di, L. Geospatial Web Services. In *Emerging Spatial Information Systems and Applications*, 1st ed.; IGI Global: Hershey, PA, USA, 2006; pp. 1–35.
7. Shukla, J.; Palmer, T.N.; Hagedorn, R.; Hoskins, B.; Kinter, J.; Marotzke, J.; Miller, M.; Slingo, J.; Shukla, J.; Palmer, T.N.; et al. Toward a New Generation of World Climate Research and Computing Facilities. *Bull. Am. Meteorol. Soc.* **2010**, *91*, 1407–1412. [CrossRef]

8. Sherretz, L.A.; Fulker, D.W. Unidata: Enabling Universities to Acquire and Analyze Scientific Data. *Bull. Am. Meteorol. Soc.* **1988**, *69*, 373–376. [CrossRef]
9. Schnase, J.L.; Duffy, D.Q.; Tamkin, G.S.; Nadeau, D.; Thompson, J.H.; Grieg, C.M.; McInerney, M.A.; Webster, W.P. MERRA Analytic Services: Meeting the Big Data challenges of climate science through cloud-enabled Climate Analytics-as-a-Service. *Comput. Environ. Urban Syst.* **2017**, *61*, 198–211. [CrossRef]
10. Khan, M.A.; Uddin, M.F.; Gupta, N. Seven V's of Big Data understanding Big Data to extract value. In Proceedings of the 2014 Zone 1 Conference of the American Society for Engineering Education, Bridgeport, CT, USA, 3–5 April 2014; pp. 1–5.
11. Habermann, T. Metadata Life Cycles, Use Cases and Hierarchies. *Geosciences* **2018**, *8*, 179. [CrossRef]
12. Greenberg, J. Metadata and the World Wide Web. *Encycl. Libr. Inf. Sci.* **2003**, *3*, 1876–1888.
13. Li, S.; Dragicevic, S.; Castro, F.A.; Sester, M.; Winter, S.; Coltekin, A.; Pettit, C.; Jiang, B.; Haworth, J.; Stein, A.; et al. Geospatial big data handling theory and methods: A review and research challenges. *ISPRS J. Photogramm. Remote Sens.* **2016**, *115*, 119–133. [CrossRef]
14. Bernard, L.; Mäs, S.; Müller, M.; Henzen, C.; Brauner, J. Scientific geodata infrastructures: Challenges, approaches and directions. *Int. J. Digit. Earth* **2014**, *7*, 613–633. [CrossRef]
15. Domenico, B.; Caron, J.; Davis, E.; Kambic, R.; Nativi, S. *Thematic Real-Time Environmental Distributed Data Services (THREDDS): Incorporating Interactive Analysis Tools into NSDL*; Multimedia Research Group, University of Southampton: Southampton, UK, 1997; Volume 2.
16. John Caron, U.; Davis, E. UNIDATA's THREDDS data server. In Proceedings of the 22nd International Conference on Interactive Information Processing Systems for Meteorology, Oceanography, and Hydrology, Atlanta, GA, USA, 27 January–3 February 2006.
17. Sun, Z.; Di, L.; Hao, H.; Wu, X.; Tong, D.Q.; Zhang, C.; Virgei, C.; Fang, H.; Yu, E.; Tan, X.; et al. CyberConnector: A service-oriented system for automatically tailoring multisource Earth observation data to feed Earth science models. *Earth Sci. Inform.* **2018**, *11*, 1–17. [CrossRef]
18. Di, L.; Sun, Z.; Yu, E.; Song, J.; Tong, D.; Huang, H.; Wu, X.; Domenico, B. Coupling of Earth science models and earth observations through OGC interoperability specifications. In Proceedings of the 2016 IEEE International Geoscience and Remote Sensing Symposium (IGARSS), Beijing, China, 10–15 July 2016; pp. 3602–3605.
19. Di, L.; Sun, Z.; Zhang, C. Facilitating the Easy Use of Earth Observation Data in Earth system Models through CyberConnector. In Proceedings of the AGU Fall Meeting, New Orleans, LA, USA, 11–15 December 2017. Abstract #IN21D-0072.
20. Sun, Z.; Di, L. CyberConnector COVALI: Enabling inter-comparison and validation of Earth science models. In Proceedings of the AGU Fall Meeting, Washington, DC, USA, 10–14 December 2018. Abstract #IN23B-0780.
21. Schellnhuber, H.J. 'Earth system' analysis and the second Copernican revolution. *Nature* **1999**, *402*, C19–C23. [CrossRef]
22. Calvin, K.; Bond-Lamberty, B. Integrated human-earth system modeling—State of the science and future directions. *Environ. Res. Lett.* **2018**, *13*, 063006. [CrossRef]
23. Hurrell, J.W.; Holland, M.M.; Gent, P.R.; Ghan, S.; Kay, J.E.; Kushner, P.J.; Lamarque, J.-F.; Large, W.G.; Lawrence, D.; Lindsay, K.; et al. The Community Earth system Model: A Framework for Collaborative Research. *Bull. Am. Meteorol. Soc.* **2013**, *94*, 1339–1360. [CrossRef]
24. Reid, W.V.; Bréchignac, C.; Tseh Lee, Y. Earth system research priorities. *Science* **2009**, *325*, 245. [CrossRef] [PubMed]
25. Lovelock, J. Gaia: The living Earth. *Nature* **2003**, *426*, 769–770. [CrossRef]
26. Holm, P.; Goodsite, M.E.; Cloetingh, S.; Agnoletti, M.; Moldan, B.; Lang, D.J.; Leemans, R.; Moeller, J.O.; Buendía, M.P.; Pohl, W.; et al. Collaboration between the natural, social and human sciences in Global Change Research. *Environ. Sci. Policy* **2013**, *28*, 25–35. [CrossRef]
27. Burnett, K.; Ng, K.B.; Park, S. A comparison of the two traditions of metadata development. *J. Am. Soc. Inf. Sci.* **1999**, *50*, 1209–1217. [CrossRef]
28. Di, L.; Moe, K.L.; Yu, G. Metadata requirements analysis for the emerging Sensor Web. *Int. J. Digit. Earth* **2009**, *2*, 3–17. [CrossRef]
29. Di, L.; Schlesinger, B.M.; Kobler, B.U.S. *Federal Geographic Data Committee (FGDC) Content Standard for Digital Geospatial Metadata*; Federal Geographic Data Committee: Reston, VA, USA, 2000.
30. Yue, P.; Gong, J.; Di, L. Augmenting geospatial data provenance through metadata tracking in geospatial service chaining. *Comput. Geosci.* **2010**, *36*, 270–281. [CrossRef]

31. Di, L.; Kobler, B. NASA Standards for Earth Remote Sensing Data. *Int. Arch. Photogramm. Remote Sens.* **2000**, *33*, 147–155.
32. Yue, P.; Sun, Z.; Gong, J.; Di, L.; Lu, X. A provenance framework for Web geoprocessing workflows. In Proceedings of the 2011 IEEE International Geoscience and Remote Sensing Symposium, Vancouver, BC, Canada, 24–29 July 2011; pp. 3811–3814.
33. Sun, Z.; Yue, P.; Di, L. GeoPWTManager: A task-oriented web geoprocessing system. *Comput. Geosci.* **2012**, *47*, 34–45. [CrossRef]
34. Sun, Z.; Yue, P.; Lu, X.; Zhai, X.; Hu, L. A Task Ontology Driven Approach for Live Geoprocessing in a Service-Oriented Environment. *Trans. GIS* **2012**, *16*, 867–884. [CrossRef]
35. Sun, Z.; Peng, C.; Deng, M.; Chen, A.; Yue, P.; Fang, H.; Di, L. Automation of Customized and Near-Real-Time Vegetation Condition Index Generation Through Cyberinfrastructure-Based Geoprocessing Workflows. *IEEE J. Sel. Top. Appl. Earth Obs. Remote Sens.* **2014**, *7*, 4512–4522. [CrossRef]
36. Tan, X.; Di, L.; Deng, M.; Huang, F.; Ye, X.; Sha, Z.; Sun, Z.; Gong, W.; Shao, Y.; Huang, C. Agent-as-a-service-based geospatial service aggregation in the cloud: A case study of flood response. *Environ. Model. Softw.* **2016**, *84*, 210–225. [CrossRef]
37. Sun, Z.; Di, L.; Zhang, C.; Fang, H.; Yu, E.; Lin, L.; Tang, J.; Tan, X.; Liu, Z.; Jiang, L.; et al. Building robust geospatial web services for agricultural information extraction and sharing. In Proceedings of the 2017 6th International Conference on Agro-Geoinformatics, Fairfax, VA, USA, 7–10 August 2017; pp. 1–4.
38. Tan, X.; Di, L.; Deng, M.; Chen, A.; Sun, Z.; Huang, C.; Shao, Y.; Ye, X. Agent-and Cloud-Supported Geospatial Service Aggregation for Flood Response. *ISPRS Ann Photogramm. Remote Sens. Spat. Inf. Sci.* **2015**, *2*, 13–18. [CrossRef]
39. Jiang, L.; Sun, Z.; Qi, Q.; Zhang, A. Spatial Correlation between Traffic and Air Pollution in Beijing. *Prof. Geogr.* **2019**, *71*, 654–667. [CrossRef]
40. Liang, L.; Geng, D.; Huang, T.; Di, L.; Lin, L.; Sun, Z. VCI-based Analysis of Spatio-temporal Variations of Spring Drought in China from 1981 to 2015. In Proceedings of the 2019 8th International Conference on Agro-Geoinformatics (Agro-Geoinformatics), Stanbul, Turkey, 16–19 July 2019; pp. 1–6.
41. Zhong, S.; Di, L.; Sun, Z.; Xu, Z.; Guo, L. Investigating the Long-Term Spatial and Temporal Characteristics of Vegetative Drought in the Contiguous United States. *IEEE J. Sel. Top. Appl. Earth Obs. Remote Sens.* **2019**, *12*, 836–848. [CrossRef]
42. Zhong, S.; Xu, Z.; Sun, Z.; Yu, E.; Guo, L.; Di, L. Global vegetative drought trend and variability analysis from long-term remotely sensed data. In Proceedings of the 2019 8th International Conference on Agro-Geoinformatics (Agro-Geoinformatics), Istanbul, Turkey, 16–19 July 2019; pp. 1–6.
43. Bai, Y.; Di, L. Providing access to satellite imagery through OGC catalog service interfaces in support of the Global Earth Observation System of Systems. *Comput. Geosci.* **2011**, *37*, 435–443. [CrossRef]
44. Chen, Z.; Chen, N. Use of service middleware based on ECHO with CSW for discovery and registry of MODIS data. *Geo-Spat. Inf. Sci.* **2010**, *13*, 191–200. [CrossRef]
45. Bai, Y.; Di, L.; Wei, Y. A taxonomy of geospatial services for global service discovery and interoperability. *Comput. Geosci.* **2009**, *35*, 783–790. [CrossRef]
46. Chen, N.; Di, L.; Yu, G.; Gong, J.; Wei, Y. Use of ebRIM-based CSW with sensor observation services for registry and discovery of remote-sensing observations. *Comput. Geosci.* **2009**, *35*, 360–372. [CrossRef]
47. Di, L.; Yu, G.; Shao, Y.; Bai, Y.; Deng, M.; McDonald, K.R. Persistent WCS and CSW services of GOES data for GEOSS. In Proceedings of the 2010 IEEE International Geoscience and Remote Sensing Symposium, Honolulu, HI, USA, 25–30 July 2010; pp. 1699–1702.
48. Hu, C.; Di, L.; Yang, W. The research of interoperability in spatial catalogue service between CSW and THREDDS. In Proceedings of the 2009 17th International Conference on Geoinformatics, Fairfax, VA, USA, 12–14 August 2009; pp. 1–5.
49. Bai, Y.; Di, L.; Chen, A.; Liu, Y.; Wei, Y. Towards a Geospatial Catalogue Federation Service. *Photogramm. Eng. Remote Sens.* **2007**, *73*, 699–708. [CrossRef]
50. Spéry, L.; Claramunt, C.; Libourel, T. A Spatio-Temporal Model for the Manipulation of Lineage Metadata. *Geoinformatica* **2001**, *5*, 51–70. [CrossRef]
51. She, J.; Feng, X.; Liu, B.; Xiao, P.; Wang, P. *Conceptual Data Modeling on the Evolution of the Spatiotemporal Object*; Chen, J., Pu, Y., Eds.; International Society for Optics and Photonics: The Hague, The Netherlands, 2007; Volume 6753, p. 67530H.

52. Simmhan, Y.L.; Plale, B.; Gannon, D. A survey of data provenance in e-science. *ACM SIGMOD Rec.* **2005**, *34*, 31–36. [CrossRef]
53. Sun, Z.; Yue, P.; Hu, L.; Gong, J.; Zhang, L.; Lu, X. GeoPWProv: Interleaving Map and Faceted Metadata for Provenance Visualization and Navigation. *IEEE Trans. Geosci. Remote Sens.* **2013**, *51*, 5131–5136.
54. Di, L.; Shao, Y.; Kang, L. Implementation of Geospatial Data Provenance in a Web Service Workflow Environment with ISO 19115 and ISO 19115-2 Lineage Model. *IEEE Trans. Geosci. Remote Sens.* **2013**, *51*, 5082–5089.
55. West, L.A.; Hess, T.J. Metadata as a knowledge management tool: Supporting intelligent agent and end user access to spatial data. *Decis. Support Syst.* **2002**, *32*, 247–264. [CrossRef]
56. Nogueras, J.; Zarazaga, F.J.; Muro, R.P. Interoperability between metadata standards. In *Geographic Information Metadata for Spatial Data Infrastructures*; Springer: Berlin/Heidelberg, Germany, 2005; pp. 89–127.
57. Zhao, P. *Geospatial Web Services: Advances in Information Interoperability: Advances in Information Interoperability*; IGI Global: Hershey, PA, USA, 2010.
58. Haslhofer, B.; Klas, W. A survey of techniques for achieving metadata interoperability. *ACM Comput. Surv.* **2010**, *42*, 7. [CrossRef]
59. Wei, Y.; Di, L.; Zhao, B.; Liao, G.; Chen, A. Transformation of HDF-EOS metadata from the ECS model to ISO 19115-based XML. *Comput. Geosci.* **2007**, *33*, 238–247. [CrossRef]
60. Di, L. The development of remote-sensing related standards at FGDC, OGC, and ISO TC 211. In Proceedings of the 2003 IEEE International Geoscience and Remote Sensing Symposium (IGARSS 2003), Toulouse, France, 21–25 July 2003; Volume 1, pp. 643–647.
61. Di, L. Distributed geospatial information services-architectures, standards, and research issues. *Int Arch. Photogramm. Remote Sens. Spat. Inf. Sci.* **2004**, *35 Pt 2*.
62. ISO. *ISO 19115: Geographic Information—Metadata*; ISO: Geneva, Switzerland, 2013.
63. Bhattacharya, A.; Culler, D.E.; Ortiz, J.; Hong, D.; Whitehouse, K.; Culler, D. *Enabling Portable Building Applications through Automated Metadata Transformation*; University of California at Berkeley: Berkeley, CA, USA, 2014.
64. Nogueras-Iso, J.; Zarazaga-Soria, F.J.J.; Be´jarbe´jar, R.; Lvarez, P.J.A.; Muro-Medrano, P.R.R.; Béjar, R.; Álvarez, P.J.; Muro-Medrano, P.R.R. OGC Catalog Services: A key element for the development of Spatial Data Infrastructures. *Comput. Geosci.* **2005**, *31*, 199–209. [CrossRef]
65. Sun, Z.; Di, L.; Gaigalas, J. SUIS: Simplify the use of geospatial web services in environmental modelling. *Environ. Model. Softw.* **2019**, *119*, 228–241. [CrossRef]
66. Singh, G.; Bharathi, S.; Chervenak, A.; Deelman, E.; Kesselman, C.; Manohar, M.; Patil, S.; Pearlman, L. A Metadata Catalog Service for Data Intensive Applications. In Proceedings of the 2003 ACM/IEEE Conference on Supercomputing, Phoenix, AZ, USA, 15–21 November 2003.
67. Tan, X.; Di, L.; Deng, M.; Fu, J.; Shao, G.; Gao, M.; Sun, Z.; Ye, X.; Sha, Z.; Jin, B. Building an Elastic Parallel OGC Web Processing Service on a Cloud-Based Cluster: A Case Study of Remote Sensing Data Processing Service. *Sustainability* **2015**, *7*, 14245–14258. [CrossRef]
68. Desai, K.; Devulapalli, V.; Agrawal Asst, S.; Kathiria Asst, P.; Patel Professor, A. Web Crawler: Review of Different Types of Web Crawler, Its Issues, Applications and Research Opportunities. *Int. J. Adv. Res. Comput. Sci.* **2017**, *8*, 1199–1202.
69. Li, W.; Wang, S.; Bhatia, V. PolarHub: A large-scale web crawling engine for OGC service discovery in cyberinfrastructure. *Comput Environ Urban Syst.* **2016**, *59*, 195–207. [CrossRef]
70. Pallickara, S.L.; Pallickara, S.; Zupanski, M.; Sullivan, S. Efficient Metadata Generation to Enable Interactive Data Discovery over Large-scale Scientific Data Collections. In Proceedings of the 2010 IEEE Second International Conference on Cloud Computing Technology and Science, Indianapolis, IN, USA, 30 November–3 December 2010.
71. Lopez, L.A.; Khalsa, S.J.S.; Duerr, R.; Tayachow, A.; Mingo, E. The BCube Crawler: Web Scale Data and Service Discovery for EarthCube. In Proceedings of the AGU Fall Meeting, San Francisco, CA, USA, 15–19 December 2014. Abstracts IN51C-06.
72. Khalsa, S.J.S. Data and Metadata Brokering–Theory and Practice from the BCube Project. *Data Sci. J.* **2017**, *16*, 1–8. [CrossRef]
73. Song, J.; Di, L. Near-Real-Time OGC Catalogue Service for Geoscience Big Data. *ISPRS Int. J. Geo-Inf.* **2017**, *6*, 337. [CrossRef]

74. Unidata THREDDs Client Catalog Spec 1.0.7. Available online: https://www.unidata.ucar.edu/software/tds/current/catalog/InvCatalogSpec.html (accessed on 26 August 2019).
75. Unidata THREDDS Support [THREDDS #BIA-775104]: Unidata THREDDs Metadata Structure and Volume. Juozasgaigalas@gmail.com. Gmail. Available online: https://mail.google.com/mail/u/0/#search/Unidata+THREDDs+metadata+structure+and+volume/FMfcgxvwzcCgSZmpPZsQFqdjLlCkPNfm (accessed on 26 August 2019).
76. Ansari, S.; Del Greco, S.; Kearns, E.; Brown, O.; Wilkins, S.; Ramamurthy, M.; Weber, J.; May, R.; Sundwall, J.; Layton, J.; et al. Unlocking the Potential of NEXRAD Data through NOAA's Big Data Partnership. *Bull. Am. Meteorol. Soc.* **2018**, *99*, 189–204. [CrossRef]
77. Theodoridis, Y.; Sellis, T.; Papadopoulos, A.N.; Manolopoulos, Y. Specifications for Efficient Indexing in Spatiotemporal Databases. In Proceedings of the Tenth International Conference on Scientific and Statistical Database Management, Capri, Italy, 3 July 1998.
78. Zhang, C.; Di, L.; Sun, Z.; Lin, L.; Yu, E.G.; Gaigalas, J. Exploring cloud-based Web Processing Service: A case study on the implementation of CMAQ as a Service. *Environ. Model. Softw.* **2019**, *113*, 29–41. [CrossRef]
79. Aronson, E.; Ferrini, V.; Gomez, B. *Geoscience 2020: Cyberinfrastructure to Reveal the Past, Comprehend the Present, and Envision the Future*; National Science Foundation: Alexandria, VA, USA, 2015.
80. Heiss, W.H.; McGrew, D.L.; Sirmans, D. Nexrad: Next generation weather radar (WSR-88D). *Microw. J.* **1990**, *33*, 79–89.
81. Bromwich, D.H.; Wilson, A.B.; Bai, L.; Liu, Z.; Barlage, M.; Shih, C.-F.; Maldonado, S.; Hines, K.M.; Wang, S.-H.; Woollen, J.; et al. The Arctic System Reanalysis, Version 2. *Bull. Am. Meteorol. Soc.* **2018**, *99*, 805–828. [CrossRef]

9

A Task-Oriented Knowledge Base for Geospatial Problem-Solving

Can Zhuang [1], Zhong Xie [1,2,*], Kai Ma [1], Mingqiang Guo [1] and Liang Wu [1,2]

1 School of Information Engineering, China University of Geosciences, Wuhan 430074, China; zhuangcan@cug.edu.cn (C.Z.); makai@cug.edu.cn (K.M.); guomingqiang@mapgis.com (M.G.); wuliang@cug.edu.cn (L.W.)

2 National Engineering Research Center for GIS, Wuhan 430074, China

* Correspondence: xiezhong@cug.edu.cn

Abstract: In recent years, the rapid development of cloud computing and web technologies has led to a significant advancement to chain geospatial information services (GI services) in order to solve complex geospatial problems. However, the construction of a problem-solving workflow requires considerable expertise for end-users. Currently, few studies design a knowledge base to capture and share geospatial problem-solving knowledge. This paper abstracts a geospatial problem as a task that can be further decomposed into multiple subtasks. The task distinguishes three distinct granularities: Geooperator, Atomic Task, and Composite Task. A task model is presented to define the outline of problem solution at a conceptual level that closely reflects the processes for problem-solving. A task-oriented knowledge base that leverages an ontology-based approach is built to capture and share task knowledge. This knowledge base provides the potential for reusing task knowledge when faced with a similar problem. Conclusively, the details of implementation are described through using a meteorological early-warning analysis as an example.

Keywords: task; workflow; geospatial problem-solving; knowledge base

---

1. Introduction

In recent years, with the rapid development of cloud computing and web technologies, an increasing number of geospatial information resources (GIRs), e.g., geospatial data, geospatial analysis functions, models, applications, etc., have been encapsulated into a wide variety of geographic information services (GIServices) [1] which are accessible to general public users over the web [2,3]. For example, a web service toolkit, named GeoPW [4], provides a set of geoprocessing services, which are used to fulfill data processing and spatial analysis tasks over distributed information infrastructures [5]. In the geospatial community, The Open Geospatial Consortium (OGC) established a series of standard interface specifications, such as Web Feature Service (WFS), Web Map Service (WMS), Web Coverage Service (WCS), and Web Processing Service (WPS), which further improve the interoperability and web-based sharing of GIServices [5–7].

In the geospatial application domain, geospatial problems usually relate to heterogeneous data and several computational processes [8]. The capabilities of a single GIService are limited and cannot be effectively conducted because of the complexity of geospatial problems [4]. In the last decade, workflow-based approaches have evolved to a major way to address complex geospatial problems [9]. Currently, with the assistance of standard interface specifications, GIServices published by different organizations can be chained as a geoprocessing workflow that can describe the execution order of problem-solving steps and enhance the power of atomic GIServices to fulfill complex geoprocessing tasks [10–13]. In general, geospatial problems require comparatively deep expertise, which therefore need experts to contribute their problem-solving knowledge by means of a conceptual workflow.

In previous studies, there are already some investigations in the formalization of workflow [7,14, 15] and semantic interoperability for GIServices [16–18]. Additionally, a number of studies have employed the task concept to facilitate the expression of user requirements at a semantic level [14,19,20]. In fact, many geospatial problems can share similar conceptual workflows. Therefore, the conceptual workflows can be formalized into a knowledge base, which can facilitate future users to solve the similar problems.

In this paper, we focus on using ontologies in association with a task-oriented approach to construct a knowledge base to enhance geospatial problem-solving. It is generally believed that ontology is the foundation and a significant part of the semantic web. Ontology provides unified terms to improve the semantic interoperability of domain knowledge [21]. A task is introduced as a reusable component to model the sequence of inference steps involved in the process of solving certain kinds of geospatial problems at a conceptual level. The knowledge base can store conceptual workflows that are considered to be a priori knowledge accumulated from past experience of domain experts [22], which can enable problem-solving knowledge reusable [23]. A geospatial problem is abstracted as a task, and the knowledge for the task is considered as a problem solution. Under many circumstances, tasks need to decompose into simpler tasks, each of which can be solved by one or a set of functions [24]. As the smaller task is simpler than the overall task, the complexity of the task is reduced significantly [22]. Hence, we further divide the task into three distinct granularities: (1) a geooperator, which is basic processing functionality; (2) an atomic task, which is indecomposable; and (3) a composite task, which is decomposed into multiple subtasks.

The main work of this paper includes the following: (1) Concepts: the task concept is introduced as a reusable component for geospatial problem-solving and is used to reflect users' requirements; (2) Model: a task model is proposed to simulate problem-solving processes; (3) Knowledge base: an ontological knowledge base is designed, that comprises several interoperable ontologies to capture and share problem-solving knowledge; and (4) Implementation: taking the meteorological early-warning (MEW) analysis, for example, we describe the details of the implementation conclusively. We focus on geospatial problem solution as a task that is composed of conceptual geoprocessing operations not in connection with any concrete services. The instantiation and execution of a task, the low-level interaction with operations (such as accessing input data), and the validation of the processing chain are not in the scope of this article.

The remainder of this paper is organized as follows. Section 2 offers related work on the task-based approach and geospatial problem-solving. Section 3 proposes the task concept and task model to describe problem-solving processes. Section 4 presents a task-oriented knowledge base, some core ontologies are described in detail. Section 5 introduces the detailed information of implementation. Finally, conclusions and future work are given in Section 6.

2. Related Work

2.1. The Task-Based Approach

The notion of the task was proposed by Albrecht in the field of geographic information system as early as in the 1990s [25] and has been used in many studies. However, there is still no unified definition of a task [19]. In general, the task concept reflects user requirements and describes all actions or operations to solve a specific problem. Some studies have been performed using a task-based approach. We summarize and classify them as follows:

1. The task-based language. A task ontology language based on the OWL (Web Ontology Language), named OWL-T, has been proposed to define task templates to formalize user demands and business processes at a high-level abstraction, which is used for the task of a trip plan [26]. Hu et al. [19] extended the task-oriented approach to the OGC Sensor Web domain. A Task Model Language, called TaskML, is a language for modeling tasks. The significant features of TaskML are Task Trigger, Task Priority, and Task QoS.

2. The task ontology approach. Sun et al. [27] proposed a task ontology-driven approach for the geospatial domain to realize live geoprocessing in a service-oriented environment, which includes three steps: task model generation, process model instantiation, and workflow execution. A case study of flood analysis is used to illustrate the effect and role of the task. Liu [28] proposed a task ontology model for domain-independent dialogue management and created a dialogue manager that is task-independent. Park et al. [29] presented a task ontology based on travelers' perspectives using tasks, activities, relations, and properties. A prototype system was developed using task-oriented menus.

3. A task-based approach for geospatial data acquisition. Wiegand and García [21] proposed a task-based approach to advance geospatial data source retrieval. More concretely, they designed a conceptual model that combines ontologies of tasks, data sources, metadata, and places and uses the Jess rule engine and Protégé tool to provide automatic processing for data retrieval. Qiu et al. [30] proposed a task-oriented approach for efficient disaster data management that performed mapping from emergency tasks to data sources and calculated the correlation between the data set and a generic task. A flood emergency example illustrates the use of this approach.

2.2. Geospatial Problem-Solving

Currently, geoprocessing service technology is widely employed to solve specific geospatial problems in distributed information infrastructures. Much research has been devoted to utilizing or facilitating geoprocessing services to support problem solving. Mikita [31] published a geoprocessing service for forest owners to optimize clearcut size and shape during the process of forest recovery. Müller [18] proposed a hierarchical framework to identify the semantic and syntactic properties of geoprocessing services with four levels of granularity, which is conducive to service retrieval, service comparison, service invocation.

In most cases, a single geoprocessing service is not enough to solve the complex geospatial problem. Therefore, geoprocessing workflow technology provides a solution. The integration of geoprocessing services has become a popular research topic, and a series of tools and architectures were developed to support geoprocessing service chaining. For example, an open source geoprocessing workflow tool, named GeoJModelBuilder, is able to integrate interoperable geoprocessing services and compose them into a workflow [6,7]. A RichWPS orchestration engine in combination with a DSL (Domain Specific Language) is used to orchestrate WPS processes and publish the composition as a WPS process for further composition [32]. In addition, there are many popular workflow management systems to facilitate the integration of geoprocessing services, such as Taverna [33,34], Triana [35], Kepler [36], jABC [9,37]. However, they only simplify the workflow construction process at the syntactical level, and building a workflow composed of services for geospatial problem-solving is still challenging for end-users.

Recently, more studies have focused on semantic and automatic workflow composition for geospatial problem-solving. Farnaghi and Mansourian [12] proposed an automatic composition solution using the AI (Artificial Intelligence) planning algorithm and SAWSDL (Semantic Annotations for Web Service Description Language) to improve the disaster management process. Al-Areqi et al. [10] applied a constraints-driven synthesis method to implement the semiautomatic composition of a workflow for analysis of the impacts of sea-level rise. Samadzadegan et al. [38] designed a framework for an automatic workflow for fire detection early warning based on OGC services. Arul and Prakash presented a unified composition algorithm that adds a new phase called Validation and Optimization to automatic web service composition and generated a scalable composition process according to the dynamic change of user requirements [39].

3. Task as a Reusable Problem-Solving Component

3.1. An Application Scenario

In this section, we demonstrate an example that uses a workflow composed of distributed data and various geoprocessing services. This example is used throughout the remainder of the paper to help understand the concept of a geospatial task. Assuming an end-user is a staff of a meteorological disaster monitoring department, he needs to predict the probability of the occurrence of geological disasters in a certain region in the next day. The ideal result is a thematic map of the early-warning region that uses different colors to represent different early-warning levels.

To achieve the early-warning results, the most common approach is to formulate a geographic processing workflow that can generate an early-warning result map. As shown in Figure 1, the elliptical shape represents a data node, and the rectangular shape represents a data processing node. First, it uses geological hazard point data, influence factor data and early-warning unit data as input data to calculate the potential degree index of early-warning units respectively. Similarly, it can obtain effective rainfall data. Then, the potential distribution map and effective rainfall data from the previous step with forecast rainfall data go through early-warning analysis calculations to achieve the early-warning result map.

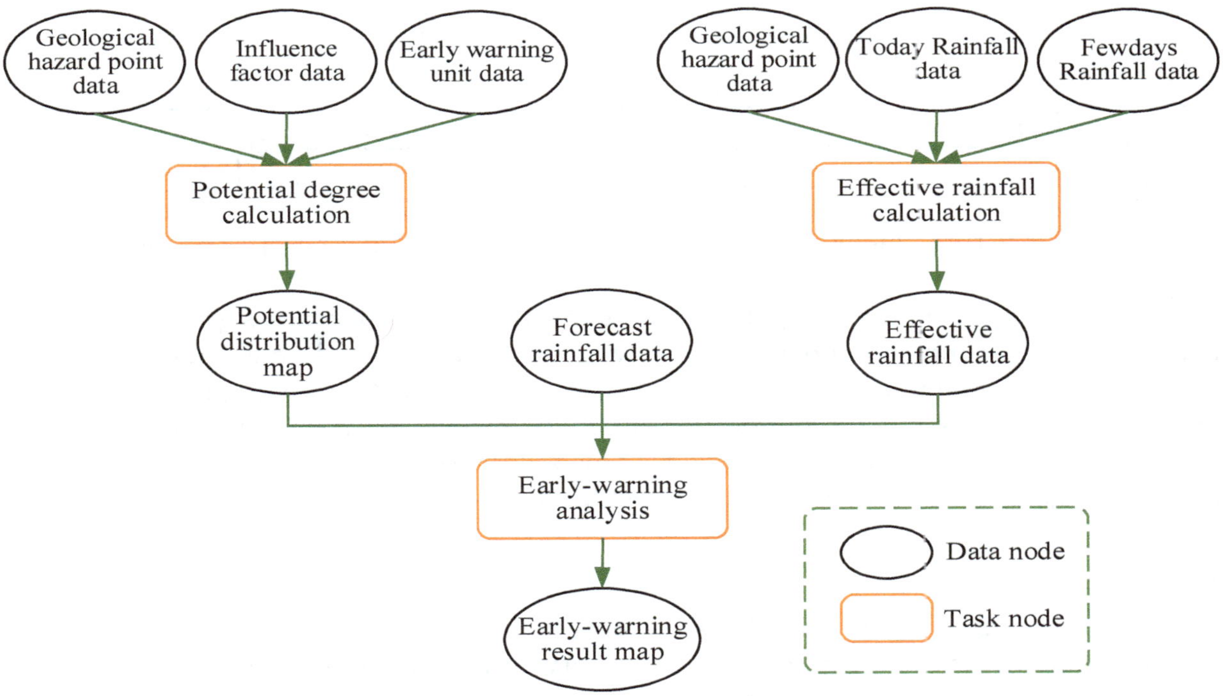

Figure 1. Sequences of the meteorological early-warning (MEW) process.

For the aforementioned application, the entire workflow can be considered a task. GIS domain experts with professional knowledge are able to analyze the technological procedures and abstract them in the form of conceptualization, which are then used to describe the skeleton knowledge of the problem-solving process. The MEW task, which was previously performed manually and had the requirements of GIS skills and knowledge of business processes, can now be executed automatically.

3.2. Task and Task Model

In this paper, the task concept is proposed to reflect user requirements, which can be accomplished by one or more geoprocessing services. A geospatial problem is abstracted as a task that denotes a high-level business goal, and users execute a sequence of processes to achieve the goal. Tasks are different from operations or services, as tasks focus on what users want to solve, while operations or services mainly focus on the implementation of geoprocessing computations.

A complex problem can consist of multiple problem-solving processes with different requirements, which makes it difficult to define the solution as a single task [22]. Hence, a complex task can be decomposed into several smaller tasks, each of which can be solved in a relatively independent way by one or more geoprocessing services and then combined together into a complete solution [24]. The granularity of the task plays an important role during the problem-solving process. As shown in Figure 2, there are three distinct granularities: (1) a geooperator as elementary functionality for an atomic task, (2) an atomic task as a building block for a composite task, and (3) a composite task as a building block for a complex geospatial task. Consequently, the task is a reusable component for construction of the problem-solving workflow.

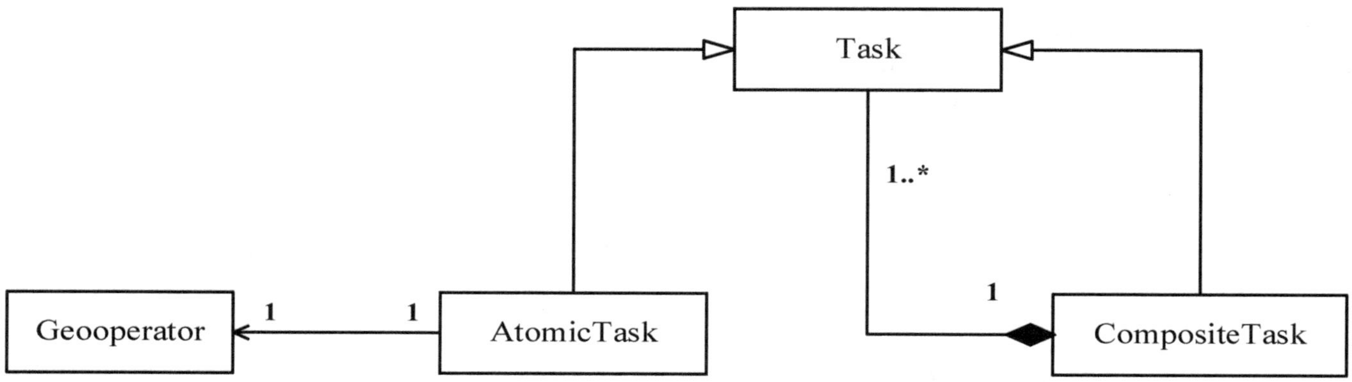

Figure 2. Relationships between Task, AtomicTask, CompositeTask, and Geooperator.

The process property of the geospatial task is expressed by a task process graph (TPG), which is used to capture the execution order of problem-solving steps and closely describe how a task should be achieved. Each TPG contains a set of edges that compose an acyclic directed graph structure. An edge denotes a workflow of two tasks. The directions of edges decide the dependencies between the tasks. The combination of TPG and task composes a task model that provides an approach to allow users to specify complex geospatial problems at an abstract level.

3.3. Geooperator

Geospatial problem-solving knowledge is represented at a conceptualized level that requires categorization and formalization of geoprocessing services. Geooperators are developed mostly for improving the discoverability and exchangeability of geoprocessing functionality and providing an approach to formalize well-defined geoprocessing functionality [40]. In Brauner's work, geooperators are categorized in terms of multiple different perspectives, such as geodata, legacy GIS, pragmatic, formal or technical perspectives [41]. An overview of perspectives and top-level categories identified by Brauner is shown in Figure 3a, and elements described by the geooperator are given in Figure 3b, which can facilitate our work. The former is used to define the subclasses of Geooperator class in the GIS operation ontology without further modification; the latter is partially transformed into data properties and object properties of the Geooperator class.

In this paper, geooperators are introduced to provide a conceptualization for geoprocessing services (such as a geospatial analysis or transformation service) that are encapsulated as standard web services (e.g., WPS) for providing geoprocessing functionalities on the web. From an object-oriented perspective, geooperators act as wrappers for existing geoprocessing services and subsequently serve as building blocks for elementary geoprocessing tasks.

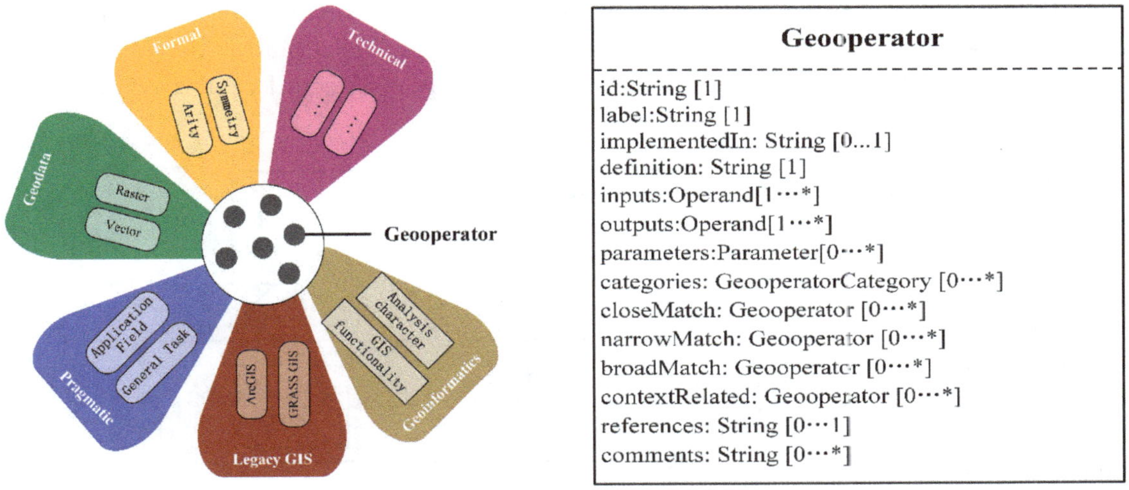

Figure 3. (**a**) Different perspectives on Geooperator [41] (**b**) Description elements of Geooperator [17].

3.4. Formal Definition

Definition 1 (Task). *A task can be defined as a quadruple:*

$$T = (PT, OP, PA, C), \tag{1}$$

where PT specifies the type of task, OP is spatial inputs and outputs (e.g., spatial datasets), PA is a set of non-spatial parameters of a task and C consists of the precondition and result that generally constrains the thematic and geometric attributes of input or output data for geoprocessing tasks [42].

Definition 2 (Task Process Graph). *A task process graph defines the basic structure of task decomposition* [43], *which is an acyclic directed graph defined as follows:*

$$TPG = (V, E), \tag{2}$$

where V is a finite set of n vertices $\{v_1, v_2, v_3, \ldots, v_n\}$, *and each node* $v \in V$ *represents a task* t_v. *E is a finite set of directed edges* $\{e_{vi,vj}\}$. *Each edge* $e_{vi,vj} \in E$ *can be characterized by a tuple* $(p_{vi,vj}, c_{ij})$. $p_{vi,vj} = \langle v_i, v_j \rangle$ *is an ordered pair that represents execution precedence between task* t_{vi} *and task* t_{vj}*; in other words,* t_{vi} *is ahead of* t_{vj} *in the sequence of task decomposition that can alsobe denoted by* $v_i \leq v_j$. c_{ij} *represents the control flow connector between two tasks, which includes sequence, branching, loop, and so forth.*

Definition 3 (Task Model). *A task model is defined by a 2-tuple as follows:*

$$TModel = (t, tpg), \tag{3}$$

where $t \in T$ *is a task instance, and tpg denotes a task process graph associated with t that defines the decomposition structure. If tpg only contains a geooperator, we consider this task to be an atomic task; otherwise, it is a composite task.*

Definition 4 (Task Decomposition). *Following the definition of the task model, we can further accomplish the task decomposition. Given a task process graph tpg = (V, E), assuming* $v \in V$, $t_v \in T$, *v associates with* t_v. *If* t_v *has a corresponding model* $tmodel_v = (t_v, tpg_v)$, *then the decomposition of the task can be defined by*

$$tpg' = Decompose(v, tpg, tmodel_v), \tag{4}$$

where tpg′ is a new task process graph obtained replacing node v with tpg_v *in* $tmodel_v$.

Taking the workflow mentioned in Section 3.1 as an example, Figure 4 depicts the task decomposition procedure. The node "Early-warning analysis" is replaced by a task process graph, which is defined in a task model, where the edges previously connected with this node are revised.

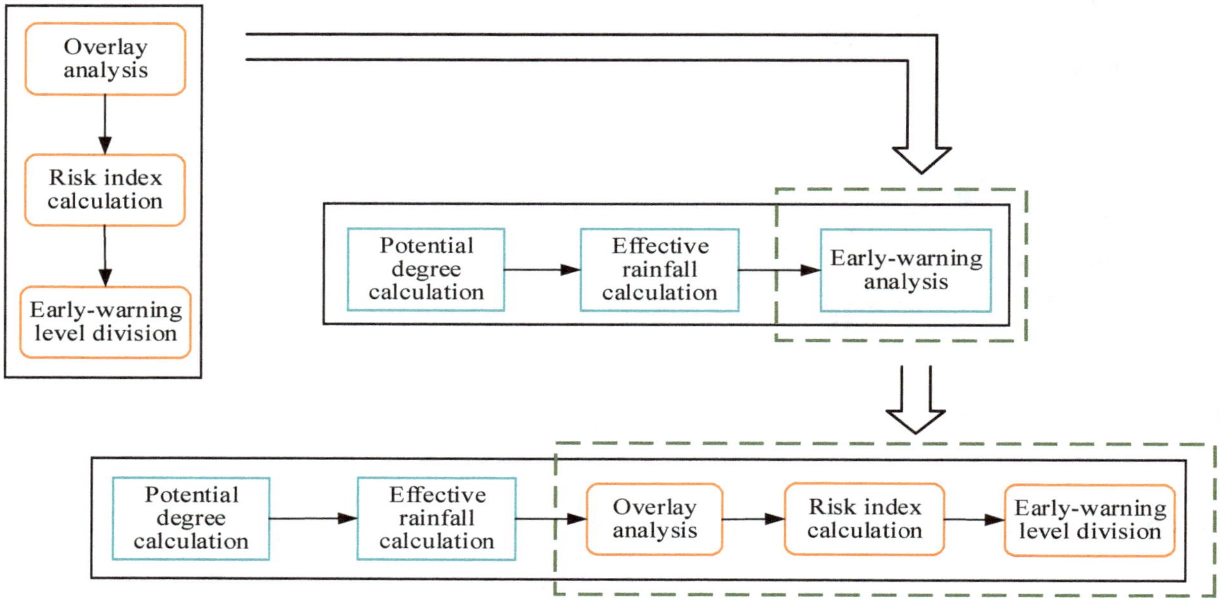

Figure 4. An example of task decomposition.

4. A Task-Oriented Knowledge Base

This section presents the knowledge base that adapts the ontology-based approach and provides comprehensive knowledge to support the geoprocessing task. To build the knowledge base, a set of ontologies are needed to capture knowledge related to the problem-solving solution. The use of ontologies makes the semantic meaning of problem-solving procedures explicit and further facilitates users to obtain the problem solution [44]. Formalizing the knowledge base will assist both GIS non-specialist users and specialists in automating problem solving, allowing reuse and sharing of solutions [21]. Accordingly, we deem that the knowledge base is valuable.

4.1. Background on Ontologies

It is widely known that ontology provides a formal language to standardize and share the semantics of various kinds of domain knowledge. The word ontology was first used as a philosophical concept and address the nature of existence, and it was subsequently introduced into the information domain by researchers. Currently, one of the most prevalent definitions of ontology is "Ontology is an explicit specification of a conceptualization", which was proposed by Gruber in 1993 [45]. Based on this definition, ontology is essentially a taxonomy of the objective world and a knowledge representation model. Meanwhile, ontology also supports non-taxonomic relations.

According to Perez [46], knowledge in ontologies is formalized by five kinds of modeling primitives: concepts, relations, functions, axioms, and instances. From a mathematical point of view, ontology can be formally expressed by an Equation as follows:

$$O = \{C, R, F, A, I\}, \quad (5)$$

where C is a set whose elements are called concepts; R is a set of relations between concepts, $R \subseteq C \times C$; F is a special relation in which the former $n-1$ elements can uniquely determine the n-th element, and it can be defined as follows: $F: C_1 \times C_2 \times \ldots \times C_{n-1} \rightarrow C_n$; A represents a geographic axiom, that is, a collection of assertions in a logical form that are always true; and I stands for instances of concepts.

In the process of building ontology, instances represent objects that can be anything in a domain, and concepts are a set of objects that are mapped to classes. The relations between concepts are realized by properties that are classified into two types: an object property and a data property [47]. An object property specifies the relations between two classes, and it connects two individuals from different classes. A data property defines the relations between individuals and data values, which is similar to an inherent attribute of an object.

4.2. Ontologies at the Heart of the Knowledge Base

To realize the capability to represent the knowledge of the problem-solving process, the knowledge base provides a set of ontologies as follows: Task Ontology, Process Ontology, GIS Operation Ontology, Interface Ontology, Data Type Ontology, GIS Data Ontology, and GIService Type Ontology. These ontologies are combined to provide support for all facets of problem-solving, each of which plays a key role in building a rich, dynamic and flexible task-oriented knowledge base. Figure 5 shows the delineations of the definitions of ontologies and how they relate to each other. Several important ontologies are discussed in detail in the following section.

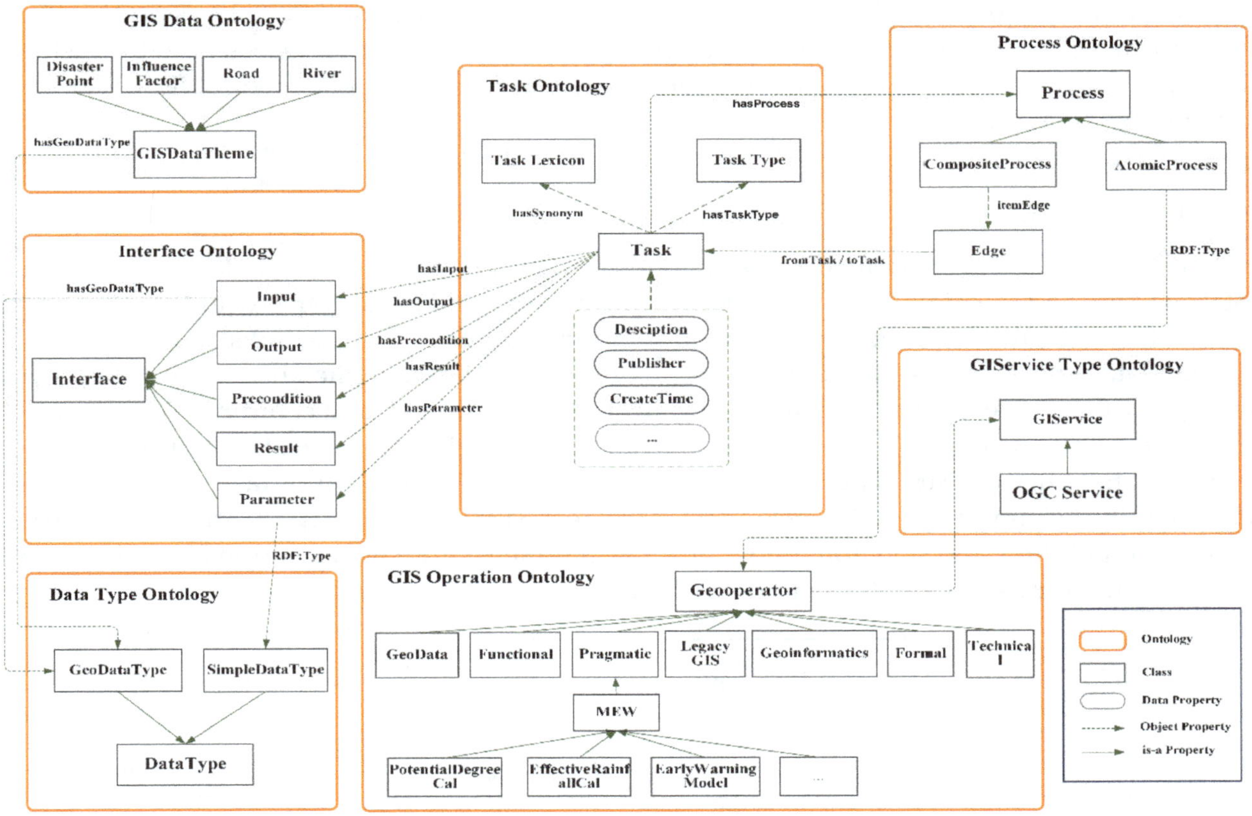

Figure 5. The relationships of ontologies in the knowledge base.

4.2.1. Task Ontology

Task Ontology is the core for supporting problem solving, that defines the Task class to represent a geospatial problem. Its property relations are composed of object properties and data properties. The data properties mainly describe the metadata information of task instances, such as Description, Publisher, Create Time, and so on. The object properties include: hasSynonym, hasTaskType, hasProcess, hasInput, hasOutput, etc.

The Task class refers to the Task Lexicon class through the hasSynonym property for semantic annotations of tasks to provide the words and phrases describing tasks, on the basis of which end users externalize their own expression of the target problem in natural language. This can broaden the scope of keyword queries and dispose synonyms to support natural language retrieval. The Task Type class describes the categorization of tasks on the basis of functionalities that tasks can implement. The MEW analysis in the example mentioned above is a sort of geospatial task. The Task Type class is linked to the Task class for semantic reference to state the type of task individuals through the predefined hasTaskType property. Each individual of the Task class has at least one conceptual solution which is denoted in the Process ontology. The interfaces of the Task class are defined in the Interface Ontology, which will be described in detail in the following section.

4.2.2. Process Ontology

Process Ontology is used to define problem-solving processes at a conceptual level for a certain type of task, that is not associated with any concrete services. The AtomicProcess and CompositeProcess classes are created as subclasses of the Process class to classify the process individuals according to the number of processes involved. The atomic process directly refers to the Geooperator class in the GIS Operation Ontology using the RDF:Type property; however, the composite process is an edge set that contains some edges. Each edge denotes the sequence of two task nodes that are semantically annotated to the Task class using the fromTask and toTask properties. A series of edges form a directed graph that is called a task process graph that describes how the task works. In this paper, we only consider the linear sequence between two tasks; other control flow logics will be included in future work.

4.2.3. Data Type Ontology

Data Type Ontology is defined to describe the data types that are divided into two categories: SimpleDataType and GeoDataType, as illustrated in Figure 6. SimpleDataType includes some primitive data types in some programming language or description language such as xml:string and xml:float in XML. GeoDataType is an abstract representation of geospatial data, which has some data properties shared by any type of geospatial data, including attribute, data format, and coordinate reference system (CRS). Based on abstract specifications of the International Standard Organization (ISO) for vector [48] and raster data [49], GeoDataType is differentiated as VectorDataType and RasterDataType, each of which has unique characteristics. In vector data, each geospatial feature must identify a geometric type, such as point, polyline, and polygon following OGC Simple Feature Specification [50]. The resolution and band number must be identified in raster data.

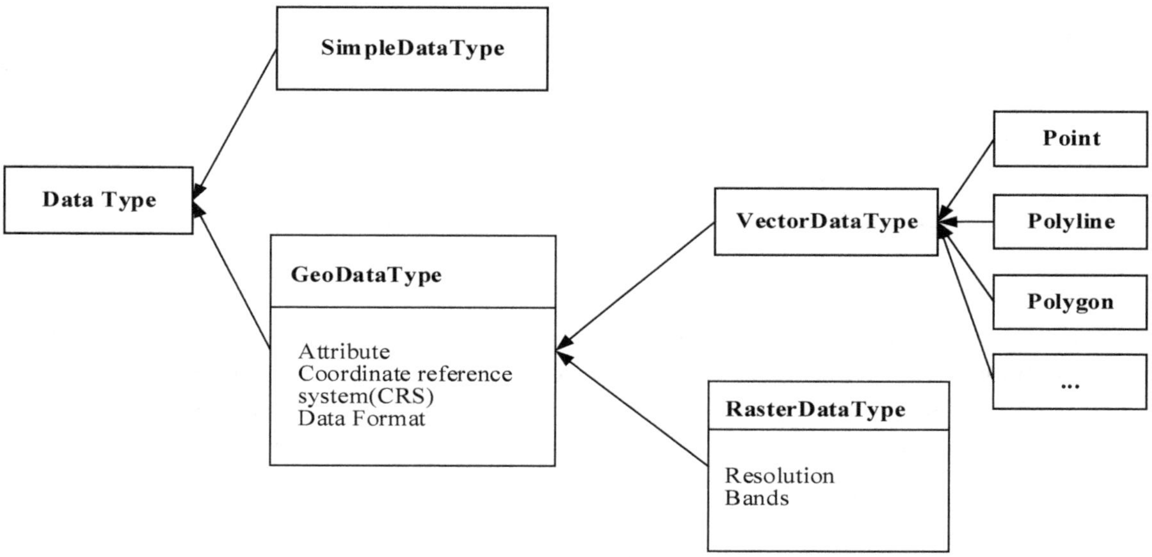

Figure 6. Data type specifications.

4.2.4. GIS Operation Ontology

In GIS Operation Ontology, the Geooperator class is employed to conceptualize geoprocessing functionalities. The notion of Geooperator has been introduced in the previous section. The geooperators are used as building blocks for the conceptual workflow of geospatial problem-solving. This ontology of the knowledge base is based on work by Hofer [42] who translated the SKOS (Simple Knowledge Organization System) thesaurus provided by Brauner [41] into an OWL ontology and included an additional concept that is known as a functional concept. The SKOS thesaurus contains 40 geooperators. This ontology can be extended by extra categories, if necessary. The categories of the Pragmatic perspective originate from the general task, and are task-oriented categories. Users can further integrate new categories based on practical application. Therefore, in this paper, an additional category named MEW is integrated into the Pragmatic perspective of the geooperator, and subcategories or geooperators can be created for a further description of geoprocessing operations. Based on this classification, geoprocessing services that perform geospatial functionalities are thought of as individuals of the Geooperator class.

4.2.5. Interface Ontology

As introduced in the previous section, tasks are used as reusable components to accomplish the composition of problem-solving processes. The composition requires an evaluation of the correspondence of interfaces. The knowledge base needs to include sufficient information of interfaces to satisfy the needs of the composition. An interface requires the description of operands that contain inputs and outputs, constraints that contain a precondition and result, and non-spatial parameters. Consequently, as illustrated in Figure 7, the Interface class consists of the subclasses Input, Output, Parameter, Precondition, and Result. GeoDataType in Data Type Ontology is used to specify operands of interfaces, whereas non-spatial parameters can refer to SimpleDataType which includes conventional data types. The Precondition class focuses on the thematic and geometric properties of the input to ensure the correct function of the operation [42]. The Postcondition class defines the expected result of the output.

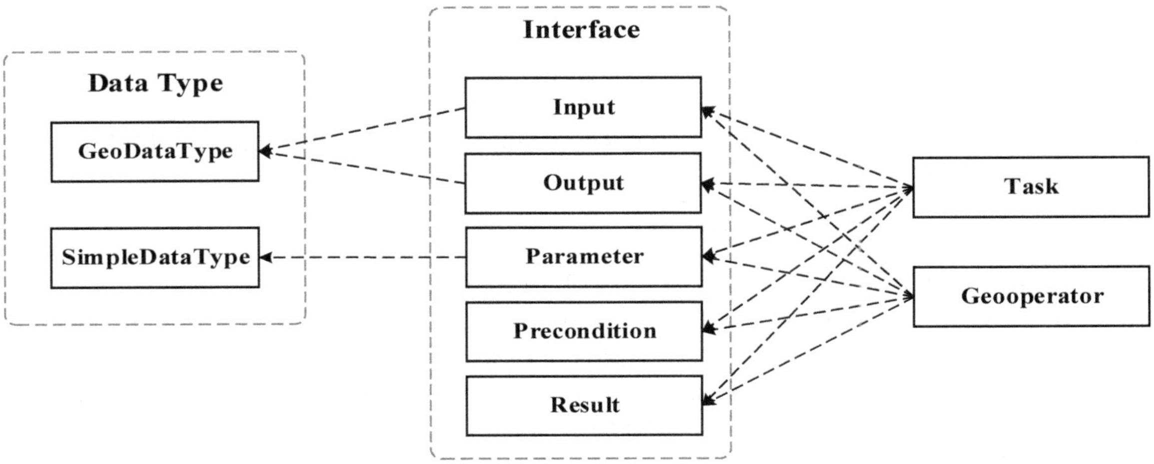

Figure 7. Interface for annotating Task and Geooperator, and Data Type for specifying Interface.

Similarly, we extend the interface properties of geooperators using the Interface Ontology which presently does not involve the related interface specifications.

5. Implementation

In Section 3, we introduce an application scenario that is a geospatial problem-solving process in the context of MEW. We take this example to demonstrate the benefits of the ontology-based knowledge base for tasks during the process of geospatial problem-solving. The implementation includes three parts: creation of ontologies, representation of knowledge, and task instances.

5.1. Creation of Ontologies

Based on the proposed architecture of the task-oriented knowledgebase described in Section 4.2, we build different abstract ontologies to represent the hierarchy and relationships of the concepts using Protégé 5.2.0 which is an OWL ontology development platform that allows creation and query of ontologies [21]. In general, an ontology is composed of the following components: concepts and properties of each concept, relationships or constraints between concepts, and instances of concepts [28]. Figure 8a presents all concepts or classes defined in the ontological knowledge base. All object properties that represent the relationships between classes are shown in Figure 8b; they include hasTaskType, hasSynonym, hasProcess, etc. The abstract ontologies can be instantiated for specific tasks. In this paper, the task instances for meteorological early-warning are implemented, which are detailed in the next section.

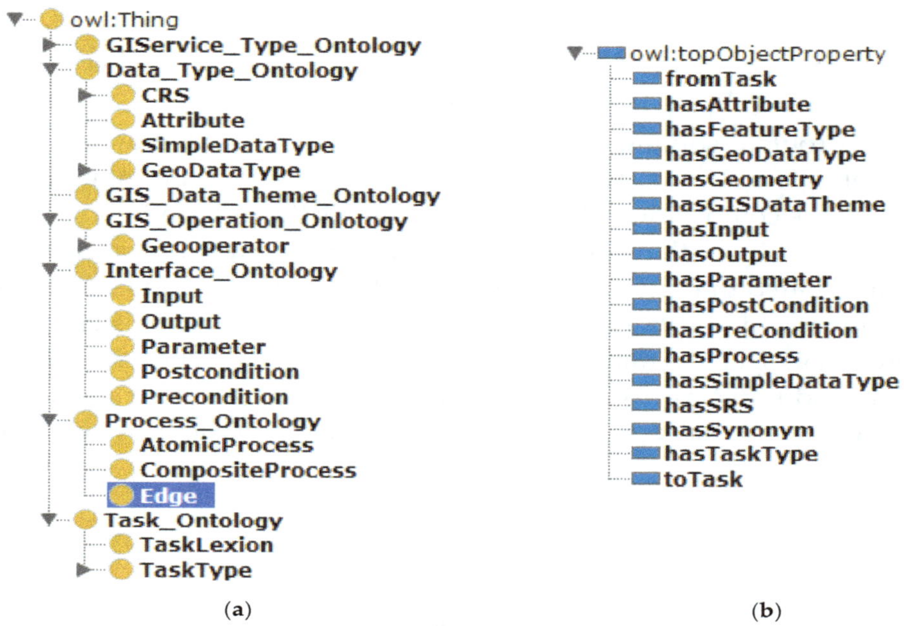

Figure 8. An excerpt of ontologies where (**a**) depicts the classes of ontologies, and (**b**) illustrates the object properties between classes.

5.2. Representation of Ontology Knowledge

Once the components of an ontology are developed, the ontology can be represented by ontology description language, such as Resource Description Framework (RDF) and Web Ontology Language (OWL). RDF is built upon XML, which uses triples of object, property, and value to describe resources. OWL is a W3C-recommended standard semantic markup language being developed by the Semantic Web community, which is an extension of RDF [15,21]. In this paper, we use OWL as a standard and machine-readable language to represent the knowledge of ontologies, which is presented as an OWL file.

Meanwhile, we use property restrictions including hasValue and quantifier restrictions to limit associations between different classes [15]. The hasValue restriction specifies that the individuals of a class have a given value. Nevertheless, the quantifier restriction limits the individuals of a class using an existential restriction (∃, owl:someValuesFrom) or a universal restriction(∀, owl:allValuesFrom). The former states that values for the restricted property have at least one instance of class, which is defined by existential restriction; however, the latter states that all values for the restricted relationship must be a type of instance. For example, an MEW analysis task only needs effective rainfall data, forecast rainfall data, and potential degree data that can be restricted with the following formal statement: ∀ hasInput (Effective_Rainfall_Data ∪ Forecast_Rainfall_Data ∪ Potential_Degree_Data). This statement defines a universal restriction on the "hasInput" property between the Task class and the Input class (Figure 5). The OWL notation using the "owl:allValuesFrom" restriction is shown in Figure 9.

```
<owl:Class rdf:ID = "Meterological Early-warning Task " >
        <rdfs:subClassof>
                <owl:Restriction>
                        <owl:onProperty rdf:resource = "hasInput " />
                        <owl:allValuesFrom rdf:resource = "#Effective_Rainfall_Data " />
                        <owl:allValuesFrom rdf:resource = "#Forecast_Rainfall_Data " />
                        <owl:allValuesFrom rdf:resource = "#Potential_Degree_Data " />
                </owl:Restriction>
        </rdfs:subClassof>
</owl:Class>
```

Figure 9. Snippets of owl notation using a universal restriction.

5.3. Task Instances

The specific task instances can be represented using classes and properties defined in the ontologies. Using the meteorological early-warning mentioned in Section 3.1 as an example, the tasks involved in the MEW example are listed in Table 1, in which there are two composite tasks (e.g., EWATask) and six atomic tasks (e.g., ERCTask, and FQTask). We use ERCTask as an example of an atomic task instance, which is used to calculate the effective rainfall. Figure 10 shows the individuals and properties involved in the ERCTask instance. The process of an atomic task is an individual of AtomicProcess, while those of composite tasks are not. We list the namespace declaration of ontologies and the syntax of class, subclass, and property definitions using OWL, as shown below Figure 10.

Table 1. Tasks involved in the MEW example

Task Type	Abbreviation	SubTask	Description
PotentialDegreeCalTask	PDCTask	FQTask FWCTask PDITask	Calculate potential degree index from multiple influence factor data
EffectiveRainfallCalTask	ERCTask		Calculate effective rainfall
EarlyWarningAnalysisTask	EWATask	OATask HRITask EWLTask	Generate a forecast map according to an early warning model
FactorQuantificationTask	FQTask		Quantify the factor data according to a certainty factor model
FactorWeightCalTask	FWCTask		Calculate factor weight
PotentialDegreeIndexCalTask	PDITask		Calculate potential degree index
OverlayAnalysisTask	OATask		Overlay the input data.
HazardRiskIndexCalTask	HRITask		Calculate the hazard risk index
EarlyWarningLevelTask	EWLTask		Divide early-warning level according to the risk index

Differing from the atomic task, the process of the composite task is composed of multiple edge individuals, each of which describes the data flow between two task instances. A set of edges compose a process graph that denotes how the task works. For example, Figure 11 shows the task instance of a composite task called EWATask. The process individual "process:EWAProcess" contains two edge individuals: "process:EWAEdge1" and "process:EWAEdge2". The former connects the two task instances: "task:QATask" and "task:HRITask", and the latter connects "task:HRITask" and "task:EWLTask". These edge individuals are linked to process individuals with the itemEdge property.

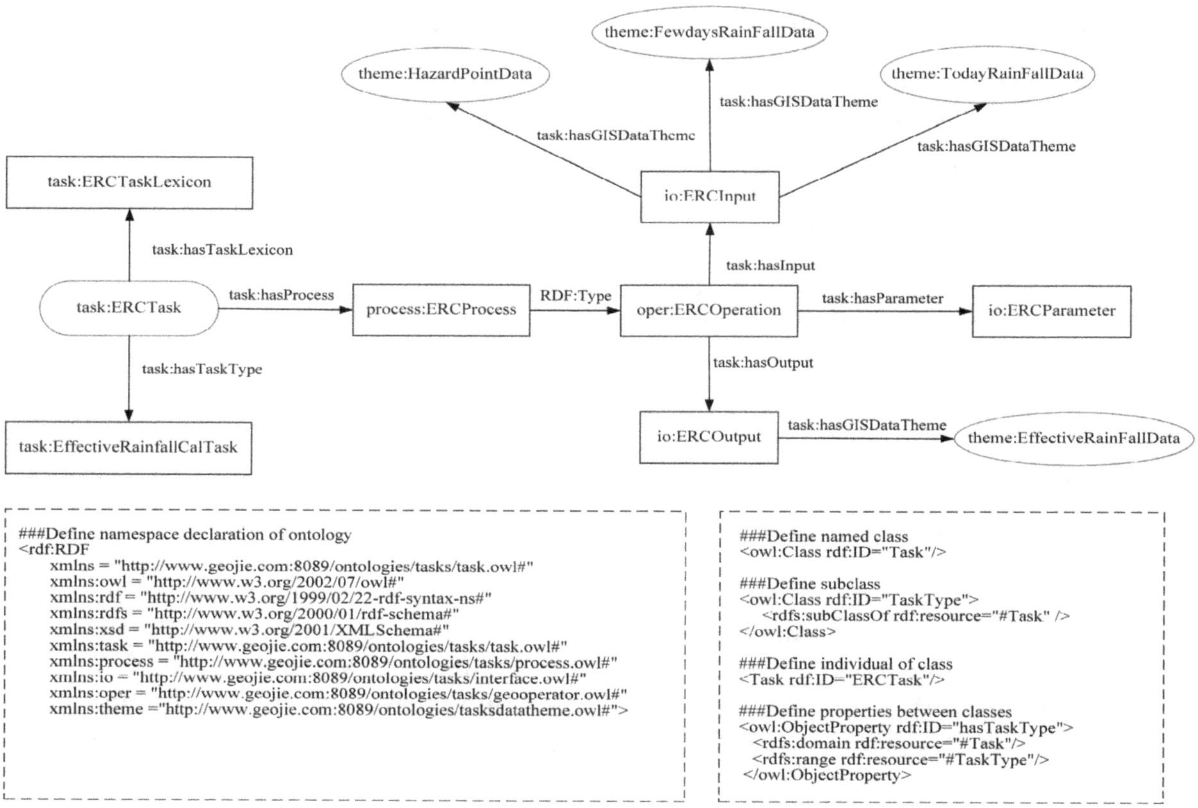

Figure 10. The task instance of an atomic task (EffectiveRainfallCalTask).

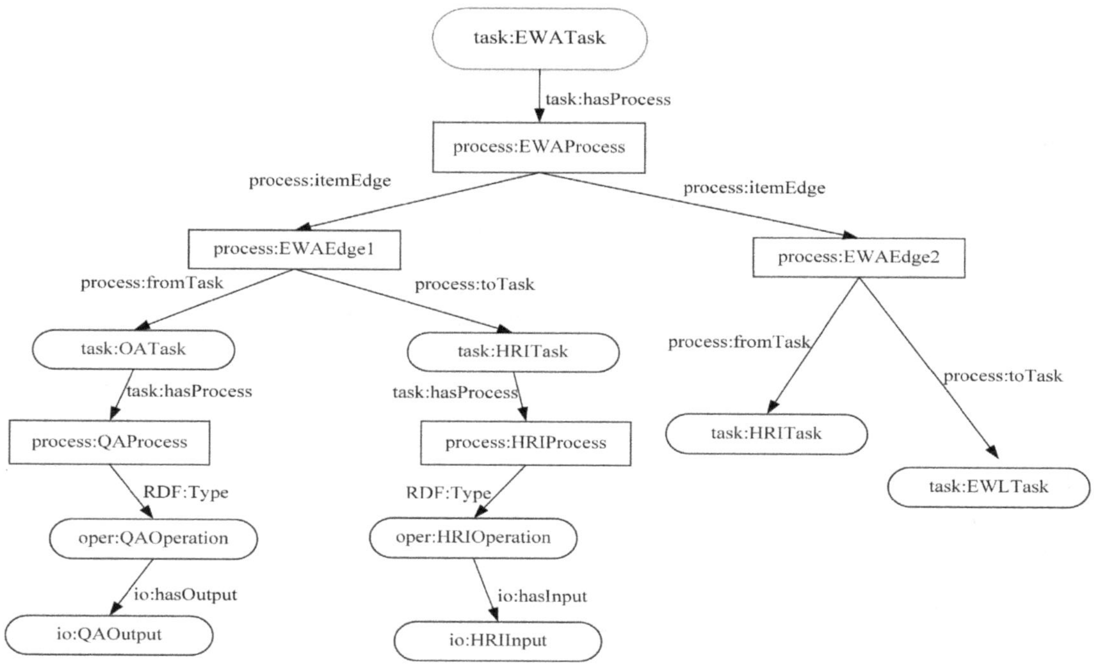

Figure 11. The task instance of a composite task (EarlyWarningAnalysisTask).

5.4. Prototype

A prototype system based on the realized ontology representation and formalized task instances was developed to facilitate users to solve complex geospatial problems. The implemented prototype, leverages a number of web techniques, such as Ajax, XML, JSON, EasyUI, GoJS, OpenLayers, Apache Axis2, and so on. Ajax is used for asynchronous data exchange between the client and server sides. XML and JSON are data exchange formats. EasyUI and GoJS are client UI frameworks, and GoJS is employed to draw a flowchart. OpenLayers is a JavaScript class library package for WebGIS client development, which is used to achieve map data access. Apache Axis2 is used to provide the Web Service interface. The Java API package Apache Jena [51], a Semantic Web framework for Java, is used to parse the ontology file, access ontology definitions, and infer knowledge [52]. The Apache Tomcat server was employed as a web container. The prototype system can be accessed using Microsoft IE or a Google browser in a Windows operating system

The MEW analysis in Henan, China is used as an example to utilize the knowledge base to support geospatial problem solving. First, we define formal semantics in the ontology-based knowledge base by creating task instances using an ontology editor. The task instance is named EWATask (Figure 11) which can decompose into three subtasks (OATask, HRITask, and EWLTask), The ontology files are generated using OWL format language which is mentioned in the previous section.

Second, the web services, including three data access services (Potential_Degree_Data, Effective_Rainfall_Data, and Forecast_Rainfall_Data) and three geoprocessing services (wps_overlay, wps _riskIndex, and wps_ewLevel), are published with the support of MapGIS IGServer [53]. The details of data access services are shown in Table 2, the geoprocessing services follow the WPS specification, and the workflow model for EWATask is shown in Figure 12.

Table 2. Data access services involved in EWATask.

Data Name	Service Type	SRS	Geometry
Potential_Degree_Data	WFS	Xi'an 80	Polygon
Effective_Rainfall_Data	WFS	Xi'an 80	Polygon
Forecast_Rainfall_Data	WFS	Xi'an 80	Polygon

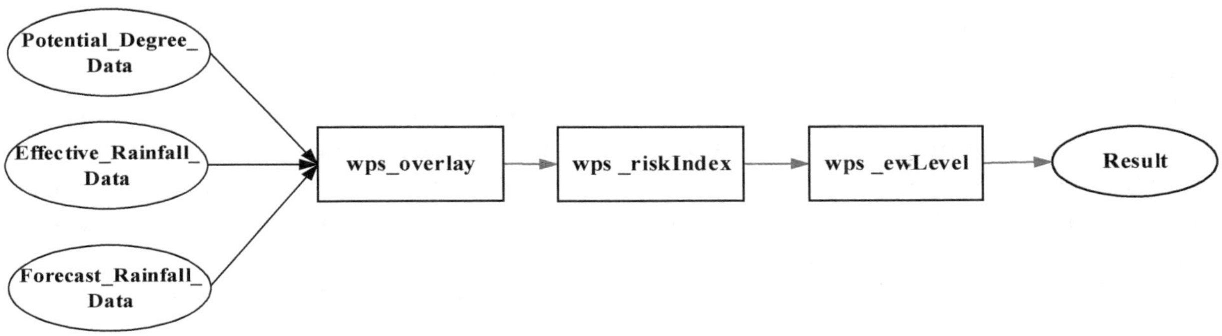

Figure 12. The workflow model of EWATask. Data access services are represented in elliptical shapes, and geoprocessing services are represented in rectangular shapes.

Finally, the prototype system provides an intuitive and easy-to-use graphical user interface (GUI). The end-users can access the GUI of the prototype system using a web browser. As shown in Figure 13, in the left panel, there is a tree structure showing the task lists that are parsed from the knowledge base. The user selects and clicks on a tasknode; then, the process model of the task will be displayed in the form of a flowchart in the right panel (step 1). Next, right-click on a process node and select the menu "service binding" (step 2). A service binding window pops up and allows end-users to bind the appropriate service and input the related parameters manually (step 3). Repeat this step for each process node. Finally, execute the task and get the result map (step 4). For instance, Mr. Wang took Henan province in China as a forecast area for the risk analysis and forecasted the possibility of the occurrence of geological hazards in the next 24 h. He clicked the EarlyWarnAnalysisTask node in the prototype system, the process graph of which was shown in the right panel (Figure 13). Following this workflow, he bound the appropriate geoprocessing services (wps_overlay, wps_riskIndex, and wps_ewLevel) that were invoked with a linear sequence (wps_overlay → wps_riskIndex → wps_ewLevel). According to the forecast results, an early-warning result map as shown in the lower right of Figure 13, was obtained, which uses different colors to represent different early-warning levels.

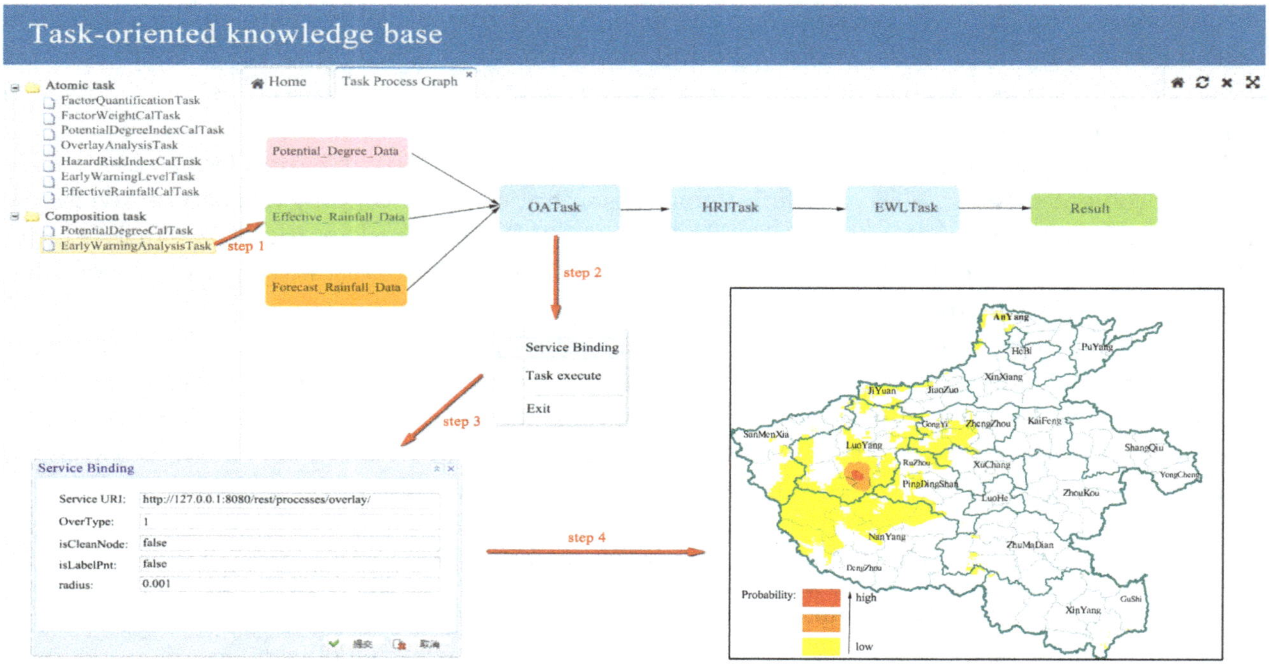

Figure 13. The graphical user interface of the prototype system.

6. Conclusions and Future Work

This paper proposes a task model and abstracts a geospatial problem as a task that can be used as a reusable component for problem-solving. A task-oriented knowledge base is built to capture sharable and reusable geospatial problem-solving knowledge. In the knowledge base, we combine multiple ontologies (e.g., Task Ontology, Process Ontology, and GIS Operation Ontology) to provide assistance for all facets of problem-solving. This knowledge base is not tightly-coupled with any specific workflow language. The required knowledge about problem-solving is stored in the knowledge base which employs ontology and task-oriented approach to achieve the formalization and reusability of tasks.

This knowledge base is tailored for domain experts to create and share their professional geospatial problem-solving knowledge. For the end-users, a user-friendly interface is needed to submit a geospatial problem and query the problem solution. An approach that has the capabilities of parsing natural language input will be developed in future work. This approach would allow users to input free-text to submit problem requirements.

In this paper, we only concentrate on using ontologies to describe a conceptual workflow that is composed of a linear sequence of GIS functionalities. We do not present an algorithm to instantiate into a concrete service chain and execute this workflow. The approach of knowledge transformation, instantiation and execution of a workflow will be implemented in future work.

Author Contributions: C.Z. designed the knowledge base, implemented the prototype system, and wrote the paper. K.M. and M.G. deployed and performed the prototype system. L.W. contributed the materials and tools. Z.X. conceived the early ideas of this work, reviewed the paper, and provided some suggestions and feedback.

Acknowledgments: We acknowledge the anonymous reviewers for their valuable comments and suggestions to improve this paper.

References

1. Yue, P.; Baumann, P.; Bugbee, K.; Jiang, L. Towards intelligent giservices. *Earth Sci. Inform.* **2015**, *8*, 463–481. [CrossRef]
2. Li, X.; Song, J.; Huang, B. A scientific workflow management system architecture and its scheduling based on cloud service platform for manufacturing big data analytics. *Int. J. Adv. Manuf. Technol.* **2016**, *84*, 119–131. [CrossRef]
3. Yang, Z.L.; Cao, J.; Hu, K.; Gui, Z.P.; Wu, H.Y.; You, L. Developing a cloud-based online geospatial information sharing and geoprocessing platform to facilitate collaborative education and research. In Proceedings of the ISPRS—International Archives of the Photogrammetry, Remote Sensing and Spatial Information Sciences, 2016 XXIII ISPRS Congress, Prague, Czech Republic, 12–19 July 2016; Volume XLI-B6, pp. 3–7.
4. Peng, Y.; Gong, J.; Di, L.; Jie, Y.; Sun, L.; Sun, Z.; Qian, W. GeoPW: Laying blocks for the geospatial processing web. *Trans. GIS* **2010**, *14*, 755–772.
5. Qi, K.; Gui, Z.; Li, Z.; Guo, W.; Wu, H.; Gong, J. An extension mechanism to verify, constrain and enhance geoprocessing workflows invocation. *Trans. GIS* **2016**, *20*, 240–258. [CrossRef]
6. Zhang, M.; Bu, X.; Yue, P. Geojmodelbuilder: An open source geoprocessing workflow tool. *Open Geospat. Data Softw. Stand.* **2017**, *2*, 8. [CrossRef]
7. Yue, P.; Zhang, M.; Tan, Z. A geoprocessing workflow system for environmental monitoring and integrated modelling. *Environ. Model. Softw.* **2015**, *69*, 128–140. [CrossRef]
8. Yue, P.; Di, L.; Yang, W.; Yu, G.; Zhao, P. Semantics-based automatic composition of geospatial web service chains. *Comput. Geosci.* **2007**, *33*, 649–665. [CrossRef]
9. Lamprecht, A.L.; Steffen, B.; Margaria, T. Scientific workflows with the jabc framework. *Int. J. Softw. Tools Technol. Transf.* **2016**, *18*, 629–651. [CrossRef]
10. Al-Areqi, S.; Lamprecht, A.L.; Margaria, T. Constraints-driven automatic geospatial service composition: Workflows for the analysis of sea-level rise impacts. In Proceedings of the International Conference on Computational Science and Its Applications, Beijing, China, 4–7 July 2016; pp. 134–150.
11. Kliment, T.; Bordogna, G.; Frigerio, L.; Crema, A.; Boschetti, M.; Brivio, P.A.; Sterlacchini, S. Image data and metadata workflows automation in geospatial data infrastructure deployed for agricultural sector. In Proceedings of the Geoscience and Remote Sensing Symposium, Milan, Italy, 26–31 July 2015; pp. 146–149.
12. Farnaghi, M.; Mansourian, A. Disaster planning using automated composition of semantic OGC web services: A case study in sheltering. *Comput. Environ. Urban Syst.* **2013**, *41*, 204–218. [CrossRef]
13. Al-Areqi, S.; Lamprecht, A.-L.; Margaria, T. Automatic workflow composition in the geospatial domain: An application on sea-level rise impacts analysis. In Proceedings of the 19th AGILE International Conference on Geographic Information Science, Helsinki, Finland, 14–17 June 2016.
14. Sun, Z.; Yue, P.; Di, L. Geopwtmanager: A task-oriented web geoprocessing system. *Comput. Geosci.* **2012**, *47*, 34–45. [CrossRef]
15. Jung, C.T.; Sun, C.H.; Yuan, M. An ontology-enabled framework for a geospatial problem-solving environment. *Comput. Environ. Urban Syst.* **2013**, *38*, 45–57. [CrossRef]
16. Lutz, M. Ontology-based descriptions for semantic discovery and composition of geoprocessing services. *Geoinformatica* **2007**, *11*, 1–36. [CrossRef]
17. Hofer, B.; Mãs, S.; Brauner, J.; Bernard, L. Towards a knowledge base to support geoprocessing workflow development. *Int. J. Geogr. Inf. Syst.* **2016**, *31*, 694–716. [CrossRef]
18. Müller, M. Hierarchical profiling of geoprocessing services. *Comput. Geosci.* **2015**, *82*, 68–77. [CrossRef]
19. Hu, L.; Yue, P.; Zhang, M.; Gong, J.; Jiang, L.; Zhang, X. Task-oriented sensor web data processing for environmental monitoring. *Earth Sci. Inform.* **2015**, *8*, 511–525. [CrossRef]
20. Gorton, S.; Reiff-Marganiec, S. Towards a task-oriented, policy-driven business requirements specification for web services. In Proceedings of the International Conference on Business Process Management, Vienna, Austria, 5–7 September 2006; pp. 465–470.
21. Wiegand, N.; García, C. A task-based ontology approach to automate geospatial data retrieval. *Trans. GIS* **2007**, *11*, 355–376. [CrossRef]
22. Luo, J. The Semantic Geospatial Problem Solving Environment: An Enabling Technology for Geographical Problem Solving under Open, Heterogeneous Environments. Ph.D. Thesis, The Pennsylvania State University, Pennsylvania, PA, USA, 2007.

23. Jung, C.T.; Sun, C.H. Ontology-driven problem solving framework for spatial decision support systems. *Tetsu- to-Hagane.* **2010**, *47*, 512–515.
24. Vahedi, B.; Kuhn, W.; Ballatore, A. Question-based spatial computing—A case study. In *Geospatial Data in a Changing World*; Springer International Publishing: Cham, Switzerland, 2016; pp. 37–50.
25. Albrecht, J. Universal elementary GIS tasks-beyond low-level commands. In Proceedings of the Sixth International Symposium on Spatial Data Handling, Edinburgh, UK, 3–7 August 1994; pp. 209–222.
26. Tran, V.X.; Tsuji, H. Owl-t: An ontology-based task template language for modeling business processes. In Proceedings of the Acis International Conference on Software Engineering Research, Management & Applications, Busan, Korea, 20–22 August 2007; pp. 101–108.
27. Sun, Z.; Yue, P.; Lu, X.; Zhai, X.; Hu, L. A task ontology driven approach for live geoprocessing in a service-oriented environment. *Trans. GIS* **2012**, *16*, 867–884. [CrossRef]
28. Yuan, X.; Liu, G. A task ontology model for domain independent dialogue management. In Proceedings of the IEEE International Conference on Virtual Environments Human-Computer Interfaces and Measurement Systems, Tianjin, China, 2–4 July 2012; pp. 148–153.
29. Park, H.; Yoon, A.; Kwon, H.C. Task model and task ontology for intelligent tourist information service. *Int. J. U- E-Serv. Sci. Technol.* **2012**, *5*, 43–58.
30. Linyao, Q.; Zhiqiang, D.; Qing, Z. A task-oriented disaster information correlation method. In Proceedings of the 2015 International Workshop on Spatiotemporal Computing, Fairfax, VA, USA, 13–15 July 2015; Volume II-4/W2, pp. 169–176.
31. Mikita, T.; Balogh, P. Usage of geoprocessing services in precision forestry for wood volume calculation and wind risk assessment. *Acta Univ. Agric. Silvic. Mendel. Brun.* **2015**, *63*, 793–801. [CrossRef]
32. Bensmann, F.; Alcacerlabrador, D.; Ziegenhagen, D.; Roosmann, R. The richwps environment for orchestration. *ISPRS Int. J. Geo-Inf.* **2014**, *3*, 1334–1351. [CrossRef]
33. Hull, D.; Wolstencroft, K.; Stevens, R.; Goble, C.; Pocock, M.R.; Li, P.; Oinn, T. Taverna: A tool for building and running workflows of services. *Nucleic Acids Res.* **2006**, *34*, 729–732. [CrossRef] [PubMed]
34. Wolstencroft, K.; Haines, R.; Fellows, D.; Williams, A.; Withers, D.; Owen, S.; Soilandreyes, S.; Dunlop, I.; Nenadic, A.; Fisher, P. The taverna workflow suite: Designing and executing workflows of web services on the desktop, web or in the cloud. *Nucleic Acids Res.* **2013**, *41*, 557–561. [CrossRef] [PubMed]
35. Taylor, I.; Shields, M.; Wang, I.; Harrison, A. The triana workflow environment: Architecture and applications. In *Workflows e-Science*; Springer: London, UK, 2007; pp. 320–339.
36. Altintas, I.; Berkley, C.; Jaeger, E.; Jones, M. Kepler: An extensible system for design and execution of scientific workflows. In Proceedings of the International Conference on Scientific and Statistical Database Management, Santorini Island, Greece, 23 June 2004; pp. 423–424.
37. Lamprecht, A.L.; Margaria, T.; Steffen, B. *Modeling and Execution of Scientific Workflows with the jABC Framework*; Springer: Berlin/Heidelberg, Germany, 2014; pp. 14–29.
38. Samadzadegan, F.; Saber, M.; Zahmatkesh, H.; Joze Ghazi Khanlou, H. An architecture for automated fire detection early warning system based on geoprocessing service composition. In Proceedings of the SMPR 2013, Tehran, Iran, 5–8 October 2013; pp. 351–355.
39. Arul, U.; Prakash, S. A unified algorithm to automatic semantic composition using multilevel workflow orchestration. In *Cluster Computing*; Springer: New York, NY, USA, 2018; pp. 1–22.
40. Hofer, B.; Brauner, J.; Jackson, M.; Granell, C.; Rodrigues, A.; Nüst, D.; Wiemann, S. Descriptions of spatial operations—Recent approaches and community feedback. *Int. J. Spat. Data Infrastruct. Res.* **2015**, *10*, 124–137.
41. Brauner, J. Formalizations for Geooperators-Geoprocessing in Spatial Data Infrastructures. Ph.D. Thesis, Technische Universität Dresden, Dresden, Germany, 2015.
42. Hofer, B.; Papadakis, E.; Mäs, S. Coupling knowledge with GIS operations: The benefits of extended operation descriptions. *Int. J. Geo-Inf.* **2017**, *6*, 40. [CrossRef]
43. Crubézy, M.; Musen, M.A. *Ontologies in Support of Problem Solving*; Springer: Berlin/Heidelberg, Germany, 2004; pp. 321–341.
44. Zhao, P.; Di, L.; Yu, G.; Yue, P.; Wei, Y.; Yang, W. Semantic web-based geospatial knowledge transformation. *Comput. Geosci.* **2009**, *35*, 798–808. [CrossRef]
45. Gruber, T.R. A translational approach to portable ontologies. *Knowl. Acquis.* **1993**, *5*, 199–220. [CrossRef]
46. Perez, A.G.; Benjamins, V.R. Overview of knowledge sharing and reuse components: Ontologies and problem-solving methods. In Proceedings of the 16th International Joint Conference on Artificial Intelligence (IJCAI'99), Stockholm, Sweden, 31 July–6 August 1999.

47. Zhong, S.; Fang, Z.; Zhu, M.; Huang, Q. A geo-ontology-based approach to decision-making in emergency management of meteorological disasters. *Nat. Hazards* **2017**, *89*, 531–554. [CrossRef]
48. The International Organization for Standardization (ISO). *ISO 19107: Geographic Information—Spatial Schema*; The International Organization for Standardization: Geneva, Switzerland, 2003.
49. ISO. *ISO 19123: Geographic Information—Schema for Coverage Geometry and Functions*; The International Organization for Standardization: Geneva, Switzerland, 2005.
50. OGC. *OGC Abstract Specifications: Topic 5—Features*; Open Geospatial Consortium: Wayland, MA, USA, 2009; pp. 8–126.
51. Apache Jena. Available online: http://jena.apache.org/ (accessed on 29 August 2018).
52. Zhang, C.; Zhao, T.; Li, W. Automatic search of geospatial features for disaster and emergency management. *Int. J. Appl. Earth Obs. Geoinf.* **2010**, *12*, 409–418. [CrossRef]
53. Wu, X.; Liu, X.; Zhou, S. *Principle and Method of MapGIS IGServer*; Publishing House of Electronics Industry: Beijing, China, 2012.

10

Integrating Geovisual Analytics with Machine Learning for Human Mobility Pattern Discovery

Tong Zhang [1], Jianlong Wang [1], Chenrong Cui [1], Yicong Li [1], Wei He [1], Yonghua Lu [2,*] and Qinghua Qiao [3]

1. State Key Laboratory of Information Engineering in Surveying, Mapping and Remote Sensing, Wuhan University, Wuhan 430079, China; zhangt@whu.edu.cn (T.Z.); wangjianlong@whu.edu.cn (J.W.); ccr2017@whu.edu.cn (C.C.); olivelee@whu.edu.cn (Y.L.); he_wei@whu.edu.cn (W.H.)
2. Shenzhen Investigation and Research Institute Co., Ltd., Shenzhen 518026, China
3. Chinese Academy of Surveying & Mapping, Beijing 100830, China; qiaoqh@casm.ac.cn

* Correspondence: luyonghua@sziri.com

Abstract: Understanding human movement patterns is of fundamental importance in transportation planning and management. We propose to examine complex public transit travel patterns over a large-scale transit network, which is challenging since it involves thousands of transit passengers and massive data from heterogeneous sources. Additionally, efficient representation and visualization of discovered travel patterns is difficult given a large number of transit trips. To address these challenges, this study leverages advanced machine learning methods to identify time-varying mobility patterns based on smart card data and other urban data. The proposed approach delivers a comprehensive solution to pre-process, analyze, and visualize complex public transit travel patterns. This approach first fuses smart card data with other urban data to reconstruct original transit trips. We use two machine learning methods, including a clustering algorithm to extract transit corridors to represent primary mobility connections between different regions and a graph-embedding algorithm to discover hierarchical mobility community structures. We also devise compact and effective multi-scale visualization forms to represent the discovered travel behavior dynamics. An interactive web-based mapping prototype is developed to integrate advanced machine learning methods with specific visualizations to characterize transit travel behavior patterns and to enable visual exploration of transit mobility patterns at different scales and resolutions over space and time. The proposed approach is evaluated using multi-source big transit data (e.g., smart card data, transit network data, and bus trajectory data) collected in Shenzhen City, China. Evaluation of our prototype demonstrates that the proposed visual analytics approach offers a scalable and effective solution for discovering meaningful travel patterns across large metropolitan areas.

Keywords: geovisual analytics; machine learning; smart card data; transit corridor; mobility community; trip

---

1. Introduction

Monitoring human movement is of fundamental importance in transportation planning and management. To facilitate public transit planning and operational management, it is appealing to understand transit movement patterns across space and time [1]. Fortunately, recent advanced geospatial data collection technologies, such as global positioning systems, digital mapping, smart card automated fare payment systems, and wireless communication techniques, are generating a wealth of spatially and temporally varying transit data that create opportunities to discover meaningful and significant movement patterns over large metropolitan areas [2,3]. Various data mining methods have been developed to uncover transit travel behavior patterns on the basis of these heterogeneous geospatial datasets, including clustering for passenger segmentation [4,5], hazard modeling for loyalty analysis [6], trip chaining methods for destination estimation [7], and choice modeling for passenger activity analysis [8].

Over the past few years, many studies have been conducted to explore urban travel patterns using various modeling and analytical approaches based on massive human mobility data, such as optimization-based routing equilibrium models for congestion alleviation [9], clustering-based correlated analyses of mobility similarities and social relationships [10], low-level mobility pattern discovery [11], and multi-scale exploration of social fragmentation [12]. With the availability of massive human mobility data, machine learning techniques have been playing a more and more important role in gaining a deep understanding of human mobility behavior [13,14], ranging from movement pattern mining [15–17], mobility prediction [18–20], and movement mode classification [21], to lifestyle discovering and prediction [22].

Recently, many attempts to visualize massive human mobility data, including cell phone data [23,24], taxi movement data [25], and social media data [26] have been reported. Some systems have been developed to perform visual analytics on smart card data [27,28], aiming to discover salient travel patterns to improve public transit planning and management. These efforts mostly focus on novel visualization designs by aggregating individual trip information into compact visual forms. With these visualization tools, users can discover and analyze significant travel characteristics efficiently. Nevertheless, most of these methods focus on the visualization of simple and intuitive spatio-temporal movement patterns, such as place-based flow variations, inter-area flow maps, or accessibility maps. While transit planners and operational managers need to reveal complex movement patterns at different spatio-temporal scales, their intuitive tools are not adequate since they are based on simple statistical methods. This motivates us to investigate the possibility of applying machine learning techniques to identify high-level, complex transit movement patterns that support advanced transit planning and management.

It can be argued that visualization should be enhanced by advanced machine learning methods, given the overwhelming size and complexity of transit data. Over the past few years, researchers have developed visual analytics tools to support interactive exploration of spatio-temporal movement patterns using massive amounts of mobility data. Among these efforts, machine learning methods have been used for pattern discovery and analysis. For example, von Landesberger et al. [29] propose to integrate interactive spatio-temporal clustering and aggregated graph representations to discover abstracted urban movement patterns using social media and mobile phone data. We choose to discover public transit corridors and mobility communities using two state-of-the-art machine learning algorithms because they produce representative high-level, complex transit mobility patterns that are useful for transit and urban planning. Furthermore, we develop specific interactive visualization forms to facilitate the understanding of identified corridors and community structure. We argue that the combination of machine learning and geovisualization is beneficial for gaining a deep understanding of complex transit mobility patterns in large metropolitan areas. We propose to examine high-level, complex public transit travel patterns using visual analytics over large-scale transit networks, which is challenging since thousands of transit passengers and massive amounts of data from heterogeneous sources are involved. Moreover, efficient representation and visualization of discovered travel patterns is also a difficult task given large quantities of transit trips. To address these challenges, this study leverages advanced machine learning methods to identify time-varying mobility patterns based on multi-source transit big data. We also devise compact multi-scale visualization forms to represent the discovered travel behavior dynamics. A web-based prototype is developed to implement the proposed geovisual analytics approach within an integrated graphic interface, enabling in-depth analysis of multi-source massive transit data. We evaluate the prototype with realistic transit data collected in Shenzhen City, China. Our empirical usability study demonstrates that our approach can offer a scalable and effective solution for discovering meaningful travel patterns across large metropolitan areas.

In this study, we aim to identify spatially and temporally varying transit movement patterns based on massive transit data over a large public transit network. We make the following technical contributions:

(1) We develop an integrated geovisual analytics approach that integrates two advanced machine learning methods with interactive maps to characterize two types of high-level, complex transit travel behavior patterns, including a clustering algorithm to identify transit corridors and a graph-embedding algorithm to identify hierarchical mobility community structure.

(2) We design novel integrated geovisual analytics interfaces for the discovered complex transit movement patterns, including specific views to visualize identified mobility communities and corridors, allowing regular users to examine and understand these ever-changing patterns at different scales and perspectives.

2. Data

Being deployed on public transit vehicles, smart card automated fare payment systems provide an efficient way to collect large volumes of travel data at the individual level. The proposed approach utilizes smart card data (SCD) collected in Shenzhen City, China. Shenzhen City has a large bus and subway network consisting of 8 subway lines, 199 subway stations, 808 bus routes, and 6226 bus stops (Figure 1).

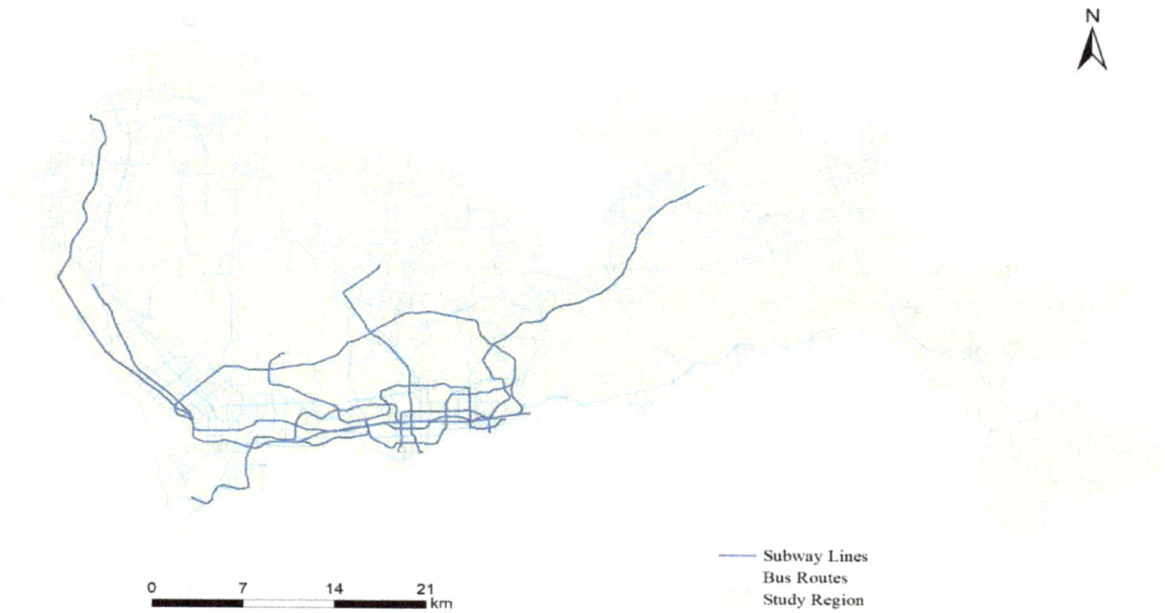

Figure 1. The study region and public transit network.

We use a week of SCD starting from 3 to 9 April 2017. The SCD used in this study holds the names of boarding and alighting stations for each subway passenger. Bus passengers are not required to tap their smart cards when alighting. Therefore the information on alighting bus stops is not recorded. In addition to the SCD, we have access to bus trajectory data, public transit network, and road network data. These three datasets are registered into the same georeference framework, i.e., World Geodetic System 1984 (WGS 84) coordinate system and Universal Transverse Mercator coordinate system (UTM) Zone 50. The public transit network dataset contains the location, identification, and schedule information of all subway lines and bus routes. Based on GPS devices installed on each bus, we can obtain bus trajectory data such as longitude and latitude, speed, and travel direction at approximate 20–60 second intervals. In addition, bus identification information including license plate numbers, number of transit lines, and company names are all saved. With more than 6 million records collected for each day, the size of the SCD set for the week amounts to 6.5 GB. Each day, the bus trajectory dataset has approximately 63–73 million GPS records.

3. Methodology

3.1. Methodology Overview

Public transit systems contain multiple components: bus stops, subway stations, bus and subway lines, bus and subway vehicles, and passengers. Most existing literature focuses on the analytics of transit lines/stops [30], schedules [31], or aggregated transit trips [32,33]. Some have explored the relationships between transit trips and points of interest [28] but have not leveraged advanced machine learning methods to analyze complex travel patterns and global mobility structures. Given a large amount of trip data, one may wish to identify significant spatio-temporal travel patterns, reveal global mobility structures, and visualize them on interactive maps. For example, questions can be raised to find interconnected road segments that contain significant transit travel demand patterns at a global scale or to delineate areas with similar transit travel characteristics. What specific spatio-temporal patterns can be discovered from these road segments and areas, and how do these patterns evolve over time? Can we jointly examine transit travel patterns from different aspects of public transit services in an integrated interactive user interface?

To answer these questions, we can define geovisual analytics tasks as follows:

(1) Global pattern discovery task 1: discover hierarchical mobility structure based on transit trip data and analyze their inter-correlations;
(2) Global pattern discovery task 2: identify significant transit corridors for any specified time intervals;
(3) Local pattern exploration task 1: explore the intrinsic information of identified individual transit corridors;
(4) Local pattern exploration task 2: examine the temporal evolution of the discovered mobility community structure;
(5) Comprehensive analysis task: design and implement linked or integrated views to visually analyze different components of public transit services (including corridors, community structures, and stops) and discovered travel patterns.

The proposed approach delivers a comprehensive solution to pre-process, analyze, and visualize complex public transit travel patterns (Figure 2). This approach first fuses SCD with other urban data to reconstruct transit trips. It then segments the study region into hierarchical areal units (i.e., public transit mobility community) according to mobility features of trips and static local features using graph embedding. Based on recovered public transit trip data, we develop a clustering-based algorithm for extracting transit corridors to represent mobility interactions between different regions. Based on detected corridors and mobility communities, we develop various visualization forms to represent these transit movement patterns on maps and other views. An interactive web-based mapping prototype is also developed to enable the visual exploration of mobility structures over space and time. Specific visualization forms are designed and implemented in the web-based prototype to facilitate the analysis of massive transit data, including map-based visualization, focus views, and auxiliary views, which will be detailed in Section 3.5. The discovered mobility community structure and transit corridors can be visualized on interactive maps. The focus views consist of four types of visualization: (1) a corridor detail view that shows detailed trip information based on a simplified schematic map for each selected corridor; (2) a community tracking view that presents evolving changes of specific communities across time; (3) a stop glyph plotting the statistical information of all trips that originate, end, or pass a specific bus stop or subway station; and (4) a corridor–community correlation view illustrating the spatio-temporal correlations between transit corridors and mobility communities using parallel coordinate plots.

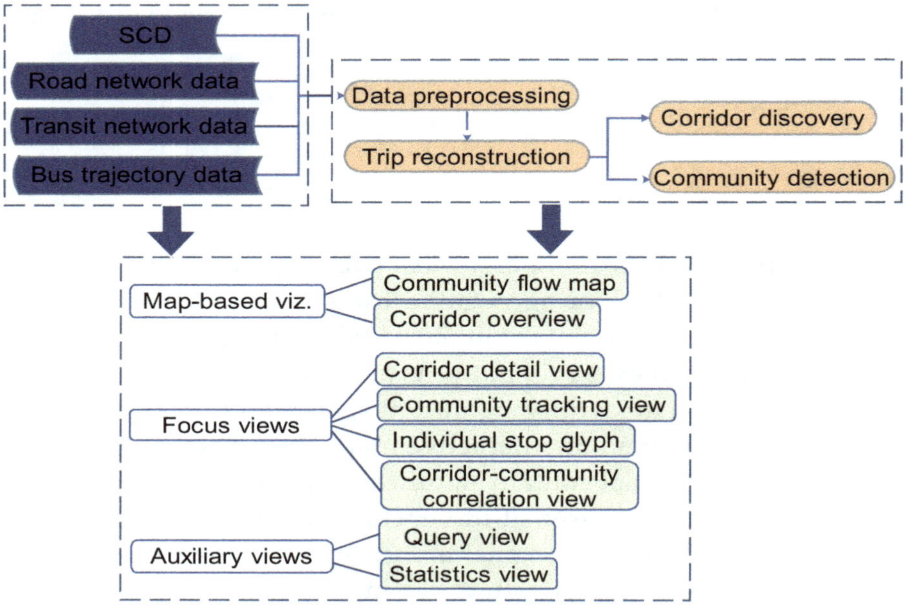

Figure 2. Overview of the proposed geovisual analytic approach. SCD—smart card data.

The trip reconstruction and corridor discovery methods have been described in a separate paper [34]. Below, we briefly introduce the two methods. Implementation details are reported in Section 4.

3.2. Data Pre-Processing and Trip Reconstruction

Following the procedure developed in Zhang et al. [34], we performed data pre-processing for the original datasets and reconstructed transit trips, which were used in subsequent geovisual analytics tasks. For a passenger, a public transit trip consists of multiple consecutively linked trip legs with a specific travel purpose [35]. Our visual analytics approach is based on trips rather than trip legs because trips are better at revealing realistic travel demands and behavior patterns. In this subsection, we briefly describe the data pre-processing and trip reconstruction steps.

After removing erroneous SCD and bus trajectory records, we corrected inconsistent stops names and locations between heterogeneous datasets based on the method developed by [36]. All of the datasets were imported into Microsoft SQL Server databases in which spatial indices were constructed based on transit network data to accelerate data query and trip reconstruction. The original SCD was divided into subway-based and bus-based datasets by each date since subway-based records have full boarding and alighting information and an alighting stop estimation procedure was expected for bus-based SCD.

First, we needed to estimate boarding and alighting stops for bus-based trip legs. For each bus-based SCD record, the boarding stop can be identified by matching the license plate number from the SCD and the bus trajectory dataset to find GPS sampling points that are close to boarding time, which were matched to the transit network to find the most probable boarding stop. Then, we proceeded to estimate alighting stops: (1) alighting stops can be easily derived by searching for the closest stop to the next boarding stop if a passenger makes another boarding during the same day; (2) if the current trip leg is the final one of the day, the first boarding stop of the next day is used to estimate the possible alighting stop of this last trip leg; (3) otherwise, we can search other dates or other similar passengers to make estimations. Upon the availability of both boarding and alighting stops, complete bus trip legs could be recovered. These bus trip legs were then connected with subway legs to reconstruct a complete trip if these trip legs taken by the same individuals were within the 30 min threshold.

3.3. Extracting Transit Corridors

The concept of "transit corridors" has been widely adopted and put into real-world planning and management practices [37]. We define a corridor as a directional linear road segment consisting of multiple transit stops with significant numbers of passengers. Note that corridors may contain multiple branches and may overlap with other corridors. Based on massive transit trip data, time-varying transit corridors can be extracted to represent the most significant travel demand patterns across space and time. The corridor-extracting algorithm is developed on the basis of public transit trips, each of which is characterized by one single travel purpose. Each trip may consist of multiple legs, and each leg corresponds to one smart card transaction. We proposed a share-flow clustering algorithm [34]. The algorithm was based on the concept of "accumulated transit flow", which calculates the number of stops each passenger passes after boarding. If two stops have a large shared "accumulated transit flow", then the downstream stop is "directly transit-flow accessible" from its preceding adjacent stop. Starting from stops with a large amount of "accumulated transit flow", the algorithm iteratively evaluates adjacent stops along the travel direction. If this next stop is directly transit-flow accessible from the current stop, the road segment between the current and the next stop will be linked to the current corridor. After the initial growth of corridors, a pruning and merging process is performed to remove short and non-significant corridor candidates. By this clustering algorithm, linear corridors can be discovered dynamically for any specific time interval. The algorithm can be described in the following steps:

(1) Network modeling. The public transit network can be modeled as a directed graph and is mapped to the road network $\boldsymbol{G}$ **(*V*, *E*)**, where $\boldsymbol{V}$ denotes the sets of road intersections $\boldsymbol{V_r}$ and transit stops $\boldsymbol{V_t}$ ($\boldsymbol{V_t}$ have been projected into road links), $\boldsymbol{E}$ denotes road segments between road intersections and transit stops. We extract a small set of connected segments $\boldsymbol{E_c}$ whose end nodes have a large shared "accumulated transit flow" and identify them as transit corridors.

(2) Computing accumulated transit flow. For each node v in $\boldsymbol{V_t}$, the number of passengers who board at v or before it is recorded as n_v. For each passenger, the number of stops she has passed after boarding is recorded until she exits from the vehicle. Then for each v, this number is used as the "accumulated transit flow" $at(v)$.

(3) Corridor initialization. We choose nodes with a significant number of accumulated transit flows as seeds to grow corridors.

(4) Corridor expansion. The seed nodes are stacked into a priority queue, ranked by its accumulated transit flow. The one with highest $at(v)$ is popped out and used as the initial seed s_0 to expand a corridor. From s_0, the algorithm searches for one adjacent stop, s_1, that meets the criterion of significant "shared accumulated transit flow" between s_0 and s_1. "Shared accumulated transit flow" is defined as $sa(0 \rightarrow 1) = [at(1) - at(0)] / at(0)$, i.e., the change ratio of accumulated transit flow for the two adjacent stops/nodes. Meanwhile, the two nodes must meet another criterion, namely, "shared transit flow", which is defined as $st(0 \rightarrow 1) = n_0 \cap n_1$. If the two nodes meet both criteria, the algorithm expands a corridor from node 0 to 1. This procedure repeats until no downstream nodes meet the two criteria. Then another seed in the queue is fetched to grow another corridor, until all seeds are popped.

(5) Corridor pruning. We need to prune short corridors (with less than 4 stops) or non-significant corridors (transit flows are less than a pre-defined threshold).

(6) Corridor merging. This final step is to merge corridors if they are already connected or overlapped.

Usually, the algorithm can extract 5–10 transit corridors for peak hours and non-peak hours during weekdays and weekends based on the trip data we have produced.

3.4. Discovering Mobility Communities

It is desirable to represent high-level urban mobility structures with multi-scale communities when dealing with overwhelming amounts of mobility data [38]. Each mobility community is featured with similar travel characteristics. The representation of a hierarchical community structure can significantly facilitate the understanding of inter-area interconnections in a city. Traditionally, the construction of community structure uses community detection algorithms developed in network science [39]. These community detection algorithms first build a graph to represent connections between nodes and then employ clustering, optimization, or statistical inference methods to divide the entire graph into groups, ensuring nodes within each group are more densely connected than external nodes [40]. However, public transit passengers usually travel long distances away from their origins, and these mobility behaviors must be accounted for when extracting transit mobility structure. This study proposes a different community definition that considers not only local trip statistics but also trip destinations and other dynamic travel characteristics such as travel frequency and transferring patterns. All of this information can be readily computed from trip data.

Instead of denoting each subway station or bus stop by a graph node, we use region partitions with similar possible boarding stations nearby. We first partition the study region into regular grid cells, each of which has a size of 100 m × 100 m. We remove grid cells located in mountainous and water areas (inaccessible by transit services). Then a vector for each cell is built to record possible boarding stations (stops) that are close to them. Finally, a heuristic algorithm is employed to merge neighboring cells with the most similar vectors. After the merging of original grid cells, we obtain 18,109 grid groups, most of which consist of 2–7 original grid cells. These grid groups are then denoted as graph nodes, whose number is much smaller than original transit stops, thereby dramatically reducing computational expenses in community detection.

Traditional community detection algorithms cannot handle our problem and are not scalable to a large public transit network. In order to handle such complex trip behavior features, we use a graph-embedding method to uncover a dynamic community from realistic SCD. Graph embedding aims to produce a compact vector representation for each node and preserves graph structure within a low dimensional space [41].

We define a directed weighted graph $\boldsymbol{G}_t(\boldsymbol{V}, \boldsymbol{E})$ for a time interval t. $\boldsymbol{V}$ is the set of grid groups, and $\boldsymbol{E}$ represents transit connection edges between nodes in $\boldsymbol{V}$. Each edge e is weighted by realistic traffic flow between its origin node and ending node during t. Based on these weights, we can construct a traffic flow matrix $\boldsymbol{F}$, where $f_{i \to j}$ denotes the number of transit passengers travelling from node i to j. We also construct an adjacency matrix $\boldsymbol{A}$ to describe local connectivity between graph nodes. The matrix $\boldsymbol{A}$ can be used to represent first-order transit connectivity proximity. The global network structure can be preserved via a high-order proximity based on traffic flow matrix $\boldsymbol{F}$. The high-order proximity is defined as the similarity between traffic connectivity structures of a pair of nodes.

Since transit travel behaviors are largely non-linear and non-stationary, we leverage deep learning methods to learn network embedding. A classical auto-encoder framework (structure deep network embedding, SDNE) is adopted to learn latent network representations [42]. The auto-encoder framework consists of an encoder and a decoder. The encoder contains multiple layers, each of which can be defined as

$$\boldsymbol{z}^{(i)} = \sigma(\boldsymbol{W}^{(i)}\boldsymbol{z}^{(i-1)} + \boldsymbol{b}^{(i)}), \tag{1}$$

where $\boldsymbol{z}^{(i)}$ denotes the hidden representation for the ith layer and $\boldsymbol{z}^0$ is the original input data $\boldsymbol{X}$, which is a n-dimensional vector. $\boldsymbol{W}^{(i)}$ and $\boldsymbol{b}^{(i)}$ are learnable parameters. $\sigma(.)$ is the non-linear activation function.

If we use K layers in the encoder, the input vector $\boldsymbol{z}^0$ would be mapped into a hidden representation $\boldsymbol{z}^{(K)}$. Correspondingly, the decoder transforms $\boldsymbol{z}^{(K)}$ back to a reconstructed vector $\boldsymbol{Y}$ after performing K layers of nonlinear transformation operations,

$$\boldsymbol{z}\prime^{(j+1)} = \sigma(\boldsymbol{W}\prime^{(j)}\boldsymbol{z}\prime^{(j)} + \boldsymbol{b}'^{(j)}), \tag{2}$$

where $\boldsymbol{z}'^{(j)}$ denotes the reconstructed data vector for the jth layer and $\boldsymbol{W}'^{(j)}$ and $\boldsymbol{b}\prime^{(j)}$ are learnable parameters. Note that $\boldsymbol{z}\prime^{(0)} = \boldsymbol{z}^{(K)}$, $\boldsymbol{Y} = \boldsymbol{z}'^{(K)}$.

The model parameters can be learned by minimizing the reconstruction error between the reconstructed vector $\boldsymbol{Y}$ and the input vector $\boldsymbol{X}$:

$$\boldsymbol{L}\,(X,\,\boldsymbol{Y}) \;=\; \sum_{i=1}^{n} \|y_i - x_i\|_2^2. \tag{3}$$

If multiple transit features are used as the input vector to the encoder (including flow, speed, destination, and travel frequency), we obtain $\boldsymbol{L}_1$. If the adjacency matrix $\boldsymbol{A}$ is used as the input, we can build

$$\boldsymbol{L}_2(\boldsymbol{A},\,\boldsymbol{Y}) \;=\; \sum_{i,j>0}^{|V|} f_{i\to j}\|y_j^{(K)} - y_i^{(K)}\|_2^2, \tag{4}$$

which preserves the high-order proximity of G.

The two reconstruction error functions can be linearly combined into a comprehensive joint loss function:

$$\boldsymbol{L}_{all} = \boldsymbol{L}_1 + \alpha \boldsymbol{L}_2. \tag{5}$$

The model is randomly initialized and optimized with a stochastic gradient descent. After model convergence, we can obtain the final embedding representation for all nodes in G. Based on learned compact node representations, we can use hierarchical clustering to generate hierarchical mobility communities.

3.5. *Visual Analytics Design*

In the literature, public transportation visualization studies mostly focus on public transit networks and represent travel statistics based on stops and routes. We propose to examine and evaluate public transit services from different perspectives, namely, hierarchical mobility communities and significant transit corridors, in addition to the public transit network. Several visual designs are proposed to facilitate this comprehensive visual analytics strategy.

Our visualization design consists of three types of views (Figure 2): (1) Map-based visualization that uses interactive maps to depict extracted mobility communities and inter-community transit flow. Detected corridors are also illustrated in the map view. (2) Focus view is designed to present detailed information on user-selected corridors, transit mobility communities, and individual transit stops. Correlations between corridors and communities can also be visualized. (3) Other auxiliary views, including a query view and a statistics view. The query view enables interactive data selection for visual analytics for any time interval. The statistics view uses statistical diagrams to present summary information on corridors.

3.5.1. Mobility Communities

Based on realistic SCD, we can extract two-level mobility community structures over Shenzhen City. After performing graph embedding for g grid groups, we can perform hierarchical clustering based on these grid groups to produce two levels of mobility communities (see Section 4 for implementation details). Low-level communities are only based on the distances between embedded vectors. Based on low-level communities, we can further produce high-level communities by accounting for spatial contiguity and cohesiveness using regionalization methods. Figure 3a illustrates a high-level mobility community structure for April 3 (a holiday). Transit flows between these high-level communities are mapped to depict an overview of aggregated transit flows across the study region. High-level community structure is favored for global travel pattern discovery and analysis. When zooming in to the low-level, detailed community structures are visualized with different colors representing different cluster types (Figure 3b). With interactive community maps presented in figures, users can perform global pattern discovery task 1 to identify the hierarchical mobility community structure and visualize inter-community interactions conveniently.

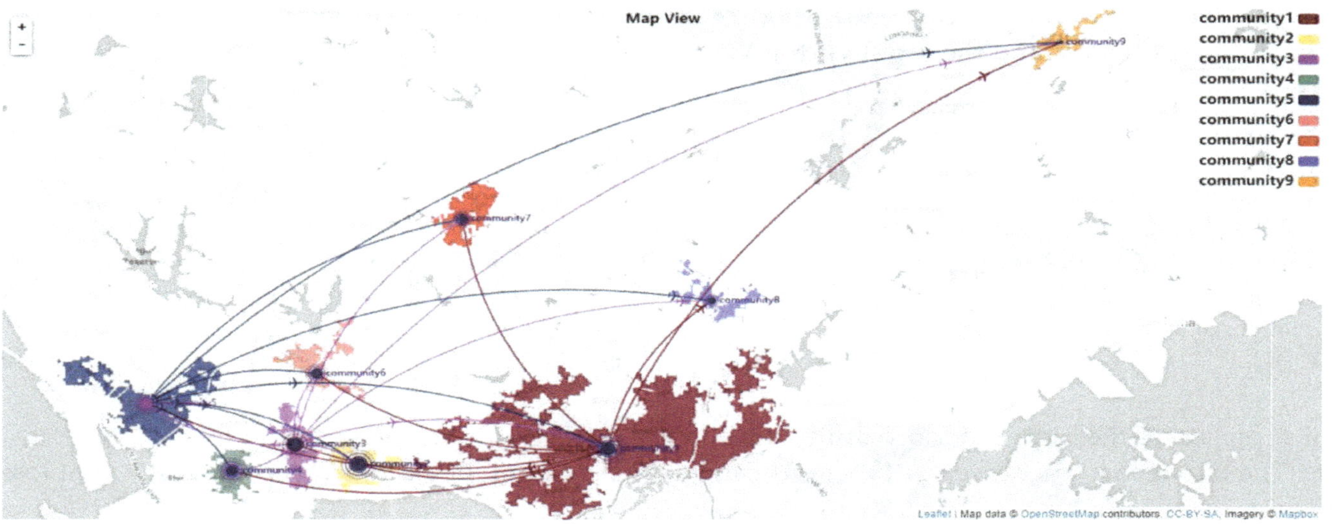

(a)

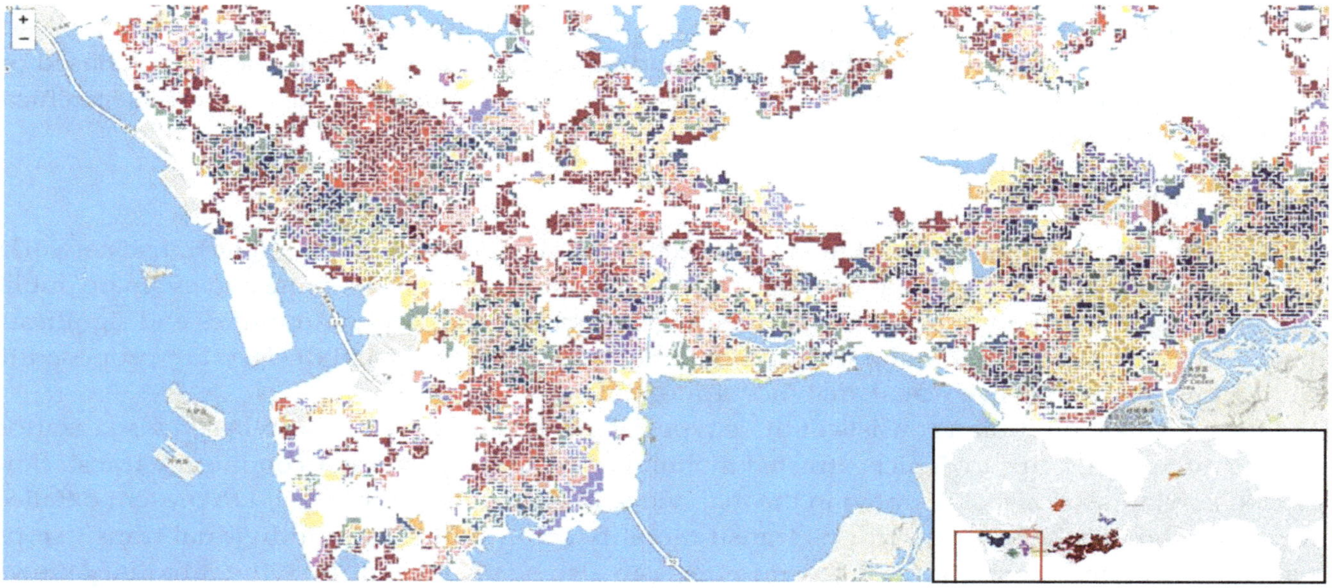

(b)

Figure 3. Transit-based mobility community structure on a holiday. (**a**) High-level community flow map; (**b**) low-level community clusters.

In the mobility community tracking view, one can select (from the map) one specific mobility community and track the temporal changes in its shape and flows between itself and other nearby communities (Figure 4). As the detected community structures evolve over time, a community may undergo different changes, including splitting into separate communities or merging with other communities. Each community is represented by a vertical bar, the height of which is proportional to the number of transit trips of the community. Ribbons connecting bars between different times represent transit flows between communities. Wide (narrow) ribbons indicate high (low) volumes of transit trips. The vertical positions of the bottom of bars also indicate the topological relations of communities from adjacent dates. Bars that are far away from each other correspond to communities that are also distant on the map. As we fix the position of the original selected bar, the overlapping relations can also be revealed by their relative vertical positions between two bars from adjacent dates. As community structure undergoes constant changes, users can conduct local pattern exploration task 2 to track the evolving trend of any chosen communities and gain a deep understanding of the mobility structure of the study region.

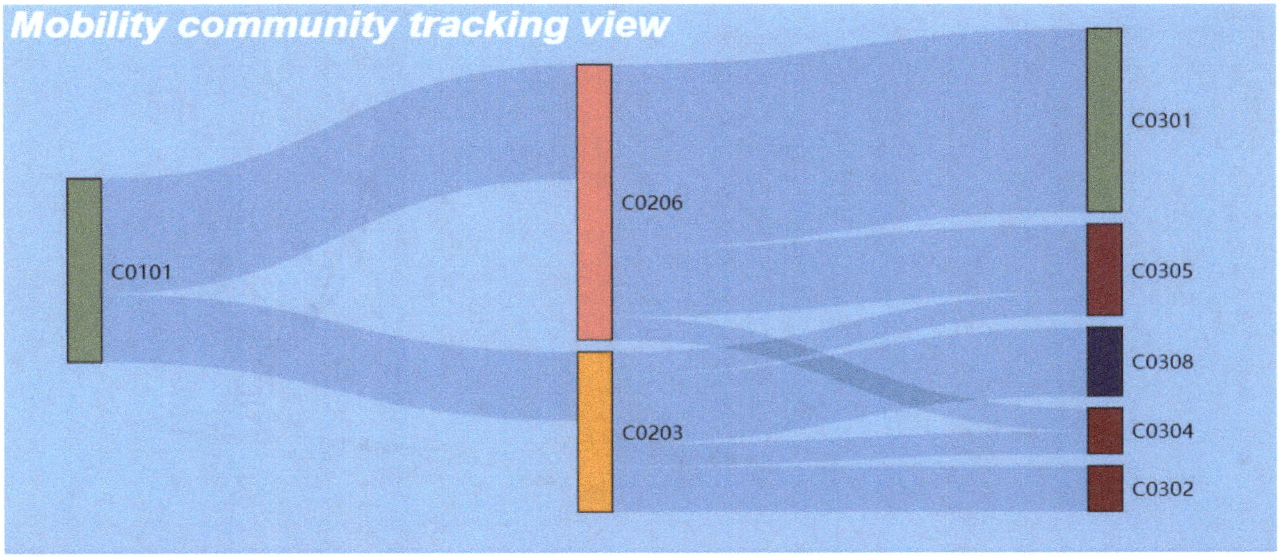

Figure 4. Mobility community tracking view. We can observe that the selected community "C0101" (indicating a detected community on April 1) has a significant number of transit trips connecting it and two communities on April 2 ("C0206" and "C0203"). These two are further connected with five communities (C0301, C0302, C0304, C0305, and C0308) on April 3. We can see that "C0101" is strongly connected with "C0206" and "C0301", as indicated by the width of ribbons between these communities.

3.5.2. Corridors

Figure 5 shows five discovered corridors on weekdays on the map. The width of corridors represents the size of transit flows. The flow direction is shown by animated particles [43]. A dedicated summary glyph in the statistics view is also designed to present a compact summary of all corridors in a radial layout, in which each segment corresponds to a corridor (Figure 6). For each corridor, the number of boardings and alightings within a corridor are divided into four categories: only trip origins fall within the corridor (destinations are outside); only trip destinations are located within the corridors (origins are not); both origins and destinations are within the corridors; and both origins/destinations are outside the corridor. Four bars are used to represent these four types of trips. Bar heights are proportional to the trip counts. The overall performance of the corridors is also represented by a line of dots in the inner circle. The performance is computed as the ratio (percentage) of on-vehicle time versus overall travel time of a transit trip. Dots close to the circle center indicate low performance. Based on this corridor overview map, users can perform global pattern discovery task 2 and examine the distribution of primary transit corridors.

Users can select and observe details of a corridor (Figure 7). The layout of a corridor is simplified in a schematic form to retain only topological connections (similar to metro maps). The above-mentioned four types of trips are visualized for the selected corridor: each ribbon that connects two adjacent stops is divided into four components, and the width represents flow counts. Key stops within a corridor are depicted as rings, with red/green representing boarding/alighting counts. Passengers boarding at a stop are further categorized into two groups: those who will be alighting within the same corridor and those who will be alighting outside the corridor. The two groups are denoted by dark and light red, respectively. Similarly, passengers who alight at a stop are divided into two groups: "boarding from at least 5 stops away" and "boarding close to the current stop". Dark- and light-green colors are used to denote the two groups. When selecting a particular corridor and observing its details in the corridor detail view, users can conduct local pattern exploration task 1 to fetch boarding, alighting, origin, and destination information in a compact visualization form.

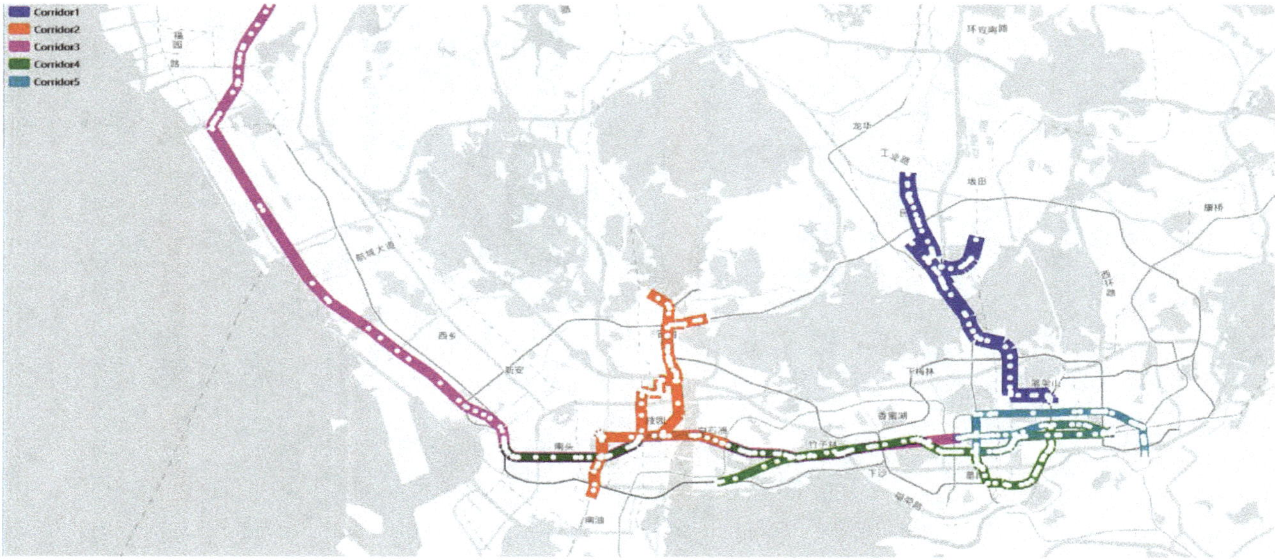

Figure 5. Public transit corridors for weekday peak hours (8:00AM–10:00AM).

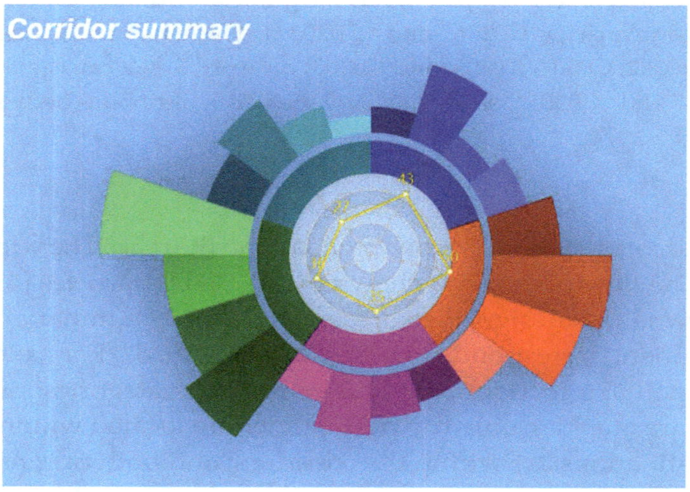

Figure 6. Corridor summary glyph.

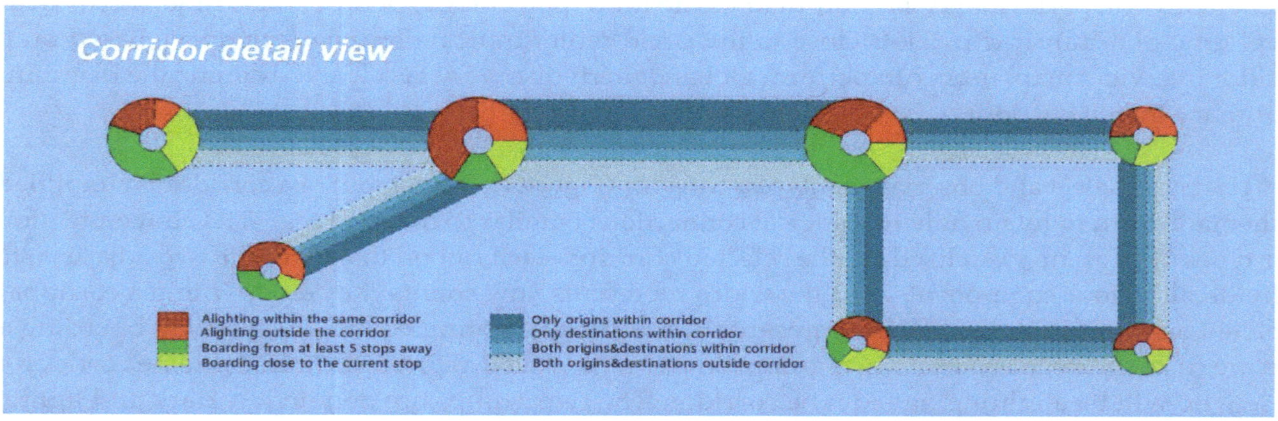

Figure 7. Corridor detail view.

3.5.3. Transit Stops

The trip information of individual transit stops is plotted in a glyph when users click on a stop on the map. Figure 8 shows that the stop glyph can visualize in-vehicle, boarding, and alighting passenger numbers. In-vehicle passengers can be further divided into boarding from distant and nearby stops. Boarding passengers consist of initial boarding and transferring passengers. Alighting passengers comprise those who finish their trips and those who transfer at this stop. One can easily tell the role played by this stop for the whole transit network: it could be an important origin stop, destination stop, or a transfer stop.

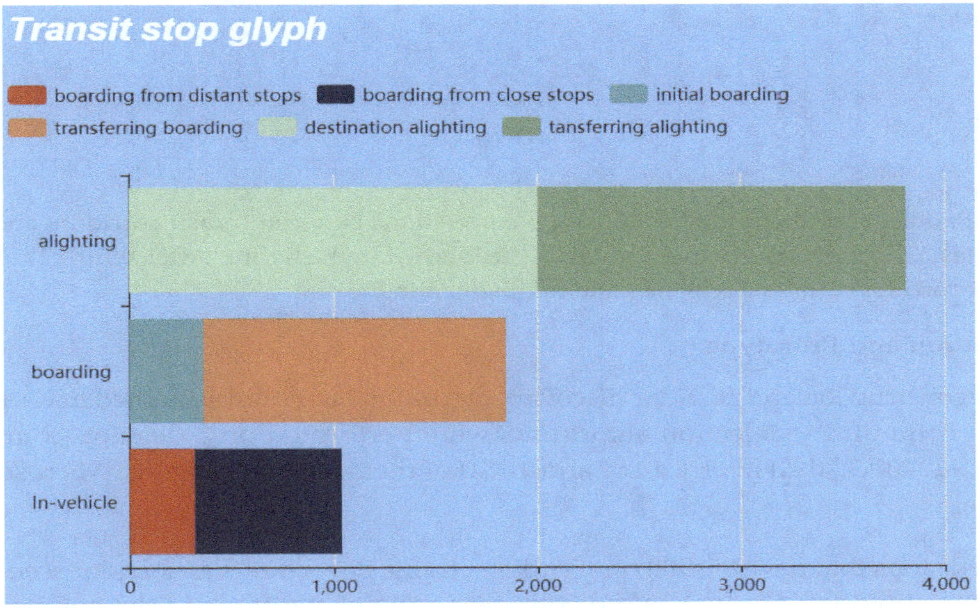

Figure 8. Transit stop glyph.

3.5.4. Correlations between Corridors and Communities

In the previous sections, we introduced our approaches to discover primary transit corridors with significant travel demands and mobility communities with similar travel patterns. Furthermore, geovisual analytics can be performed to examine correlations between these two identified time-varying mobility representations. An integrated parallel coordinate plot is designed to describe their correlations. For any pre-specified time interval, we can draw identified corridors as polylines and represent discovered communities as vertical parallel axes. For each corridor (i.e., polyline), the intersection point on an axis (i.e., a community) indicates the number of transit passengers who originate from the community towards the corridor. In this way, spatio-temporal correlations between each corridor and each community can be illustrated in a compact manner. We can easily find which community contributes the biggest portion of transit flow to a particular corridor or identify which corridor is the most correlated one for a specific community. We can also get to know the composition of any corridor or community. For example, the parallel coordinate plot can reveal whether most trips from a community are correlated to a few corridors or are evenly distributed over a number of corridors across the city. Note that these correlations are not equivalent to intersection relations between corridors and communities, which are explicit on maps. As long as the origins of constituent trips for a corridor can be traced to a community, the corridor and the community establishes a correlation. The number of these correlations represents the intensity of interactions between a corridor–community pair. For each vertical axis, a filtering box can be specified to find corridors that meet the transit flow number search range criterion. Multiple filtering boxes on different axes can be interactively designated to further identify corridors that are correlated to selected communities based on specified transit flow ranges (Figure 9). This geovisual analytics tool enables users to undertake the comprehensive analysis task to mine correlation knowledge between corridors and mobility communities.

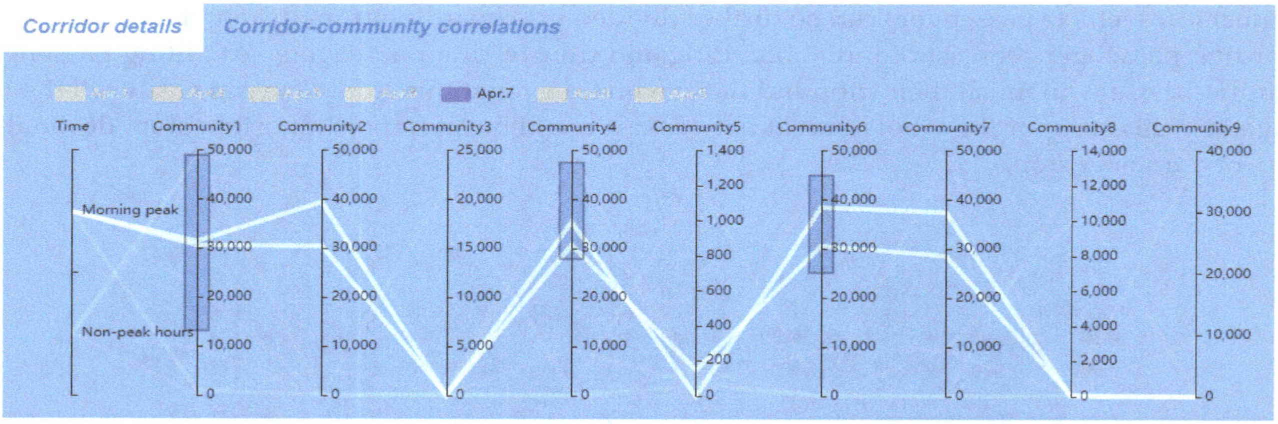

Figure 9. Parallel coordinate plot to illustrate correlations between transit corridors and mobility communities. Corridors discovered for April 7 are shown in the figure. After setting three filtering boxes, four corridors can be discovered and highlighted in the plot.

4. Implementation and Prototype

The trip reconstruction and corridor discovery algorithms were implemented in C++. The corridor extraction and community detection algorithms were performed on a desktop computer with an Intel™ Xeon E3-1240@3.70 GHz processor and 16GB of memory, running on a Microsoft Window 10 operating system.

We selected stops that have 85–90th percentile of traffic flow counts as corridor seeds. The "shared transit flow" threshold (*st*) was set as the 50th percentile of the transit flow counts. The "shared accumulated transit flow" threshold (*sa*) can be set between −15% and 25%.

Mobility community structures were extracted using a hierarchical clustering tool implemented in the SciPy package based on network embedding results produced by a structural deep network embedding (SDNE) method [27]. The graph-embedding algorithm was implemented using TensorFlow 1.14.0 by Python 3.6. The clustering was performed based on Euclidean distance. The autoencoder network contains three layers: the input layer contains 18,109 neurons, which correspond to 18,109 grid groups in the study region; the hidden layer has 2000 neurons; and the output layer produces a 128 dimensional vector as the final embedding result for each graph node. Deeper layers would lead to performance degradation, as demonstrated by our tests. The model parameter initialization was based on a Gaussian distribution (with mean $\mu = 1$ and standard deviation $\sigma = 0.01$). The weight in the joint loss function (Equation (5) was set as 0.2 since this delivered the best performance. The learning rate was set as 0.001. In order to produce cohesive and contiguous high-level communities, we applied a regionalization algorithm, REDCAP [44], based on the communities produced by SciPy.

We compared the performance of our community detection algorithm with a classical community detection algorithm developed by Newman and Leicht [45]. To evaluate the performance, we used the modularity metric proposed by Newman and Girvan [46].

As indicated by Table 1, our graph-embedding algorithm outperformed Newman and Leicht's algorithm by a large margin. Note that our case study is different from the regular community detection problem, in which higher modularity values indicate good community partitions. Since we encourage a community to have dense inter-community transit trips and sparse intra-community trips, lower modularity values are better.

Table 1. Performance comparison of community detection using the modularity metric (low-level communities).

	Weekdays	Weekends
[44]	0.0372	0.0707
Ours	0.0218	0.0229

The visual designs were implemented in a web-based prototype, which was developed with PyCharm Pro 2018.3.1 on a Windows 10 operating system. Major visualization modules were developed in JavaScript following the standards of HTML5 and CSS3. The user interfaces comprises four major components (Figure 10): (1) query view in the upper-left portion; (2) map view in the upper right; (3) statistics view in the bottom-left corner; and (4) focus views for corridors and communities in the bottom-right region. In the query view, users can specify the time range and select SCD falling within this range for analysis. The map view depicts discovered corridors and mobility community structures. Different corridors are differentiated by distinct colors. The map view embeds a Baidu Map as the background map. Flow maps can be produced to describe primary transit flows between major communities. Focus views illustrate three types of detailed displays: corridor detail view, community tracking view, and corridor–community correlation view. All of these views are dynamically linked. User interactions within any view will apply to other linked views for the same data (community, transit stop, or corridor).

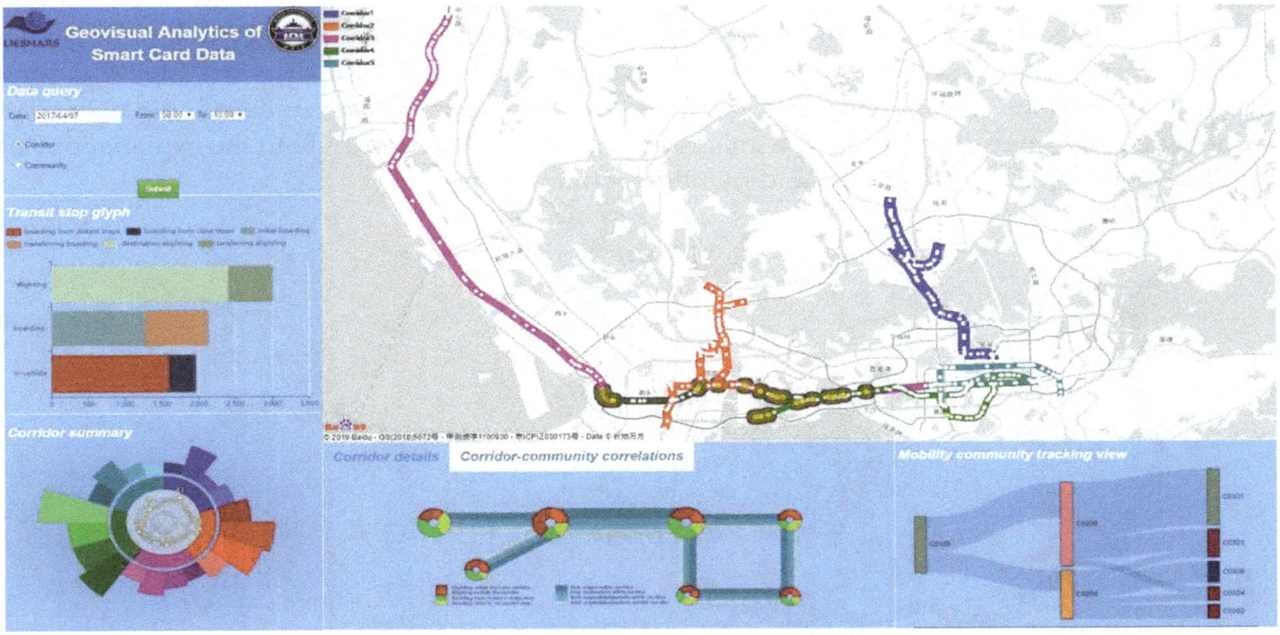

Figure 10. Web-based prototype user interface.

5. Analysis and Discussion

5.1. Geovisual Analytics Workflow and Examples

Typically, a user can first specify the time range of SCD for analysis. For example, she can focus on morning peak hours of a weekday and invoke back-end algorithms to extract mobility structure and primary transit corridors. Then, both corridors and community structures can be visualized in multiple linked views to enable further examination. The integration of flow map and corridor map with community map at two scales can help users understand the overall transit mobility structure of the city. At a glance, users can identify the major origin/destination areas and how many passengers travel between these areas. Meanwhile, the statistics view presents summary information on all corridors, which allows users to compare the extracted corridors in terms of their trip types and performance. Furthermore, users can select a corridor and visualize it in the detailed view to gain more information on its constituent trips. Users can also select any transit stop to see the decomposition of its boarding and alighting trips. The prototype also allows users to examine the evolution of any chosen mobility community in the detail view. With all these linked views, users can perform the comprehensive analysis task to discover global and local transit travel patterns across the city over time.

For example, users can discover high-level mobility communities for any date. Figure 3a presents these communities for a holiday. The identified community structure synthesizes transit travel patterns that are much easier to understand than original massive transit footprints. The largest community (No. 1) is located on the east side of the downtown area, which is served by multiple subway lines and dozens of subway stations. This community attracts a large amount of leisure-oriented trips that originate from all over the city. In the west side of downtown, three separate communities (Nos. 2, 3, and 4) can be observed, and each attracts short trips close to it. Other communities are distributed over the suburbs, which are mostly residential areas. Many passengers living in these suburb communities travel to downtown areas for leisure purposes on the holiday.

Figure 11 shows that it can beneficial to examine intrinsic travel patterns by integrating transit corridors with mobility communities. It can be observed that the most salient corridor connects community Nos. 8 and 9 and community No. 1, which have the most job opportunities in the city. Based on the flow direction of the corridor (shown by animated particles), we can find that a large number of commuters travel towards community No. 1 for work in the morning. Another corridor in the Northeast indicates that many passengers who live in remote outskirts make trips towards community No. 6, which features many industrial parks and high-technology companies. With these interactive maps, both corridor and community information can be combined to further investigate the travel origins and destinations for different times and dates, thereby deepening our understanding of evolving movement patterns across the city. These integrated maps can also contribute to the explanation of the interactions between transit corridors and mobility communities.

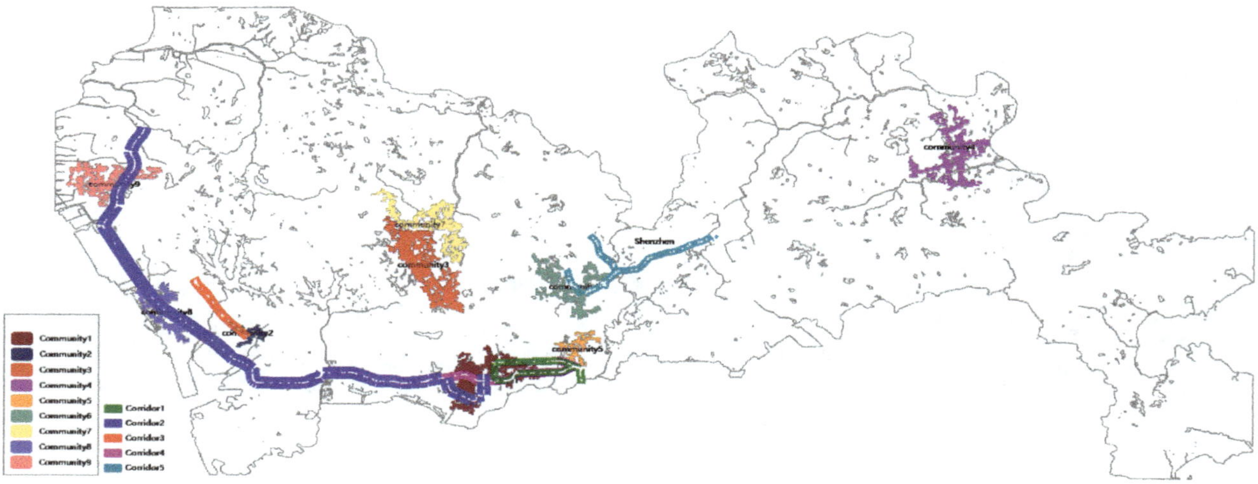

Figure 11. Discovered corridors and mobility communities for 11:00AM–1:00PM on weekdays.

5.2. User Evaluation

Twenty-three users were interviewed to obtain comments and feedback on our geovisual analytics approach based on their experience using the prototype. Sixteen of them are experts in public transportation, and among them, nine have geovisual analytics development knowledge (experienced users). The users can be classified into three groups: (1) experienced users with background knowledge (9 users); (2) non-experienced users with background knowledge (7 users); and (3) non-experienced users without background knowledge (7 users). Before allowing them to use the prototype, we introduced the proposed geovisual analytics approach and the web-based prototype. We asked users to evaluate 6 geovisual analytics tasks: (1) to discover and visualize transit corridors; (2) to extract and visualize mobility community structure; (3) to obtain transit stop summary statistics information; (4) to evaluate the corridor detailed view; (5) to evaluate the mobility community tracking view; (6) to evaluate the corridor–community correlation view. Numeric scores were obtained from questionnaires, with "0" indicating the worst user experience and "5" indicating the best. Figure 12 summarizes the scores of interviewed users.

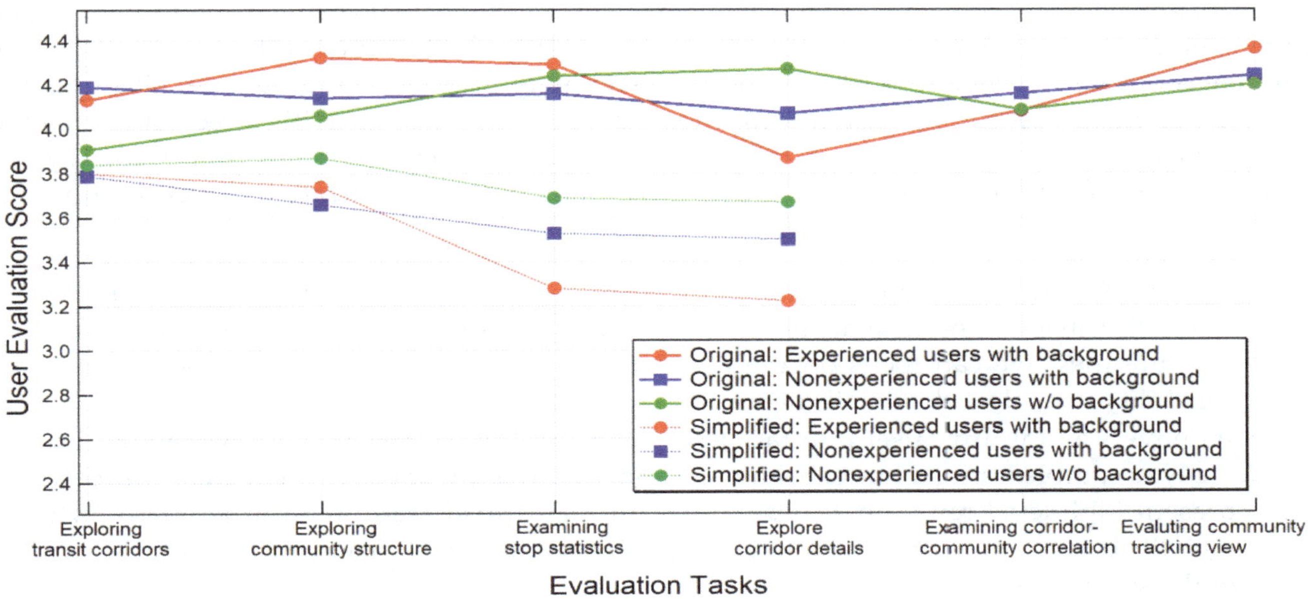

Figure 12. Average evaluation scores for 6 selected geovisual analytics tasks.

According to ratings and comments, different groups of users agreed that our integrated analytics approach and web-based prototype deliver an interesting and applicable solution for human mobility pattern discovery given massive transit data and complex transit networks. As shown in Figure 12, experienced users and users with background knowledge tended to give more positive ratings than non-experienced users or users without any background knowledge for most evaluation tasks. It may take more time for the third group users to understand the interfaces and functions of the system, thereby reducing their time to fully explore all of the views and leading to lower scores. Evaluation tasks (1) and (4) had relatively low ratings, probably due to the unfamiliar concept of transit corridors for some users. Users may have had difficulties selecting and examining particular corridors between different views, which was confirmed by subsequent feedback interviews. The corridor detail view is also not intuitive to use, according to the users' comments, since it requires users to switch their focus between the map view and the detail view frequently.

We also implemented a simplified web-based version for user evaluation. Compared to the original version, the simplified prototype only has an integrated map view to show discovered corridors and community structure. It does not implement linked views and only has limited visualizations (e.g., without animated particles to show the flow direction in corridors, without individual transit stop glyphs and corridor detail views).

With such a simplified system, we conducted user evaluation interviews with the same three groups of users. The same evaluation interview procedure was conducted to solicit their scores and feedback. Note that only four evaluation tasks were evaluated, namely, exploring transit corridors, exploring community structures, examining stop statistics, and exploring corridor details. The average evaluation scores for the simplified system are also shown in Figure 12. As we can see, these evaluation scores are significantly lower than scores obtained based on the original prototype for the evaluated four tasks.

5.3. Discussion

This study adopted the concepts of community structure and transit corridor to construct high-level aggregate mobility knowledge from massive SCD and other urban data. The results of community detection and corridor discovery algorithms were integrated into an interactive visualization interface that consists of multiple linked views to enable efficient visual analysis of spatio-temporal transit mobility patterns at multiple scales and resolutions. The map view offers an overview interface to help users preserve context information when they focus on a particular corridor, community, or stop visualizations. Specific views such as corridor detail view, mobility community tracking view, and a parallel coordinate correlations plot, along with summary glyphs (including a transit stop glyph and corridor summary glyph) complement the map view to provide intuitive geovisual analytics tools for the discovery of detailed knowledge of any specific component of the public transit system. The advantages of integrating machine learning with interactive visualization can be summarized as follows:

(1) It offers an efficient and effective method to explore a massive amount of transit trips, which is otherwise challenging to analyze and visualize. Based on discovered corridors and mobility communities, we can focus on the most significant travel patterns while still having the capability to explore the details of any stop.
(2) It delivers an intuitive user interface to combine multiple views that allows regular users to analyze complex transit travel behaviors from different perspectives. For example, corridors present high-level representations of concentrated trips based on road networks, whereas mobility communities are produced to synthesize similar travel characteristics over the partition of the study region.
(3) It is beneficial for many transit management applications, such as demand modeling, transit planning, and daily operations, since they provide an applicable approach to highlight aggregated movement patterns at multiple spatial and temporal resolutions. The prototype can also be used by regular passengers to plan their transit trips and choose their residence or work place.

For most city residents, transit travel follows a weekly rhythm: they commute to work on every weekday and enjoy their leisure time on weekends. One-week data could then be sufficient to extract typical transit movement patterns for the study area. In the literature, we can also find other researchers also using one-week public transit data (i.e., smart card data) for their studies. For example, Long and Thill [47] examined job–housing relationships in Beijing with one-week bus-based SCD. Alsger et al. [48] validated origin-destination estimation algorithms based on one-week SCD in Southeast Queensland, Australia. If we can access SCD and GPS trajectory data from other time periods, the same geovisual analytics approach can be readily applied.

6. Conclusions and Further Work

In this study, we applied two machine learning methods, including a clustering algorithm to extract transit corridors and a graph-embedding algorithm to discover mobility community structure. These high-level representations are visualized in a web-based interactive interface to allow users to examine massive SCD in a highly aggregated and efficient manner. Our prototype demonstrates that the proposed visual analytics approach can offer a scalable and effective solution for discovering meaningful travel patterns across a big metropolitan area. We plan to improve the usability of the prototype based on users' comments in the near future. It is favorable to allow users to designate algorithm configurations in the graphic user interface, which contributes to a better understanding of the underpinning machine learning algorithms, and this will be implemented in the near future.

Author Contributions: Conceptualization, T.Z.; Methodology, T.Z.; J.W.; Software, C.C.; Y.L. (Yicong Li), W.H.; Formal Analysis, C.C.; Y.L. (Yicong Li); Data Curation, C.C.; Q.Q.; Writing-Original Draft Preparation, T.Z.; Writing-Review & Editing, J.W., Q.Q.; Visualization, J.W.; C.C.; Project Administration, Y.L. (Yonghua Lu); Funding Acquisition, T.Z.; Y.L. (Yonghua Lu).

References

1. Pelletier, M.-P.; Trepanier, M.; Morency, C. Smart card data use in public transit: A literature review. *Transp. Res. Part C* **2011**, *19*, 557–568. [CrossRef]

2. Sun, L.; Axhausen, K. Understanding urban mobility patterns with a probabilistic tensor factorization framework. *Transp. Res. Part B* **2016**, *91*, 511–524. [CrossRef]
3. Zhao, J.; Qu, Q.; Zhang, F.; Xu, C.; Liu, S. Spatio-temporal analysis of passenger travel patterns in massive smart card data. *IEEE Trans. Intell. Transp. Syst.* **2017**, *18*, 3135–3146. [CrossRef]
4. El Mahrsi, M.; Come, E.; Oukhellou, L.; Verleysen, M. Clustering smart card data for urban mobility analysis. *IEEE Trans. Intell. Transp. Syst.* **2017**, *18*, 712–728. [CrossRef]
5. Kieu, L.; Ou, Y.; Cai, C. Large-scale transit market segmentation with spatial-behavioural features. *Transp. Res. Part C* **2018**, *90*, 97–113. [CrossRef]
6. Trépanier, M.; Habib, K.; Morency, C. Are transit users loyal? Revelations from a hazard model based on smart card data. *Can. J. Civ. Eng.* **2012**, *39*, 610–618. [CrossRef]
7. Li, T.; Sun, D.; Jing, P.; Yang, K. Smart card data mining of public transport destination: A literature review. *Information* **2018**, *9*, 18. [CrossRef]
8. Wang, Y.; Correia, G.; Romph, E.; Timmermans, H. Using metro smart card data to model location choice of after-work activities: An application to Shanghai. *J. Transp. Geogr.* **2017**, *63*, 40–47. [CrossRef]
9. Çolak, S.; Lima, A.; González, M.C. Understanding congested travel in urban areas. *Nat. Commun.* **2016**, *7*, 10793. [CrossRef]
10. Toole, J.L.; Herrera-Yaqüe, C.; Schneider, C.M.; González, M.C. Coupling human mobility and social ties. *J. R. Soc. Interface* **2015**, *12*, 20141128. [CrossRef]
11. Schneider, C.M.; Rudloff, C.; Bauer, D.; González, M.C. Daily travel behavior: Lessons from a week-long survey for the extraction of human mobility motifs related information. In Proceedings of the 2nd ACM SIGKDD International Workshop on Urban Computing, Chicago, IL, USA, 11 August 2013. Article No. 3.
12. Hedayatifar, L.; Bar-Yam, Y.; Morales, A.J. Social fragmentation at multiple scales. *arXiv* **2018**, arXiv:1809.07676.
13. Mazimpaka, J.; Timpf, S. Trajectory data mining: A review of methods and applications. *J. Spat. Inf. Sci.* **2016**, *13*, 61–99. [CrossRef]
14. Toch, E.; Lerner, B.; Ben-Zion, E.; Ben-Gal, I. Analyzing large-scale human mobility data: A survey of machine learning methods and applications. *Knowl. Infor. Syst.* **2019**, *58*, 501–523. [CrossRef]
15. Xie, R.; Ji, Y.; Yue, Y.; Zuo, X. Mining individual mobility patterns from mobile phone data. In Proceedings of the 2011 International Workshop on Trajectory Data Mining and Analysis, Beijing, China, 18 September 2011; pp. 37–44.
16. Khoroshevsky, F.; Lerner, B. Human mobility-pattern discovery and next-place prediction from GPS data. In *Multimodal Pattern Recognition of Social Signals in Human-Computer-Interaction*; Schwenker, F., Scherer, S., Eds.; Lecture Notes in Computer Science; Springer: Cham, Switzerland, 2017; Volume 10183, pp. 24–35.
17. Chen, X.; Shi, D.; Zhao, B.; Liu, F. Mining individual mobility patterns based on location history. In Proceedings of the IEEE First International Conference on Data Science in Cyberspace (DSC), Changsha, China, 13–16 June 2016.
18. Ouyang, X.; Zhang, C.; Zhou, P.; Jiang, H. DeepSpace: An online deep learning framework for mobile big data to understand human mobility patterns. *arXiv* **2016**, arXiv:1610.07009.
19. Kim, D.; Song, H. Method of predicting human mobility patterns using deep learning. *Neurocomputing* **2018**, *280*, 56–64. [CrossRef]
20. Wang, C.; Ma, L.; Li, R.; Durrani, T.; Zhang, H. Exploring trajectory prediction through machine learning methods. *IEEE Access* **2019**, *7*, 101441–101452. [CrossRef]
21. Chen, R.; Chen, M.; Li, W.; Wang, J.; Yao, X. Mobility modes awareness from trajectories based on clustering and a convolutional neural network. *ISPRS Int. J. Geo-Inf.* **2019**, *8*, 208. [CrossRef]
22. Ben-Zion, E.; Lerner, B. Identifying and predicting social lifestyles in people's trajectories by neural networks. *EPJ Data Sci.* **2018**, *7*, 45. [CrossRef]
23. Gonzalez, M.C. *Transportation Model in the Boston Metropolitan Area from Origin Destination Matrices Generated with Big Data*; New England University Transportation Center Year 24 Final Report (MITR24-5); Massachusetts Institute of Technology: Cambridge, MA, USA, 2016.
24. Di Lorenzo, G.; Sbodio, M.; Calabrese, F.; Berlingerio, M.; Pinelli, F.; Nair, R. AllAbroad: Visual exploration of cellphone mobility data to optimise public transport. *IEEE Trans. Vis. Comput. Graph.* **2016**, *22*, 1036–1050. [CrossRef]
25. Zhou, Z.; Yu, J.; Guo, Z.; Liu, Y. Visual exploration of urban functions via spatio-temporal taxi OD data. *J. Vis. Lang. Comput.* **2018**, *48*, 169–177. [CrossRef]

26. Kim, S.; Jeong, S.; Woo, I.; Jang, Y.; Maciejewski, R.; Ebert, D. Data flow analysis and visualization for spatiotemporal statistical data without trajectory information. *IEEE Trans. Vis. Comput. Graph.* **2017**, *24*, 1287–1300. [CrossRef] [PubMed]
27. Tao, S.; Rohde, D.; Corcoran, J. Examining the spatial-temporal dynamics of bus passenger travel behaviour using smart card data and the flow-comap. *J. Transp. Geogr.* **2014**, *41*, 21–36. [CrossRef]
28. Zeng, W.; Fu, C.-W.; Arisona, S.; Schubiger, S.; Burkhard, R.; Ma, K.-L. Visualizing the relationship between human mobility and points of interest. *IEEE Trans. Intell. Transp. Syst.* **2017**, *18*, 2271–2284. [CrossRef]
29. Von Landesberger, T.; Brodkorb, F.; Roskosch, P.; Andrienko, N.; Andrienko, G.; Kerren, A. MobilityGraphs: Visual analysis of mass mobility dynamics via spatio-temporal graphs and clustering. *IEEE Trans. Vis. Comput. Graph.* **2016**, *22*, 11–20. [CrossRef] [PubMed]
30. Sun, Y.; Shi, J.; Schonfeld, P. Identifying passenger flow characteristics and evaluating travel time reliability by visualizing AFC data: A case study of Shanghai Metro. *Public Transp.* **2016**, *8*, 341–363. [CrossRef]
31. Palomo, C.; Guo, Z.; Silva, C.; Freire, J. Visually exploring transportation schedules. *IEEE Trans. Vis. Comput. Graph.* **2016**, *22*, 170–179. [CrossRef] [PubMed]
32. Zeng, W.; Fu, C.-W.; Arisona, S.; Erath, A.; Qu, H. Visualizing mobility of public transportation system. *IEEE Trans. Vis. Comput. Graph.* **2014**, *20*, 1833–1842. [CrossRef]
33. Song, Y.; Fan, Y.; Li, X.; Ji, Y. Multidimensional visualization of transit smartcard data using space-time plots and data cubes. *Transportation* **2018**, *45*, 311–333. [CrossRef]
34. Zhang, T.; Li, Y.; Yang, H.; Cui, C.; Li, J.; Qiao, Q. Identifying primary public transit corridors using multi-source big transit data. *Int. J. Geogr. Inf. Sci.* **2019**, 1–25. [CrossRef]
35. Primerano, F.; Taylor, M.; Pitaksringkarn, L.; Tisato, P. Defining and understanding trip chaining behaviour. *Transportation* **2008**, *35*, 55–72. [CrossRef]
36. Nassir, N.; Hickman, M.; Ma, Z. Activity detection and transfer identification for public transfer fare card data. *Transportation* **2015**, *42*, 683–705. [CrossRef]
37. Carr, J.; Dixon, C.; Meyer, M. *Guidebook for Corridor-Based Statewide Transportation Planning*; Transportation Research Board: Washington, DC, USA, 2010.
38. Yildirimoglu, M.; Kim, J. Identification of communities in urban mobility networks using multi-layer graphs of network traffic. *Transp. Res. Part C* **2018**, *89*, 254–267. [CrossRef]
39. Newman, M. Communities, modules, and large-scale structure in networks. *Nat. Phys.* **2011**, *8*, 25–31. [CrossRef]
40. Fortunato, S.; Hric, D. Community detection in networks: A user guide. *Phys. Rep.* **2016**, *659*, 1–44. [CrossRef]
41. Perozzi, B.; Al-Rfou, R.; Skiena, S. Deepwalk: Online learning of social representations. In Proceedings of the 20th ACM SIGKDD International Conference on Knowledge Discovery and Data Mining, New York, NY, USA, 24–27 August 2014; pp. 701–710.
42. Wang, D.; Cui, P.; Zhu, W. Structural deep network embedding. In Proceedings of the KDD '16, 22nd ACM SIGKDD International Conference on Knowledge Discovery and Data Mining, San Francisco, CA, USA, 13–17 August 2016.
43. Scheepens, R.; Hurter, C.; van de Wetering, H.; van Wijk, J. Visualization, selection, and analysis of traffic flows. *IEEE Trans. Vis. Comput. Graph.* **2016**, *22*, 379–388. [CrossRef] [PubMed]
44. Guo, D. Regionalization with dynamically constrained agglomerative clustering and partitioning (REDCAP). *Int. J. Geogr. Inf. Sci.* **2008**, *22*, 801–823. [CrossRef]
45. Newman, E.; Leicht, E. Mixture models and exploratory analysis in networks. *Proc. Natl. Acad. Sci. USA* **2007**, *104*, 9564–9569. [CrossRef] [PubMed]
46. Newman, E.; Girvan, M. Finding and evaluating community structure in networks. *Phys. Rev. E* **2004**, *69*, 026113. [CrossRef]
47. Long, Y.; Thill, J. Combining smart card data and household travel survey to analyze jobs–housing relationships in Beijing. *Comput. Environ. Urban Syst.* **2015**, *53*, 19–35. [CrossRef]
48. Alsger, A.; Assemi, B.; Mesbah, M.; Ferreira, L. Validating and improving public transport origin-destination estimation algorithm using smart card fare data. *Transp. Res. Part C* **2016**, *68*, 490–506. [CrossRef]

11

High-Performance Overlay Analysis of Massive Geographic Polygons that Considers Shape Complexity in a Cloud Environment

Kang Zhao [1], Baoxuan Jin [2,*], Hong Fan [1], Weiwei Song [3], Sunyu Zhou [3] and Yuanyi Jiang [3]

1 State Key Laboratory for Information Engineering in Surveying, Mapping, and Remote Sensing, Wuhan University, 129 Luoyu Road, Wuhan 430079, China
2 Information Center, Department of Land and Resources of Yunnan Province, Kunming 650504, China
3 Department of Geoinformation Science, Kunming University of Science and Technology, Kunming 650504, China
* Correspondence: jbx@yngc.org

Abstract: Overlay analysis is a common task in geographic computing that is widely used in geographic information systems, computer graphics, and computer science. With the breakthroughs in Earth observation technologies, particularly the emergence of high-resolution satellite remote-sensing technology, geographic data have demonstrated explosive growth. The overlay analysis of massive and complex geographic data has become a computationally intensive task. Distributed parallel processing in a cloud environment provides an efficient solution to this problem. The cloud computing paradigm represented by Spark has become the standard for massive data processing in the industry and academia due to its large-scale and low-latency characteristics. The cloud computing paradigm has attracted further attention for the purpose of solving the overlay analysis of massive data. These studies mainly focus on how to implement parallel overlay analysis in a cloud computing paradigm but pay less attention to the impact of spatial data graphics complexity on parallel computing efficiency, especially the data skew caused by the difference in the graphic complexity. Geographic polygons often have complex graphical structures, such as many vertices, composite structures including holes and islands. When the Spark paradigm is used to solve the overlay analysis of massive geographic polygons, its calculation efficiency is closely related to factors such as data organization and algorithm design. Considering the influence of the shape complexity of polygons on the performance of overlay analysis, we design and implement a parallel processing algorithm based on the Spark paradigm in this paper. Based on the analysis of the shape complexity of polygons, the overlay analysis speed is improved via reasonable data partition, distributed spatial index, a minimum boundary rectangular filter and other optimization processes, and the high speed and parallel efficiency are maintained.

Keywords: overlay analysis; shape complexity; massive data; cloud; parallel computing

1. Introduction

Overlay analysis is a common geographic computing operation and an important spatial analysis function of geographic information systems (GIS). It is widely used in applications related to spatial computing [1,2]. This operation involves the spatial overlay analysis of different data layers and their attributes in the target area. It connects multiple spatial objects from multiple data sets, creates a new clip data set, and quantitatively analyzes the spatial range and characteristics of the interactions among different types of spatial objects [3]. The development of geospatial science has entered a new stage with the rapid popularization of the global Internet, sensor technologies, and Earth observation technologies. The transformation of a space information service from digital Earth to intelligent Earth has posed challenges, such as being data-intensive, computationally intensive, and time–space-intensive and high concurrent access [4,5]. Overlay analysis deals with massive data, for which traditional data processing algorithms and models are no longer suitable. For example, the number of land use classification patches in Yunnan Province investigated in this study is hundreds of thousands at the county level, millions at the city level, and tens of millions at the provincial level. With the development of the social economy and the progress of data acquisition technologies, the number of land use classification patches will continue to increase. Effectively calculating land use change using traditional single-computer calculation models is difficult.

The rise of parallel computing technologies, such as network clustering, grid computing, and distributed processing in recent years has gradually shifted research on high-performance GIS spatial computing from the optimization of algorithms to the parallel transformation and parallel strategy design of GIS spatial computing in a cloud computing environment [6]. Recently, MapReduce and Spark technology have been applied to overlay analysis of massive spatial data, and some results have been achieved. Nevertheless, the massive spatial data is different from the general massive Internet data. The spatial characteristics of spatial data and the complexity of a spatial analysis algorithm determine that simply copying a cloud computing programming paradigm cannot achieve high-performance geographic computing. Therefore, this study chooses the classical Hormann clipping algorithm [7] to analyze and measure the impact of the shape complexity of geographic polygons on parallel overlay analysis, and proposes a Hilbert partition method based on the shape complexity measure to solve the data skew caused by the difference of the shape complexity of polygons. In addition, through the combination of MBR (Minimum Bounding Rectangle) filtering, R-tree spatial index and other optimizations, an efficient parallel overlay analysis algorithm is designed. The experimental analysis shows that the proposed method reduces the number of polygon intersection operations, achieves better load balancing of computing tasks, and greatly improves the parallel efficiency of overlay analysis. When the computational core increases, the algorithm achieves an upward acceleration ratio, and the computational performance presents a nonlinear change.

The rest of this paper is organized as follows. Section 2 reviews the research background and related studies, including those on shape complexity and overlay analysis algorithms. In Section 3, the Hormann algorithm is improved for a parallel polygon clipping process, and the process of a parallel polygon clipping algorithm is optimized according to the shape complexity of polygons. Section 4 describes the experimental process in detail and analyzes the experimental results. Section 5 provides the conclusion drawn from this research, followed by potential future work.

2. Relevant Work

This paper discusses the related research work from two aspects: shape complexity and overlay analysis algorithm.

2.1. Shape Complexity

Many studies use abstract language to describe the shapes and complex details of geometric objects, such as "the structure of polygons with multiple holes, the number of vertices is very large, and polygons with multiple concaves." To evaluate the computational cost, the complexity and computational efficiency of geometric computing problems should be accurately measured [8]. The concept of complexity is also introduced [9–12]. Many applications related to spatial computing heavily depend on algorithms to solve geometric problems. When dealing with large-scale geographic computing problems, the evaluation of computational cost must consider the quantity of input data, the complexity of graphical objects, and the time complexity of computing models [10]. When the amount of input data and the algorithm are determined, the complexity of different graphical objects frequently leads to considerable differences in the computing efficiency.

Mandelbrot described the complexity of geometric objects from the perspective of a fractal dimension [13]. The most commonly used method is the box-counting technique [14,15]. Brinkhoff quantitatively reported the complexity of polygons from three aspects, namely, the frequency of local vibration, the amplitude of local vibration, and the deviation from the convex hull, to describe the complexity of a global shape [16]. On the basis of Brinkhoff's research, Bryson proposed a conceptual framework to discuss the query processing-oriented shape complexity measures for spatial objects [17]. Rossignac [8] analyzed shape complexity from the aspects of algebraic, topological, morphological, combinatorial, and expression complexities. Rossignac also reduced the shape complexity by using a triangular boundary representation at different scales [8]. Ying optimized graphic data transmission on the basis of shape complexity [18].

From the above discussion, we know that the complexity of graphics has different meanings and measurement methods in different professional fields, such as design complexity, visual complexity and so on. Therefore, we should consider the shape complexity from the perspective of geographic computing. Shape complexity directly affects the efficiency of spatial analysis and spatial query computation, such as the numbers of vertices and local shapes (such as the concavity) of graphics, considerably influencing the efficiency of spatial geometry calculation. These values are important indicators for evaluating the calculation cost. Fully considering the influence of graphical complexity on specific geographic computations can effectively optimize the computing efficiency of applications.

2.2. Overlay Analysis

The study on vector overlay analysis arithmetic originates from the field of computer graphics. For example, two groups of thousands of overlay polygons are often clipped in 2D and 3D graphics rendering. Subsequently, different overlay analysis algorithms have been produced. Among which, the Sutherland–Hodgman [19], Vatti [20], and Greiner–Hormann [7] algorithms are the most representative when dealing with arbitrary polygon clippings. The Sutherland–Hodgman algorithm is unsuitable for complex polygons. The Weiler–Atherton algorithm requires candidate polygons to be arranged clockwise and with no self-intersecting polygons. The Vatti algorithm does not restrict the types of clipping, and thus, self-intersecting and porous polygons can also be processed. The Hormann algorithm clips polygons by judging the entrance and exit of directional lines. This algorithm also addresses point degradation by moving small distances [21]. In addition, the Hormann algorithm can deal with self-intersecting and nonconvex polygons. The Weiler algorithm uses tree data structures, whereas the Vatti and Greiner–Hormann algorithms adopt a bilinear linked list data structure. Therefore, the Vatti and Greiner–Hormann algorithms are better than the Weiler algorithm in terms of complexity and running speed.

Subsequent researchers have implemented many improvements to the aforementioned traditional vector clipping algorithms [22–24], which simplify the calculation of vector polygon clippings. However, these studies are based on the optimization of a serial algorithm. When overlay analysis is applied to the field of geographic computing, it will deal with more complex polygons (such as polygons with holes and islands) and a larger data volume (the number of land use classification patches in a province is tens or even hundreds of millions, and a polygon may have tens of thousands of vertices). A vector clipping algorithm can be applied efficiently to computer graphics but cannot be applied efficiently to geographic computing. Moreover, many traditional geographic element clipping algorithms also exhibit poor suitability and performance degradation. With the development of computer technology and the increase in spatial data volume, traditional vector clipping algorithms frequently encounter efficiency bottlenecks when dealing with large and complex geographic data sets. Therefore, improving the overlay algorithm and using the parallel computing platform for the overlay analysis of massive data is a new research direction.

With the rapid development of MapReduce and Spark cloud computing technologies, the use of large-scale distributed storage and parallel computing technology for massive data processing and analysis has become an effective technical approach [6,25]. Recent studies have applied the MapReduce and Spark technology to the overlay analysis of massive spatial data. Wang [26] used MapReduce to improve the efficiency of overlay analysis by about 10 times by a grid partition and index. Zheng [27] built a multilevel grid index structure by combining the first-level grid with quartering based on Spark distributed computation platform. Zheng's experiments show that a grid index algorithm achieves good results when polygons are uniformly distributed; otherwise, the efficiency of the algorithm is low. Xiao [28] proves that parallel task partitioning based on polygons' spatial location achieves better load balancing than random task partitioning. In addition, SpatialHadoop [29–32] and GeoSpark [32–35] extend Hadoop and Spark to support massive spatial data computing better. Among them, Spatial Hadoop designed a set of spatial object storage model, which provides HDFS with grid, R-tree, Hilbert curve, Z curve and other indexes. In addition, it also provides a filtering function for filtering data that need not be processed. GeoSpark also adds a set of spatial object models and extends RDD (Resilient Distributed Dataset) to SRDD (Spatial Resilient Distributed Dataset) that supports spatial object storage. GeoSpark also provides filtering functions to filter data that need not be processed. The design ideas of SpatialHadoop and GeoSpark have great reference value for the research of this paper.

In summary, using a Spark cloud computing paradigm to develop high-performance geographic computing is a cheap and high-performance method. It is also one of the research hotspots in the field of high-performance geographic computing. Recent studies have implemented overlay analysis in Spark, which greatly improves the efficiency of overlay analysis. Optimizing strategies for spatial data characteristics, such as reasonable data partitioning and an excellent spatial data index, play an important role in improving the efficiency of parallel computing, and Hilbert partitioning is more suitable for parallel overlay analysis of non-uniform spatial distribution data than grid partitioning. It is noticed that the current parallel overlay analysis is based on the third-party clipping interface and ignores the impact of the shape complexity of geographic polygons on the clipping algorithm, which will cause a serious data skew.

3. Methodology

In this section, the core idea of the paper will be introduced. First, an excellent basic overlay analysis algorithm is selected for execution on each computing node. Then, according to the complexity and location of graphics, the polygons are divided reasonably, and an index based on spatial location is established to realize fast data access and load balancing of parallel computing nodes.

3.1. Basic Overlay Analysis Algorithm Running on Each Computing Node.

3.1.1. Hormann Algorithm and Improvement of Intersection Degeneration Problem.

The basic overlay analysis algorithm, which is the basic processing program for each parallel computing node, performs overlay analysis on two sets of polygons. In designing an overlay analysis algorithm, we use the Hormann algorithm, which can deal with complex structures (e.g., self-intersection and polymorphism with holes), as reference. However, the point coordinate perturbation method used by the Hormann algorithm is not the best solution for the degradation problem, which brings cumulative errors in the area statistics of massive patches. Therefore, we solve the intersection degeneration by judging the azimuth interval between intersecting correlation lines. We also use the improved algorithm as the basis of parallel overlay analysis.

In order to achieve a simplified expression of the overlay analysis of two groups of polygons, we assume that each group of polygons has only one polygon object, because the overlay analysis of two groups of layers with multiple polygons only increases the number of iterations. The processing steps of the improved Hormann algorithm are as follows:

1. Calculating the intersections of the clipped and target polygons
2. Judging the entry and exit of the intersection point by the vector line segment (judging the entry or exit point of the intersection point) and adding the entry point to the vertex sequence of the clipping result polygon

3. Comparing the azimuth intervals of the degenerated vertices of the intersection points and adding the overlapping vertices of the azimuth intervals to the vertex sequence of the clipping result polygon
4. Forming a new polygon (clipping result) in accordance with the sequence of vertices

As shown in Figure 1a, the clipped polygon P_1 and the target polygon P_2 intersect. Intersection points K_1 and K_2 can be obtained through a collinearity equation. By judging the positive and negative values of the product of vector line segments, the intersection points can be judged to enter and exit. As illustrated in the same figure, $\overrightarrow{A_1A_2} \times \overrightarrow{B_1B_2} > 0$, and thus, K_1 is the entry point relative to P_2. Moreover, $\overrightarrow{A_2A_3} \times \overrightarrow{B_1B_2} < 0$, and thus, K_2 is the exit point. The resulting polygon is composed of a sequence of vertices that consist of K_1, A_2, and K_2. As illustrated in Figure 1b–d, the entry and exit points are unsuitable for describing intersection degradation.

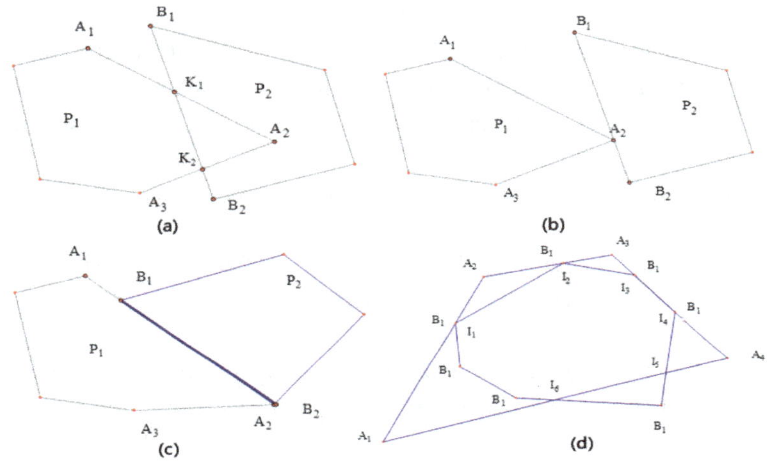

Figure 1. Polygon overlay.

Figure 2 shows how to deal with the phenomenon of intersection degeneration. In Figure 2, the dotted arrow N points toward the north, which is the starting point of the azimuth calculation. Therefore, each line segment has its own azimuth. The clipped and target polygons have an intersection point K, which is also the location of vertices A_m and B_m. The azimuth intervals of the clipped and target polygons at intersection K are $A_C(\alpha_1, \alpha_2)$ and $A_T(\alpha_1, \alpha_2)$. If $A_C(\alpha_1, \alpha_2)$ and $A_T(\alpha_1, \alpha_2)$ have overlapping parts (yellow in the figure), then both polygons overlap near the intersection point, which should be added to the vertex sequence of the resultant polygon.

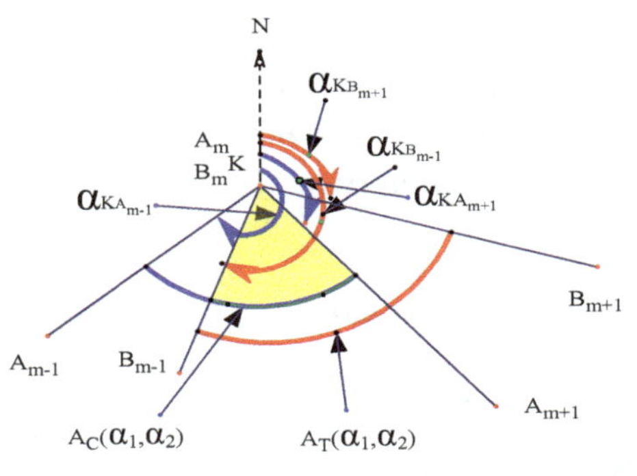

Figure 2. Diagram of azimuth interval calculation.

3.1.2. Effect of Shape Complexity on Parallel Clipping Efficiency

In parallel clipping computing, each computing node is usually assigned the same number of polygons. Generally speaking, it is difficult to ensure that complex polygons are evenly allocated to each computing node; usually, one computing node is allocated more complex polygons. Although the total number of polygons allocated by each computing node is the same, this computing node needs a long time to complete the allocated computing task, while other computing nodes will be in a waiting state. Therefore, ignoring the complexity differences of polygons will result in a situation in which each computing node cannot complete the computing task at the same time, thus the efficiency of parallel computing is reduced.

Complexity is an intuitive linguistic concept. Generally speaking, different professional fields pay different attention to shape complexity. In the field of geographic computation, shape complexity is related to specific geographic algorithms. The same shape corresponds to different geographic algorithms and may have different shape complexities.

On the other side, specifically to geographic polygons, different polygons have different morphological characteristics, such as convex, concave, self-intersection and a large number of vertices. To measure complexity, information must be compressed into one or more comparable parameter and expression models. Although the starting point and location are completely different, similar shapes may still appear. Therefore, when discussing the shape complexity of a polygon, we can neglect the spatial position and scale, and focus on the influence of polygon features on geographic computing.

Shape complexity can be defined from the perspective of geographic computing:

Definition 1. *Shape complexity is a measure of the computational intensity index of shapes participating in the calculation of geographic algorithms. Shape complexity can be measured by the number of repetitions of basic operations in a geographic algorithm caused by a shape.*

As for the overlay analysis of polygons, the most basic operation of the Hormann algorithm is to find the intersection point of two sides. Therefore, for the Hormann algorithm, the complexity of a polygon is the number of edges it possesses.

Based on these analyses, we know that shape complexity is an absolute value, which is difficult to program. Therefore, it is necessary to get a relative value by normalization to measure the graphics complexity.

Definition 2. *Given a set of polygons* $P = \{P_1, P_1, \cdots P_{n,}\}$, *the number of vertices of the polygon is* V_i, V_{min} *is the minimum number of vertices of all polygons, and* V_{max} *is the maximum number of vertices of all polygons. Then, the complexity* W_i *of the polygon* P_i *can be expressed as*

$$W_i = \frac{V_i - V_{min}}{V_{max} - V_{min}} \tag{1}$$

Since a polygon is usually represented by a sequence of vertices in a polygon storage model. The number of edges of a polygon is the same as that of vertices, so in Definition 2, we use vertices of a polygon instead of edges.

Therefore, in parallel overlay analysis, we can take shape complexity as an indicator for data partitioning. The ideal state is that the polygon complexity of each data partition is the same, at which time all computing nodes will complete the computing task at the same time.

3.2. Data Balancing and Partitioning Method that Considers Polygon Shape Complexity

3.2.1. Data Partitioning and Loading Strategy

Data partitioning is the key to accelerating a polygon clipping algorithm based on a high-performance computing platform. A complete piece of data is divided into relatively small, independent multiblock data, which provide a basis for distributed or parallel data operation. Spatial data partitioning differs from general data partitioning. In addition to balancing the amount of data, the spatial location relationship, such as spatial aggregation and proximity of data, should also be considered. Commonly used spatial data partitioning methods are meshing and filling curve partitioning [36]. Meshing is simple and considers the spatial proximity of data, but it cannot guarantee a balanced amount of data. The Hilbert curve is a classical spatial filling curve with good spatial clustering characteristics and that considers the spatial relationship and data load. Therefore, the data partitioning strategy in this study adopts the Hilbert filling curve algorithm combined with shape complexity to achieve load balancing.

In Figure 3, Hilbert partitioning divides the spatial region into 2N × 2N grids. During the iteration process, N is the order of the Hilbert curve, i.e., the number of iterations. In general, N is determined by the number of spatial objects, and the amount of spatial data requires $n < 2^{2\times N}$.

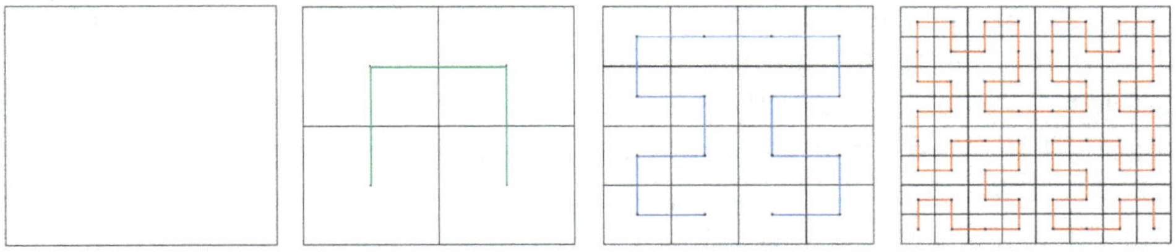

Figure 3. Hilbert partitioning and Hilbert curve generation.

Hilbert partitioning consists of the following four steps:

(1) Determine the order of the Hilbert curve, generate the Hilbert grid and the Hilbert curve, number the Hilbert curve sequentially, and obtain the Hilbert grid coding set, GHid $= \{GH_1, GH_2 \cdots GH_n\}$.
(2) Calculate the polygon MBR center point, find its corresponding mesh, and use the Hilbert coding of the mesh as the Hilbert coding of the polygon to obtain the Hilbert coding set of the polygon, PHid $= \{PH_1, PH_2 \cdots PH_n\}$.
(3) In accordance with the number of computing nodes M, divide the Hilbert coding set of the polygons into M partitions, and calculate the start–stop coding of the Hilbert coding of polygons in each partition.
(4) Merge the grids of the Hilbert partitions to obtain partition polygons PS $= \{PS_1, PS_2, \cdots, PS_M\}$.

In actual partitioning, the shapes of polygons are different because the polygons are not in an ideal uniform distribution. If only one polygon central point is strictly required for each grid, then Hilbert's order N may be extremely large, the edge length of the grid will be too small, and no polygonal MBR center may exist in many grids. Thus, Hilbert partitioning and the Hilbert curve will consume considerable computing time, and the subsequent overlay calculation will involve many cross-partition problems. Therefore, the existence of multiple polygonal MBR centers in a grid is necessary.

The order N of Hilbert grids is related to the length of the mesh edge. Grid length and order N are also determined. To obtain a reasonable order N of the Hilbert curve, we can calculate the normal distribution of the central point position of a polygon MBR, determine the optimal grid edge length, and eventually achieve balance between the order N of the Hilbert curve and the number of polygon MBR central points in each grid. The key to dividing the PHid of the Hilbert coding set of polygons is to ensure the load balance of each partition. Considering that polygon complexity may vary considerably, we cannot simply divide the Hilbert coding set PHid of polygons equally.

The shape complexity of polygon P_i is defined as W_i., $\overline{W}$ as the average complexity of all polygons, the ideal complexity of each partition as W_{ideal}, and the actual complexity as W_{actual}, then

$$W_{ideal} = \frac{\sum_{i=1}^{n} W_i}{M} \tag{2}$$

if the polygons from j to k are placed into the same partition, then,

$$W_{actual} = \sum_{i=j}^{k} W_i \tag{3}$$

$$|W_{ideal} - W_{actual}| < \overline{W} \quad (4)$$

Generally, the number of polygons in each partition is slightly different after partitioning, but the complexity of polygons in each partition is basically the same. Therefore, this strategy guarantees the load balancing of computing tasks.

3.2.2. R-tree Index Construction

R-tree is a widely adopted spatial data index method; it is used in commercial software, such as the Oracle and the SQL Server [37]. To improve the efficiency of spatial data access, an R-tree must be built. In addition, data are segmented in accordance with Hilbert data partitioning points, and the grid area of the Hilbert curve before each partitioning point is defined as a sub-index area. Moreover, the R-tree index for spatial objects is established in the sub-index area. Similarly, the mapping relationship among grid coding, polygon MBR central point coding, and sub-index area coding is established. Furthermore, the corresponding index codes on computing nodes are cached. In this experiment, we directly use the STR-tree (Sort Tile Recursive) class of the JTS (Java Topology Suite, a java software library) library to construct an R-tree index.

3.3. Process Design of Distributed Parallel Overlay Analysis

To ensure that the process is suitable for decoupling, we divide the distributed parallel overlay analysis process into six steps based on the characteristics of the algorithm: Data preprocessing, preliminary filtering, Hilbert coding, data partitioning and index building, data filtering, and overlay calculation (Figure 4).

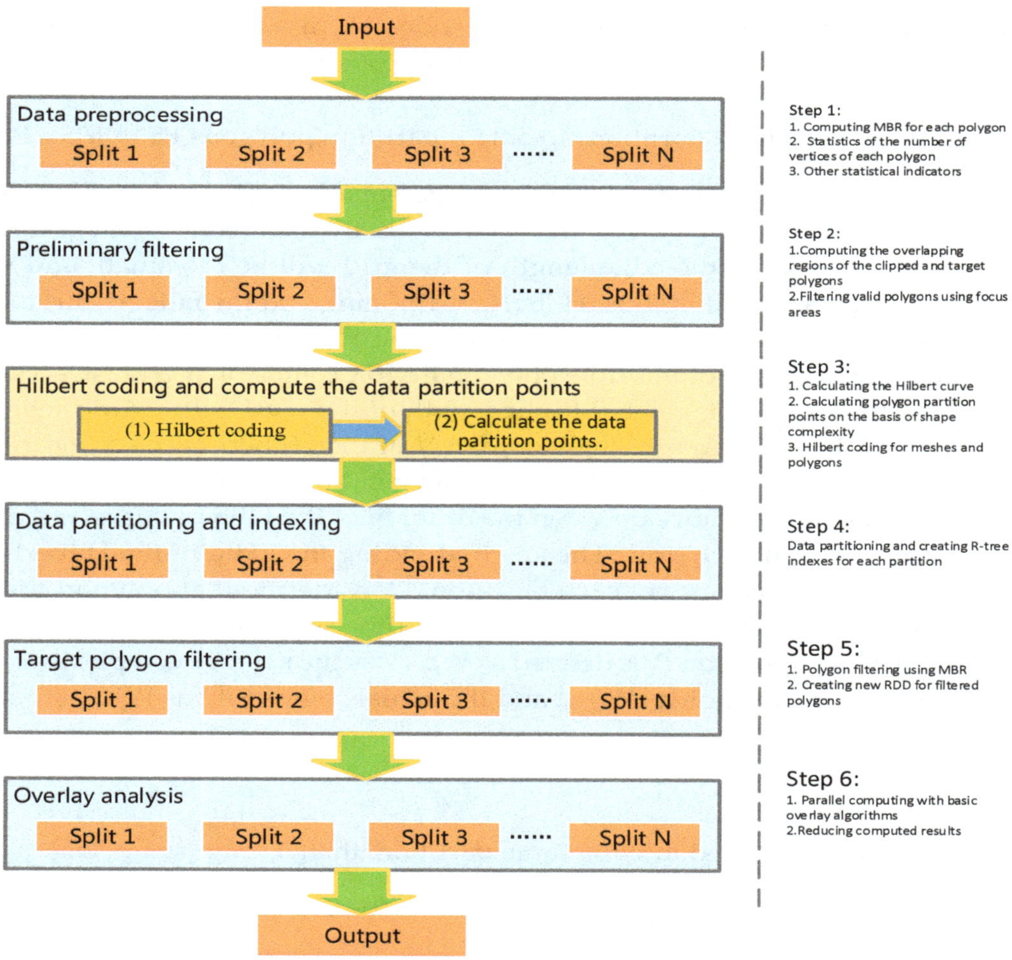

Figure 4. Parallel overlay computing flow.

(1) Data preprocessing

In the whole process, many steps need to traverse all polygons and their vertices. To reduce the number of traversals, we can conduct centralized processing in one traversal, such as calculating the MBR of a polygon, its geometric center point, the area of the MBR, and the shape complexity of the polygon, to prepare data for optimizing the processing flow. In the subsequent calculation, such information can be read directly to avoid repeated calculation. Considering the massive amount of data in preprocessing, the Spark paradigm can be used in parallel data processing. In accordance with the calculation of the number of physical nodes N, data are divided by default, and the allocated data are traversed at each computing node.

(2) Preliminary filtering

It can be determined that only the polygons in the area where the MBRs of two polygonal layers intersect need to be clipped. Therefore, filtering polygons that do not require clipping can reduce the computational cost.

(3) Hilbert coding and computation of data partition points base on polygon clip complexity

All polygons are divided by Hilbert grids in accordance with the spatial distribution position, and each grid and polygon are Hilbert coded. Then, the data partition points are calculated on the basis of shape complexity. This work is not suitable for decoupling and, therefore, cannot be executed in parallel.

(4) Data partitioning and indexing

Based on the partitioning points, the region of the Hilbert curve is regarded as a sub-index region. The STR-tree class of the JTS library is used to establish an R-tree index for each partition, and the index file is stored in each computing node.

(5) Target polygon filtering

In overlay analysis, every point of the clipped and target polygons should be traversed. Even if the two polygons are not covered, all points will be traversed, resulting in some invalid calculations. Filtering out the invalid polygons of target polygons can obviously improve the efficiency. Before overlay calculation, the target polygon without overlay analysis can be effectively eliminated by calculating whether an overlay relationship exists between the MBR of the clipped polygon and the MBR of the target polygon. The calculation method directly compares the maximum and minimum coordinates of the clipped and target polygons without using an overlay algorithm.

(6) Overlay analysis

All computing nodes use the Hormann algorithm described in Section 3.1 for parallel overlay computation. The results of each calculation node are reduced to obtain the final overlay analysis results.

3.4. *Algorithmic Analysis*

The major processes of the parallel overlay analysis conducted in this study include data preprocessing, preliminary filtering, Hilbert partitioning, R-tree index establishment, polygon MBR filtering, and polygon clipping.

In the data preprocessing, only the layer attribute data and vertex coordinate information of the polygons are included in the original data. Polygon MBR and the number of polygon vertices must be used thrice in the calculation process designed in this research. Therefore, we unified the data preprocessing, establish a new data structure and avoided repetitive calculation. Unified data preprocessing saves about half of the workload compared with separate data preprocessing.

In the Preliminary filtering, the time complexity of the MBR filtering algorithm is O(1), whereas the complexity of the overlay analysis algorithm is O(logN), where N is the number of polygon vertices. Therefore, computational complexity will be considerably reduced by filtering polygons without overlay analysis through MBR. In addition, the reduced computational complexity depends on the spatial distribution of polygons, which is an uncontrollable factor.

The time complexity of constructing the Hilbert curve is $O(N^2)$, where N is the order of the Hilbert curve. The larger N is, the longer the time that is spent on data partitioning is. However, if N is too small, then multiple polygons will correspond to the same Hilbert coding. If Hilbert partitioning strictly satisfies the condition that each mesh has only one central point of a polygon MBR, then a Hilbert grid supports a maximum of $2^{2\times N}$ polygons. Moreover, polygons in real data are generally not uniformly distributed, and no polygons exist in many Hilbert grids. Therefore, allowing an appropriate number of repetitive Hilbert-coded values is feasible. In addition, compared with grid partition, Hilbert partition can solve the problem of the uneven location of data perfectly.

R-tree is a typical spatial data index method. The time for data traversal is considerably shortened by establishing an R-tree index. The time complexity of R-tree is O(logN).

Data preprocessing, MBR filtering, R-tree index construction, and other processes are relatively time-consuming. By using multi-node parallel computing in data partitioning, the time consumed can be reduced to 1/N, where N is the number of parallel processes. In the Spark paradigm, data operations are performed in memory, and I/O operations consume minimal time. Therefore, the proposed overlay algorithm exhibits high efficiency.

4. Experimental Study

4.1. Experimental Design

To conduct overlay analysis experiments, we used the patches of land use types and the patches with a slope greater than 25 degrees in a county of Yunnan Province in 2018. There are 500,000 patches of land use types and 110,000 slope patches. These data are distributed in the area of 15,000 square kilometers. Based on this data, we constructed different data sets for the experiments. We will use different overlay analysis modes for the execution in the case of different data magnitude data, record the change of execution time, and analyze the characteristics and applicability of different overlay analysis modes.

4.1.1. Computing Equipment

Experiments were carried out using one portable computer and six X86 servers. The equipment configuration information is shown in Table 1.

Table 1. Equipment configuration.

Equipment	Num	Hardware Configuration	Operating System	Software	Remark
portable computer	1	Thinkpad T470p, 8 vcore, 16 G RAM, SSD (Solid State Drive)	Windows 10	ArcMap 10.4.1	Single computer experiment for desktop overlay analysis.
X86 Server	6	DELL R720, 24 core, 64 G RAM, HDD (Hard Disk Drive)	Centos7	Hadoop 2.7, Spark 2.3.1	Spark Computing Cluster

4.1.2. Experimental Data

(1) Clipping layer

The digital elevation model (DEM) data of a 30 m grid in the county were obtained from the Internet, and a slope map was generated by it (Figure 5). The area with a slope greater than 25 degrees was extracted, and 108,025 patches were obtained.

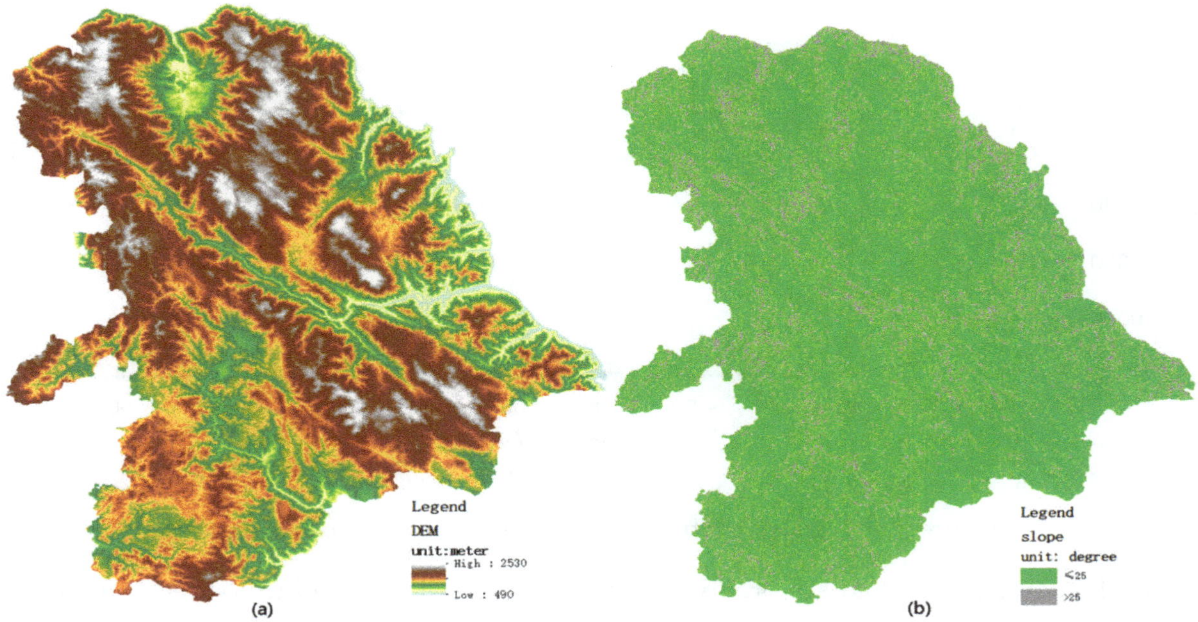

Figure 5. Extracting slope as the clipping layer from the digital elevation model (DEM).

(2) Target layer

A total of 10 groups of experimental data were obtained from 500,000 original land-type patches of the county using sparse and intensive data sets. The number of patches was 50,000, 100,000, 250,000, 500,000, 1 million, 2 million, 4 million, 6 million, 8 million and 10 million, respectively.

Through data checking, 88,000,000 vertices were found in 500,000 original terrain pattern data. Among all the polygons, the simplest polygon has four vertices, whereas the most complex polygon has 99,500 vertices. A total of 890,000 vertices were recorded in 110,000 slope patches. Among all slope patches, the simplest has 8 vertices, whereas the most complex has 5572 vertices. The statistics of the number of polygon vertices in the query and target layers are illustrated in Figures 6 and 7, respectively.

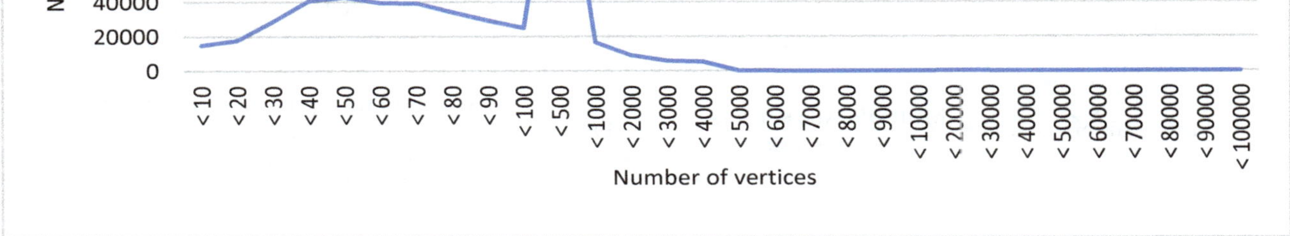

Figure 6. Polygons distribution with different number of vertices.

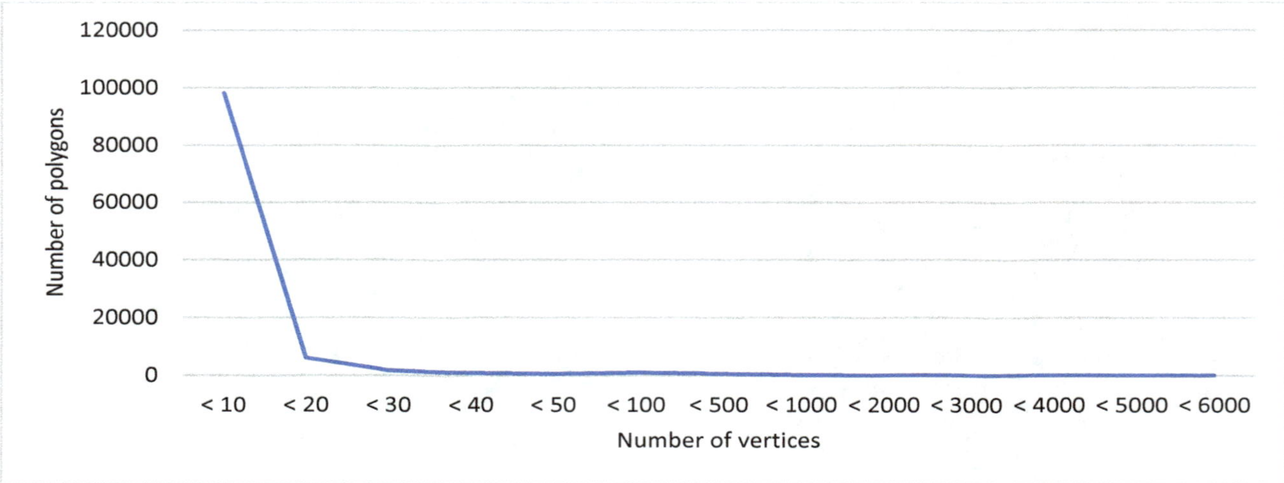

Figure 7. Polygons distribution with different number of vertices.

In parallel computing, data are organized into GeoJson format and uploaded to HDFS. HDFS data blocks are three copies, each of which is 64 MB.

4.1.3. Experimental Scene

Before describing the experimental scenario design, we first define several different modes for comparison and explain the differences of each mode (Table 2).

Table 2. Explanation of the experimental mode.

Mode Abbreviation	Equipment	Data Storage Mode	Notes
ArcMap	1 portable computer with ArcMap	Local File System	Use the clip tool of Toolbox to perform overlay analysis on the portable computer
Spark_original	Multiple X86 servers with Spark	HDFS	Directly partition the data randomly and do parallel overlay analysis without any improvement.
Spark_improved	Multiple X86 servers with Spark	HDFS	Completely implement parallel overlay analysis according to the process of Section 3.3. Hilbert partitioning method considering graph complexity
Spark_NoComlexity	Multiple X86 servers with Spark	HDFS	Except that the complexity of polygon graphics is not considered, all of them are the same as the Spark_improved mode.
Spark_MBR	Multiple X86 servers with Spark	HDFS	Based on the Spark_original model, MBR filtering is performed first, and then parallel overlay analysis is performed.
Spark_MBR_Hilbert	Multiple X86 servers with Spark	HDFS	Based on the Spark_original model, MBR filtering and a Hilbert partitioning operation are added.
Spark_MBR_Hilbert_R-tree	Multiple X86 servers with Spark	HDFS	Based on the Spark_original model, MBR filtering, Hilbert partitioning and R-tree index creation operation are added.

Among them, the ArcMap mode is a typical method used in geographic data processing. The purpose of comparing Spark_original, Spark_improved and Spark_NoComlexity modes is to determine how much the performance has been improved. The purpose of comparing Spark_MBR,

Spark_MBR_Hilbert and Spark_MBR_Hilbert_R-tree modes is to determine how much the three improved methods can improve the efficiency of parallel overlay analysis.

(1) Scene 1: Compare the performance differences of four modes: ArcMap, Spark_original, Spark_improved and Spark_NoComlexity.

Ten groups of polygons with different numbers were used for overlay analysis in four modes. We will record the completion time of the overlay analysis process and draw time-consumption curves. This experimental scenario can answer the following questions:

- How much better will Spark parallel computing improve the performance of overlay analysis compared to desktop software?
- How much better is the performance of the parallel overlay analysis algorithm proposed in this paper compared with the direct use of the spark computing paradigm?
- How much influence does the complexity difference of a geographic polygon have on parallel overlay analysis?

(2) Scene 2: Compare the performance differences of four modes: Spark_original, Spark_MBR, Spark_MBR_Hilbert and Spark_MBR_Hilbert_R-tree.

Ten groups of polygons with different numbers were used for overlay analysis in three modes. We will record the completion time of the overlay analysis process and draw time-consumption curves.

In addition to considering the influence of the shape complexity difference of a geographic polygon, three important improvements are used in our algorithm flow: (1) MBR filtering, (2) Hilbert partitioning, (3) R-tree establishment. This experimental scenario can answer: How much do the above three improvements affect the efficiency of parallel computing?

(3) Scene 3: Cluster acceleration performance testing of the proposed algorithm.

The experimental data are fixed to 10 million geographic polygons. One to six servers are used to perform overlay analysis and record the time-consumption changes of the overlay analysis algorithm in this paper. In this experimental scenario, we can see the acceleration ratio and parallel efficiency of the proposed algorithm in the Spark cluster.

4.2. Test Process and Results

4.2.1. Compare the Performance Differences of Four Modes: ArcMap, Spark_original, Spark_NoComlexity and Spark_improved

The parallel computing mode uses six computing nodes to calculate the flow before and after optimization. The experimental data are collected from 50,000, 100,000, 250,000, 500,000, 1 million, 2 million, 4 million, 6 million, 8 million, and 10 million recorded data sets. The time consumption statistics of different computing modes are illustrated in Figure 8.

As shown in Figure 9, blue, red, yellow and grey represent the time consumed by ArcMap, Spark_original, Spark_NoComlexity and Spark_improved modes. With the increase in the data volume, the four time-consumption curves show an upward trend. The changes in these curves answer three questions related to the design of the experimental scenario.

(1) When the number of polygons is less than 10 million, the efficiency of Spark_original mode is even lower than that of ArcMap mode. When the number of polygons is more than 50,000, the time-consumption of the Spark_improved mode is less than that of the ArcMap mode. When the number of polygons exceeds 1 million, ArcMap mode consumes twice as much time as the Spark_improved mode. As the amount of data increases, the time-consumption of the ArcMap mode increases dramatically, and the time-consumption curve of Spark_improved mode is still relatively flat.

(2) The efficiency of Spark_original mode is lower than that of Spark_improved mode, and the more polygons there are, the more obvious it is. This shows that the efficiency of overlay analysis using Spark directly is very low, and the algorithm optimization must be carried out according to the characteristics of spatial data and geographical calculation.

(3) By comparing the time-consumption curves, Spark_improved takes almost half as much time as Spark_NoComlexity, which is better than I thought. I think it may be related to my experimental data: in Section 4.1.2, I have found that there are many polygons with high shape complexity in the experimental data. Maybe many big polygons are partitioned into the same computational partition, which leads to data skew.

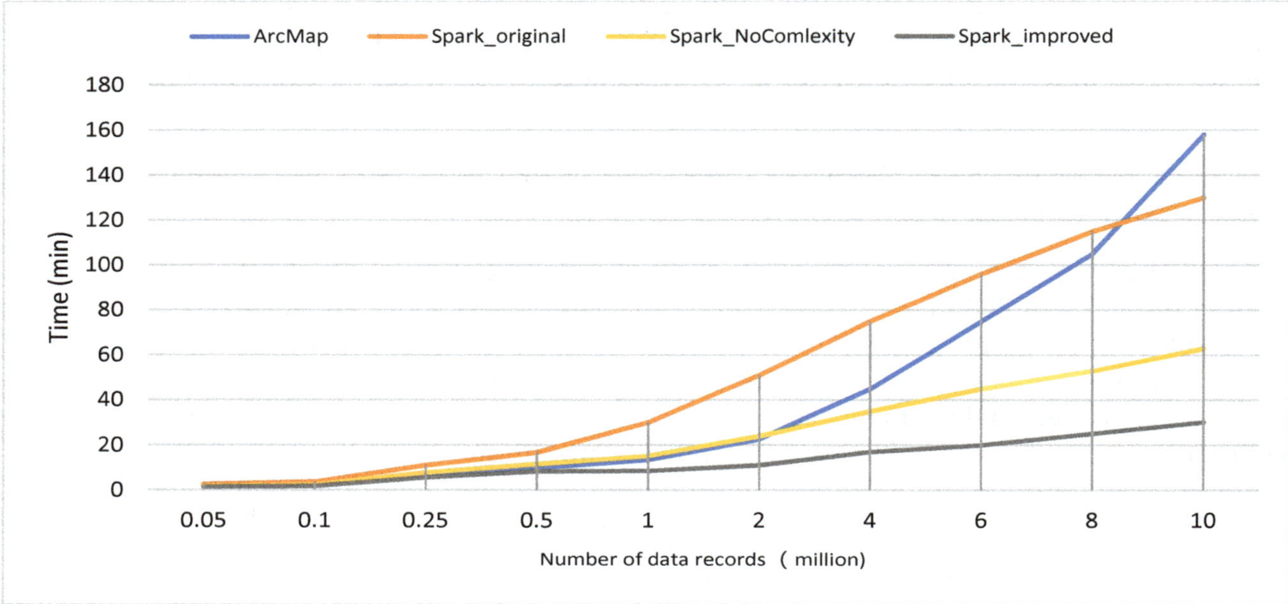

Figure 8. Time consumption statistical graphs of different computing modes.

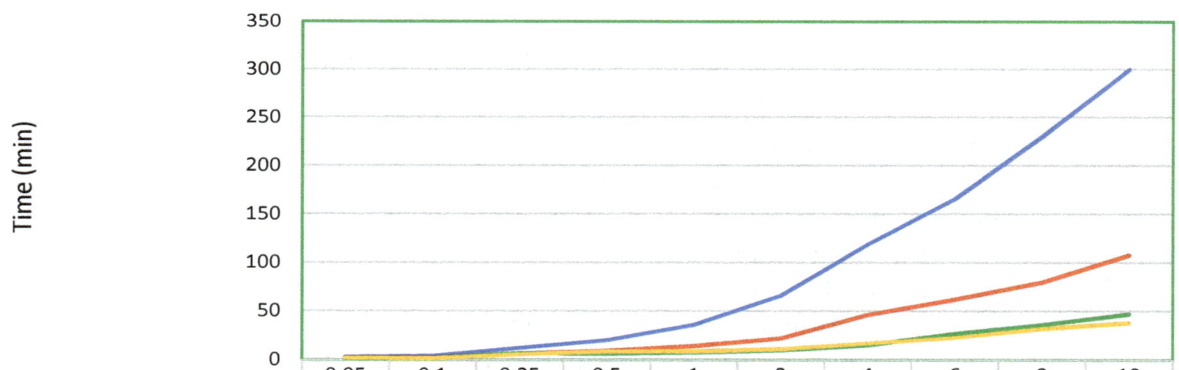

	0.05	0.1	0.25	0.5	1	2	4	6	8	10
Spark_original	2.6	4	12	20	36	66	119	166	230	300
Spark_MBR	1.9	2	6	8.9	14	22	46	62	80	108
Spark_MBR_Hilbert	1.3	1	7	6.3	7.5	9.9	15	27	36	47
Spark_MBR_Hilbert_R-tree	1.7	1	6	8.2	8.5	11	17	23	32	38

Number of data records (million)

Figure 9. Time consumption comparison of different optimization strategies.

4.2.2. Compare the Performance Differences of Four Modes: Spark_original, Spark_MBR, Spark_MBR_Hilbert and Spark_MBR_Hilbert_R-tree

As shown in Figure 9:

(1) After only adopting the MBR filtering strategy, the efficiency of overlay computation is increased by two to four times. Therefore, this strategy filters a large number of invalid overlay computations. Specific efficiency improvement is related to the size, shape, and spatial distribution of polygons in the target and clipped layers.

(2) The Hilbert partitioning algorithm based on polygon graphic complexity is used to allocate the data of each computing node. When the amount of data reaches millions, the computing performance can be doubled. As the data amount increases, the computational performance advantage becomes more evident. The experimental data verify that the spatial aggregation characteristics of Hilbert partitioning that considers polygon complexity can considerably improve spatial analysis algorithms.

(3) Index construction can generally improve the efficiency of data access, but index construction itself can result in a certain amount of computational overhead. After adding the R-tree index strategy based on the first two steps, the overlay calculation time of each order of magnitude increases slightly when the amount of data is less than 5 million. When the amount of data exceeds 5 million, the overlay calculation time decreases compared with the case without the R-tree index. Therefore, the data access time saved after the R-tree index is established offsets the time consumed by the index itself.

4.2.3. Cluster Acceleration Performance Testing of the Proposed Algorithm

The experimental data are unified using 10 million polygons, and then one server is added at a time. As the number of servers increases, the time consumed in parallel computing decreases considerably (Figure 10). However, the trend of running time decreases as the number of nodes increases.

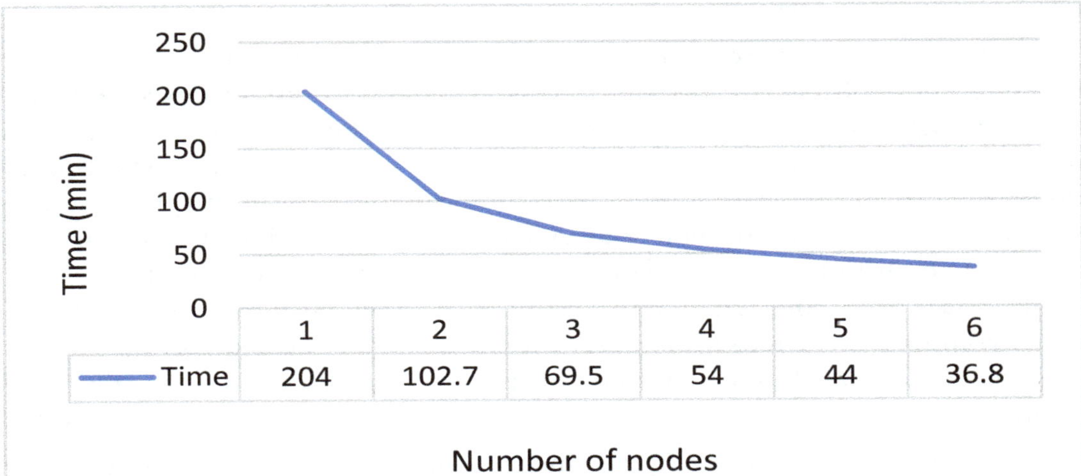

Figure 10. Average running time of different numbers of nodes.

As shown in Figure 11, as the number of servers increases, the acceleration ratio decreases slightly but is nearly linear. In addition, Figure 12 illustrates that with the increase in the number of servers, the parallel efficiency gradually decreases, and finally stabilizes to more than 90%. This is a good result if we consider that the increase in the number of servers will increase the system synchronization and network overhead.

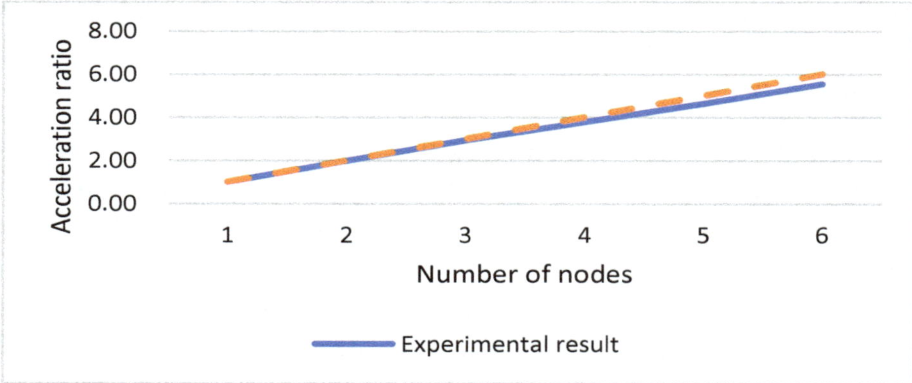

Figure 11. Acceleration ratio of different numbers of nodes.

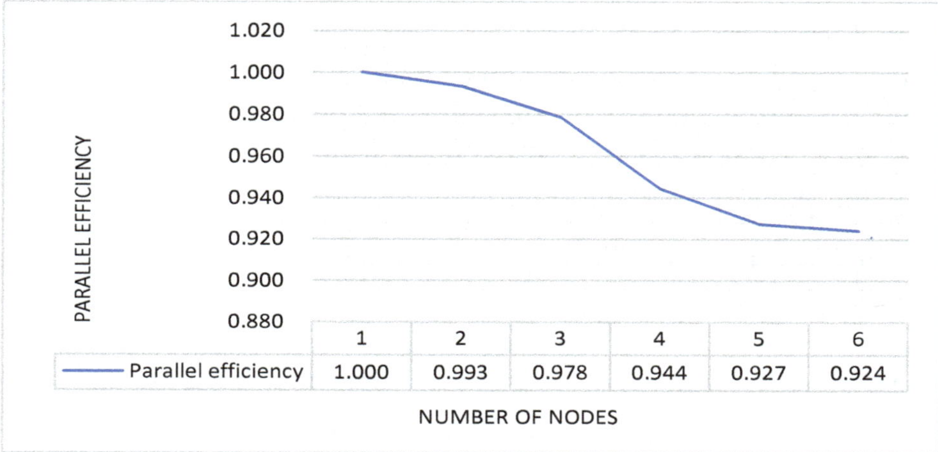

Figure 12. Parallel efficiency of different number of nodes.

4.3. Analysis of Experimental Results

Figure 9 shows that a single computer with ArcMap Soft achieves high efficiency in the overlay analysis of small data volume by adopting a reasonable algorithm and excellent multithreading processing technology. In addition, SDD also plays an important role. However, the performance of ArcMap sharply declines when the number of data records reaches millions. Spark distributed parallel computing can effectively solve such problems, but the simple transplantation of the overlay analysis algorithm into the Spark framework is not a reasonable solution. The actual geographic data are often unevenly distributed, and the complexity of polygon graphics varies greatly, which will lead to a serious data skew, and which will seriously affect the performance of parallel computing. When the data volume reaches tens of millions, the performance of our algorithm improves by more than 10 times via Hilbert partitioning based on the polygon graphic complexity and the R-tree index. In addition, the performance advantage becomes more evident as the data volume increases.

When the amount of data is constant, the time-consumption of parallel overlay analysis decreases with the increase in the number of servers. However, the decreasing trend of running time declines as the number of nodes increases. Figure 11 shows that the acceleration ratio is nearly linear. Figure 12 illustrates that parallel efficiency is still over 90% and remains stable when the number of servers increases to six, which means that higher computing efficiency can be maintained when the computing cluster expands. Therefore, it is an effective method to add physical nodes in massive data overlay analysis.

In addition, the proposed overlay analysis algorithm also has some problems to be improved, such as: (1) Big polygons will span multiple data partitions, which will lead to repeated participation of the polygon in overlay analysis on multiple servers. (2) In the current algorithm process, the R-tree index is created temporarily, which leads to the creation of the index repeatedly for each overlay analysis.

5. Conclusions

In high-performance parallel overlay analysis, the differences in shape complexity of a polygon can lead to serious data skew. In this paper, we measure the shape complexity of polygons from the perspective of geographic computing and design a high-performance parallel overlay analysis algorithm considering the shape complexity of polygons. The analysis of the algorithm shows that the algorithm reduces invalid overlay calculation by MBR filtering, achieves load balancing by use of a Hilbert partition based on the polygon shape complexity, and improves data access speed using the R-tree index. Experiments show that this is a high-performance method and can maintain high speed-up and parallel efficiency in computing cluster expansion.

In future studies, we will study the impact of the spatial distribution of graphics, spatial data storage and indexing methods on the efficiency of overlay analysis. We will also optimize spatial index storage through distributed memory database technology to further improve the efficiency of parallel overlay analysis.

Author Contributions: Kang Zhao proposed the research ideas and technical lines. Baoxuan Jin and Hong Fan provided research guidance. Weiwei Song shared valuable opinions. Sunyu Zhou and Yuanyi Jiang helped complete the programming. Zhao Kang completed the thesis.

Acknowledgments: The authors are grateful to Fan Yang and Liying Li for providing their experimental data and to Lifeng Hou for giving valuable suggestions.

References

1. Wang, S.; Zhong, E.; Lu, H.; Guo, H.; Long, L. An effective algorithm for lines and polygons overlay analysis using uniform spatial grid indexing. In Proceedings of the 2015 2nd IEEE International Conference on Spatial Data Mining and Geographical Knowledge Services (ICSDM), Fuzhou, China, 8–10 July 2015; pp. 175–179.
2. Puri, S.; Prasad, S.K. Efficient parallel and distributed algorithms for GIS polygonal overlay processing. In Proceedings of the 2013 IEEE International Symposium on Parallel & Distributed Processing, Workshops and Phd Forum, Cambridge, MA, USA, 20–24 May 2013; pp. 2238–2241.
3. van Kreveld, M.; Nievergelt, J.; Roos, T.; Widmayer, P. *Algorithmic Foundations of Geographic Information Systems*; Springer: New York, NY, USA, 1997; Volume 1340.
4. Li, Q.; Li, D. Big data GIS. *Geomat. Inf. Sci. Wuhan Univ.* **2014**, *39*, 641–644.
5. Li, D.R.; Cao, J.J.; Yuan, Y. Big data in smart cities. *Sci. China Inf. Sci.* **2015**, *58*, 108101. [CrossRef]
6. Yang, C.; Goodchild, M.; Huang, Q.; Nebert, D.; Raskin, R.; Xu, Y.; Bambacus, M.; Fay, D. Spatial cloud computing: How can the geospatial sciences use and help shape cloud computing? *Int. J. Digit. Earth* **2011**, *4*, 305–329. [CrossRef]
7. Greiner, G.; Hormann, K. Efficient clipping of arbitrary polygons. *ACM Trans. Graph. (TOG)* **1998**, *17*, 71–83. [CrossRef]
8. Rossignac, J. Shape complexity. *Vis. Comput.* **2005**, *21*, 985–996. [CrossRef]
9. Day, H. Evaluations of subjective complexity, pleasingness and interestingness for a series of random polygons varying in complexity. *Percept. Psychophys.* **1967**, *2*, 281–286. [CrossRef]
10. Tilove, R.B. Line/polygon classification: A study of the complexity of geometric computation. *IEEE Comput. Graph. Appl.* **1981**, *1*, 75–88. [CrossRef]
11. Chen, Y.; Sundaram, H. Estimating complexity of 2D shapes. In Proceedings of the 2005 IEEE 7th Workshop on Multimedia Signal Processing, Shanghai, China, 30 October–2 November 2005; pp. 1–4.
12. Huang, C.-W.; Shih, T.-Y. On the complexity of point-in-polygon algorithms. *Comput. Geosci.* **1997**, *23*, 109–118. [CrossRef]
13. Mandelbrot, B.B. *The Fractal Geometry of Nature*; WH Freeman: New York, NY, USA, 1982; Volume 1.
14. Peitgen, H.-O.; Jürgens, H.; Saupe, D. *Chaos and Fractals: New Frontiers of Science*; Springer Science & Business Media: New York, NY, USA, 1992.

15. Faloutsos, C.; Kamel, I. Beyond uniformity and independence: Analysis of R-trees using the concept of fractal dimension. In Proceedings of the Thirteenth ACM SIGACT-SIGMOD-SIGART Symposium on Principles of Database Systems, Minneapolis, MN, USA, 24–27 May 1994; pp. 4–13.
16. Brinkhoff, T.; Kriegel, H.-P.; Schneider, R.; Braun, A. Measuring the Complexity of Polygonal Objects. In Proceedings of the ACM-GIS, Baltimore, MD, USA, 1–2 December 1995; p. 109.
17. Bryson, N.; Mobolurin, A. Towards modeling the query processing relevant shape complexity of 2D polygonal spatial objects. *Inf. Softw. Technol.* **2000**, *42*, 357–365. [CrossRef]
18. Ying, F.; Mooney, P.; Corcoran, P.; Winstanley, A.C. A model for progressive transmission of spatial data based on shape complexity. *Sigspat. Spec.* **2010**, *2*, 25–30. [CrossRef]
19. Weiler, K.; Atherton, P. Hidden surface removal using polygon area sorting. *ACM SIGGRAPH Comput. Graph.* **1977**, *11*, 214–222. [CrossRef]
20. Vatti, B.R. A generic solution to polygon clipping. *Commun. ACM* **1992**, *35*, 56–63. [CrossRef]
21. Wang, H.; Chong, S. A high efficient polygon clipping algorithm for dealing with intersection degradation. *J. Southeast Univ.* **2016**, *4*, 702–707.
22. Zhang, S.Q.; Zhang, C.; Yang, D.H.; Zhang, J.Y.; Pan, X.; Jiang, C.L. Overlay of Polygon Objects and Its Parallel Computational Strategies Using Simple Data Model. *Geogr. Geo-Inf. Sci.* **2013**, *29*, 43–46.
23. Chen, Z.; Ma, L.; Liang, W. Polygon Overlay Analysis Algorithm Based on Monotone Chain and STR Tree in the Simple Feature Model. In Proceedings of the 2010 International Conference on Electrical & Control Engineering, Wuhan, China, 25–27 June 2010.
24. Wang, J. An Efficient Algorithm for Complex Polygon Clipping. *Geomat. Inf. Sci. Wuhan Univ.* **2010**, *35*, 369–372.
25. Guest, M. An overview of vector and parallel processors in scientific computation. *J. Comput. Phys. Commun.* **1989**, *57*, 560. [CrossRef]
26. Wang, Y.; Liu, Z.; Liao, H.; Li, C. Improving the performance of GIS polygon overlay computation with MapReduce for spatial big data processing. *Clust. Comput.* **2015**, *18*, 507–516. [CrossRef]
27. Zheng, Z.; Luo, C.; Ye, W.; Ning, J. Spark-Based Iterative Spatial Overlay Analysis Method. In Proceedings of the 2017 International Conference on Electronic Industry and Automation (EIA 2017), Suzhou, China, 23–25 June 2017.
28. Xiao, Z.; Qiu, Q.; Fang, J.; Cui, S. A vector map overlay algorithm based on distributed queue. In Proceedings of the 2017 IEEE International Geoscience and Remote Sensing Symposium (IGARSS), Fort Worth, TX, USA, 23–28 July 2017; pp. 6098–6101.
29. Eldawy, A.; Mokbel, M.F. A demonstration of spatialhadoop: An efficient mapreduce framework for spatial data. *Proc. VLDB Endow.* **2013**, *6*, 1230–1233. [CrossRef]
30. Eldawy, A.; Alarabi, L.; Mokbel, M.F. Spatial partitioning techniques in SpatialHadoop. *Proc. VLDB Endow.* **2015**, *8*, 1602–1605. [CrossRef]
31. Eldawy, A.; Mokbel, M.F. Spatialhadoop: A mapreduce framework for spatial data. In Proceedings of the 2015 IEEE 31st International Conference on Data Engineering, Seoul, Korea, 13–17 April 2015; pp. 1352–1363.
32. Lenka, R.K.; Barik, R.K.; Gupta, N.; Ali, S.M.; Rath, A.; Dubey, H. Comparative analysis of SpatialHadoop and GeoSpark for geospatial big data analytics. In Proceedings of the 2016 2nd International Conference on Contemporary Computing and Informatics (IC3I), Noida, India, 14–17 December 2016; pp. 484–488.
33. Yu, J.; Wu, J.; Sarwat, M. GeoSpark: A cluster computing framework for processing large-scale spatial data. In Proceedings of the 23rd SIGSPATIAL International Conference on Advances in Geographic Information Systems, Seattle, WA, USA, 3–6 November 2015; pp. 1–4.
34. Yu, J.; Wu, J.; Sarwat, M. A demonstration of GeoSpark: A cluster computing framework for processing big spatial data. In Proceedings of the 2016 IEEE 32nd International Conference on Data Engineering (ICDE), Helsinki, Finland, 16–20 May 2016; pp. 1410–1413.
35. Yu, J.; Zhang, Z.; Sarwat, M. Spatial data management in apache spark: The geospark perspective and beyond. *Geoinformatica* **2019**, *23*, 37–78. [CrossRef]
36. Luitjens, J.; Berzins, M.; Henderson, T. Parallel space-filling curve generation through sorting: Research Articles. *Concurr. Comput. Pract. Exp.* **2010**, *19*, 1387–1402. [CrossRef]
37. Kim, K.-C.; Yun, S.-W. MR-Tree: A cache-conscious main memory spatial index structure for mobile GIS. In Proceedings of the International Workshop on Web and Wireless Geographical Information Systems, Goyang, Korea, 26–27 November 2004; pp. 167–180.

12

Social Media Big Data Mining and Spatio-Temporal Analysis on Public Emotions for Disaster Mitigation

Tengfei Yang [1,2], Jibo Xie [1,*], Guoqing Li [1], Naixia Mou [3], Zhenyu Li [3], Chuanzhao Tian [1,2] and Jing Zhao [1,2]

1 Institute of Remote Sensing and Digital Earth, Chinese Academy of Sciences, Beijing 100094, China; yangtf@radi.ac.cn (T.Y.); ligq@radi.ac.cn (G.L.); tiancz@radi.ac.cn (C.T.); zhaojing01@radi.ac.cn (J.Z.)
2 University of Chinese Academy of Sciences, Beijing 100094, China
3 College of Geomatics, Shandong University of Science and Technology, Qingdao 266590, China; mounx@lreis.ac.cn (N.M.); lizy1@radi.ac.cn (Z.L.)
* Correspondence: xiejb@radi.ac.cn

Abstract: Social media contains a lot of geographic information and has been one of the more important data sources for hazard mitigation. Compared with the traditional means of disaster-related geographic information collection methods, social media has the characteristics of real-time information provision and low cost. Due to the development of big data mining technologies, it is now easier to extract useful disaster-related geographic information from social media big data. Additionally, many researchers have used related technology to study social media for disaster mitigation. However, few researchers have considered the extraction of public emotions (especially fine-grained emotions) as an attribute of disaster-related geographic information to aid in disaster mitigation. Combined with the powerful spatio-temporal analysis capabilities of geographical information systems (GISs), the public emotional information contained in social media could help us to understand disasters in more detail than can be obtained from traditional methods. However, the social media data is quite complex and fragmented, both in terms of format and semantics, especially for Chinese social media. Therefore, a more efficient algorithm is needed. In this paper, we consider the earthquake that happened in Ya'an, China in 2013 as a case study and introduce the deep learning method to extract fine-grained public emotional information from Chinese social media big data to assist in disaster analysis. By combining this with other geographic information data (such population density distribution data, POI (point of interest) data, etc.), we can further assist in the assessment of affected populations, explore emotional movement law, and optimize disaster mitigation strategies.

Keywords: social media; big data; fine-grained emotion classification; spatio-temporal analysis; hazard mitigation

---

1. Introduction

With the popularity of mobile devices and the development of the network infrastructure, social media has quickly integrated into people's lives. People can easily share what they see and hear, and even what they feel and think with social media. They are like "mobile sensors" [1] to collect information around them constantly. This provides a new way to acquire disaster-related data. Compared with traditional disaster information collection methods, social media has the characteristics of real-time information provision and low cost. Furthermore, these data contain a lot of geographic information (such as location, time, and other attribute information), which is very important for disaster mitigation. Therefore, many researchers have noticed the importance of social media in disaster mitigation. They have studied disasters from the perspectives of event extraction [2,3], user trajectory rules [4] and data fusion [5], etc., and achieved good results. However, few researchers have considered the public emotional information contained in social media (especially fine-grained emotions) as an attribute of disaster-related geographic information to aid in disaster mitigation. When disasters occur, public emotions often express the public's attitude towards disaster, needs during disaster, and feedback on disaster relief, etc. These are very helpful to understand the progress of the disaster quickly and effectively improve the efficiency of rescue. However, there is still a lack of an effective framework to quickly collect, process, and use this emotional information. There are three problems involved: (1) How can the fine-grained public emotional categories be divided during the disaster? (2) Social media has a huge user base. We take Sina micro-blog, a Chinese social media, as an example. According to statistics, as of Q3 2018, Chinese social media platform Sina micro-blog had over 431 million active monthly users [6]. When disasters occur, this will generate a lot of disaster-related data. As such, how can the fine-grained emotional information contained in these data be extracted more accurately? (3) When these fine-grained emotions are extracted, how can they be regarded as an attribute of disaster-related geographic information to assist disaster mitigation? In this paper, we used a Sina micro-blog and took an earthquake disaster as an example to describe how the framework we built extracted fine-grained public emotions and used them to serve disaster mitigation.

Unlike most emotion analysis studies (they usually divide emotions into three categories: positive, neutral, and negative), we divide the public emotions during the disaster into more dimensions, because the use of multiple dimensions of emotion in the disaster context can allow more details of the disaster to be described. Additionally, studies have illustrated the importance of multidimensional emotional information in disasters. Ekman, et al [7] showed the differences between anger, disgust, fear, and sadness in terms of antecedent events and likely behavioral responses. Oliver Gruebner et al [8] analyzed how to apply multiple dimensions of negative emotion (including anger, fear, sadness, surprise, confusion, disgust) to survey disaster mental health. Existing psychological studies [9–11] also mention the fine-grained division of emotions in a disaster. Therefore, based on these previous studies and the corpus used in this paper, we subdivide the negative emotions into anger, anxiousness, fear, and sadness.

The commonly used methods for emotion classification include rule-based algorithms and traditional machine learning models [12]. Rule-based algorithms mainly uses given emotional lexicons and corresponding grammatical rules to calculate the emotional intensity of the text [13,14].

This method relies on a large number of manual operations, such as manual development of search rules and a large-scale emotional lexicon [15], which determines the accuracy of the method. Additionally, this method is weak in dealing with stop words and new words. It is also hard to add some slang and Internet buzzwords to the emotional lexicon in time, such as "喜大普奔" (great satisfaction), "狂顶" (very supportive), etc., which often appear in social media. Traditional machine learning models, such as naive Bayes [16], maximum entropy, and support vector machine [17] do not rely on emotional lexicons or search rules.

They only need to manually annotate the training set. However, the traditional machine learning method is based on the bag-of-words model, which ignores the semantic relations in text. In other words, it does not consider the order of words in a sentence, which can easily cause misclassification of emotions. For example, the sentences "Although the earthquake is terrible, we are safe and sound" and "Although we are safe and sound, the earthquake is terrible" contain the same words, they express different emotions. Moreover, for traditional machine learning models, the input is the feature words extracted from the text after segmentation.

The definition of feature words has a significant impact on the model's efficiency [15]. We selected the deep learning method to extract public emotion from social media. Compared with the rule-based method, deep learning does not depend on any emotional lexicons. Therefore, it is not affected by new and unknown words. Unlike traditional machine learning, deep learning uses word vector models to replace the bag-of-words model, which can make good use of semantic information in sentences. Much research has indicated that the performance of deep learning [18,19] in natural language processing (NLP) tasks is better than of traditional machine learning.

Furthermore, we used extracted fine-grained public emotions and combined them with traditional geographic information data (population density distribution data, point of interest (POI) data, etc.), and the powerful spatial analysis functions of a GIS (geographic information system) to assist disaster relief. Combining public emotional information could produce the following benefits: (1) It could improve the accuracy and efficiency of disaster assessment.

For example, with the help of the powerful spatial analysis functions of a GIS, traditional geographic information data (such as population distribution, traffic distribution, etc.), and emotional distribution data can be combined to assess the affected population in real-time. People who express negative emotions are generally considered to be more affected by disasters. (2) It could help to reduce disaster-related losses. For example, disasters, especially sudden disasters (such as earthquakes, volcanic eruptions, etc.), can easily cause disaster-related mental health problems, such as post-traumatic stress disorder (PTSD) and depression [20–22].

Traditional monitoring has difficulty obtaining information about emotions of the public in the disaster area (despite the existence of a questionnaire, its real-time performance is poor). If information about the public's emotions and corresponding spatio-temporal distribution is known, the disaster reduction department can take corresponding psychological rescue measures to reduce the occurrence of disaster-related mental health problems. In addition, extreme disasters have the characteristics of inevitability and unpredictability [23].

People will express different emotions at different stages [24] and have different responses to try to overcome them [9]. For example, anxious people are more sensitive to the negative side of event-related information and can be easily influenced by rumors [25]. Therefore, through understanding the distribution of anxious people, we can release the correct disaster information at appropriate times to prevent rumors being intrusive to anxious people.

(3) Learning more about the causes of emotion could help us to optimize emergency decisions. By using different emotion categories, we can explore different emotional causes, such as why angry emotions are predominant in a certain area and more anxious emotions are predominant in another area, and why the emotion categories change in some places over time.

By understanding the causes of emotion, the disaster reduction departments can carry out targeted countermeasures. In the spatio-temporal analysis of public emotion information, the framework in this paper includes assessing the affected population in real-time, exploring an emotional movement law, and monitoring the causes of emotional change.

The structure of this paper is as follows: Section 2 describes the data acquisition, parsing, processing, and emotion classification method used in this paper. Section 3 presents the role of public emotion in assisting disaster reduction with a case study. Section 4 shows the evaluation of the experimental indicators. Section 5 concludes the paper.

2. Framework to Analyze Public Emotion from Social Media Big Data

The framework to analyze the role of public emotions in disaster mitigation proposed in this paper includes five major phases: data acquisition and processing, the construction of a word vector list, model training, emotion classification, and spatio-temporal analysis of public emotions (as shown in Figure 1).

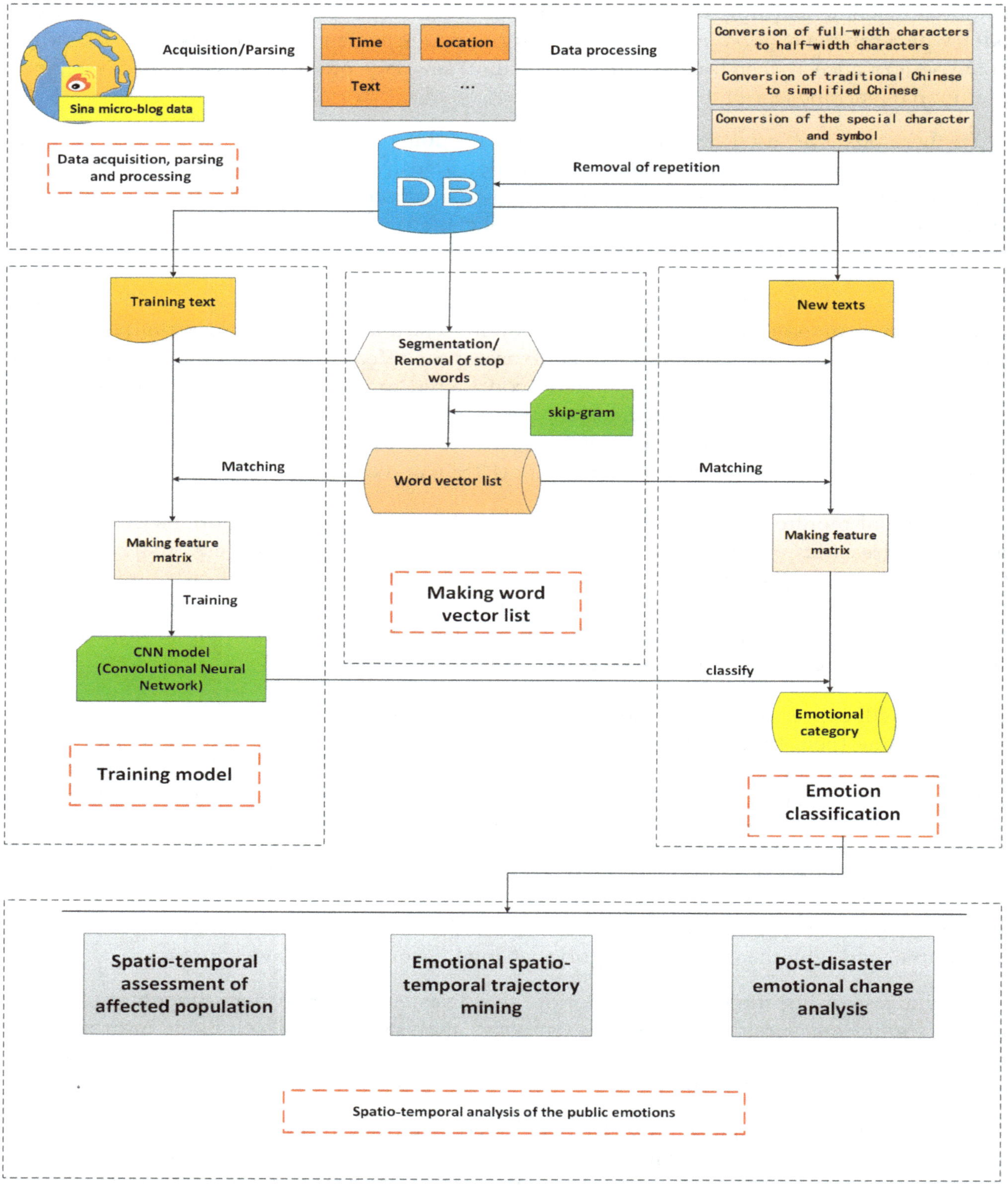

Figure 1. Framework of the automatic emotion classification and disaster analysis.

2.1. *Social Media Data Acquisition and Parsing*

We used an earthquake that happened in Ya'an, Sichuan, China, at 08:02 h on April 20, 2013, as the case study. According to the report by the China Seismograph Network (http://news.ceic.ac.cn/CC20130420080246.html), the magnitude of this earthquake was 7.0 and its focal depth was 13 kilometers. The epicenter of this earthquake was located at 30.30° N, 103.00° E, which caused about 1.52 million people to be affected in an area of 12,500 square kilometers.

In this paper, social media data was acquired from the Sina micro-blog from the region surrounding the epicenter with a radius of 200 km, which was severely damaged by the earthquake. The affected cities included Ya'an, Meishan, Ganzi, Leshan, Ziyang, Deyang, Chengdu, Aba, Zigong, Mianyang, and Neijiang, as shown in Figure 2. The time span of social media data was from April 20 until April 26, 2017. Social media platforms usually provide an interface or API (Application Programming Interface) that allows developers to retrieve social media data. However, the retrieval of data in this way has great limitations; for example, you cannot set the time-span and topics, etc. Therefore, in this paper, we used Sina micro-blog's advanced search capability to get data by using time-span, city names, and event-related key words.

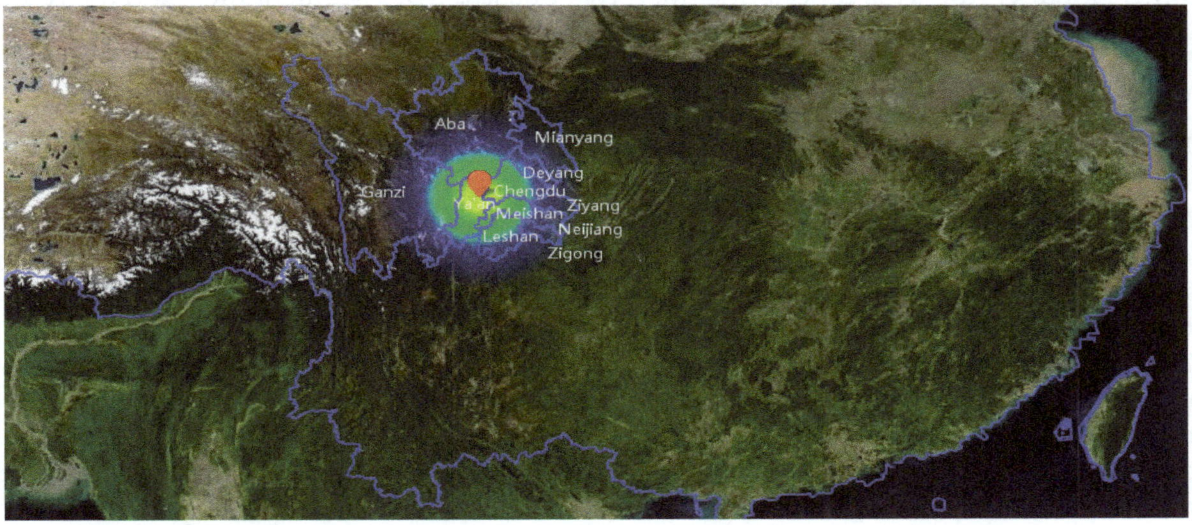

Figure 2. The study area of the 2013 Ya'an earthquake that was used in this paper.

The data format was initially hypertext markup language (HTML). We parsed the data into a structured data format including fields such as "time," "location," "text," etc. Among them, location was represented by the address and the accuracy of them were different. We take Chengdu as an example. Some addresses were described in more detail, such as "East Gate of Sichuan University," "Sishengci North Street," etc. Some addresses were roughly described, such as "Funan New District." There were also some texts that did not have address information. The reason for this is that people have different usage habits (some people do not want to share their location information). We used the API provided by Baidu (http://lbsyun.baidu.com/index.php?title=webapi/guide/webservice-geocoding) to convert these addresses to latitude and longitude. Among them, for line data, such as "Sishengci North Street," we took its midpoint coordinates to represent it. For surface data, such as "Wangjiang Campus of Sichuan University" and even "Funan New District," we extracted the central point coordinate to represent them respectively. We did not assign coordinates to those texts that did not have address information, including those with rough addresses. They were just labeled "Chengdu."

2.2. Social Media Data Processing

In the subsequence processing steps, we mainly dealt with the text data. The main text processing steps included the conversion of full-width characters to half-width characters and from traditional Chinese to simplified Chinese, as well as recognition of the special characters and symbols. The aim of the first two steps was mainly to improve the computational efficiency of the model. The third step aimed to recognize special characters and symbols, such as "(>_<)", "☹", which are deleted and ignored by many common natural language processing (NLP) tools. However, for an emotional analysis, these special characters and symbols have emotional meaning, for example, "(>_<)" and "(>_<)>" can express troubled emotions. Therefore, in this paper, we interpreted them into text that could be processed by NLP. Some special characters and symbols could be translated into text by the micro-blog platform. For example, 😭 could be translated into "tear" (泪). However, others that could not be decoded by the micro-blog platform, such as (>_<) and 💔, were interpreted according to the web's "list of emoticons," which includes the emotional implications of all kinds of emoticons through a large amount of published literature. For example, (>_<) can be translated into "troubled" (焦虑) and 💔 can be translated into "sad" (伤心).

Finally, after eliminating duplications, there were 39341 data records stored in our database.

2.3. Constructing the Word Vector List

In this paper, we first converted each word from the previously processed texts into a multidimensional vector. This process included two phases: word segmentation and the removal of stop words, and the construction of a word vector list.

2.3.1. Word Segmentation and the Removal of Stop Words

Unlike the English language, there is no space separation between Chinese words. Therefore, we needed to segment Chinese text to get separate words. Additionally, the Chinese micro-blog is more colloquial, which brings great challenges to word segmentation. We compared many different Chinese word segmentation tools, such as "Stanford NLP", "ANSJ", "NLPIR (Natural Language Processing & Information Retrieval Sharing Platform)", and so on. We found that "NLPIR" had the best performance in terms of the accuracy and speed of word segmentation.

There are many meaningless words in text after word segmentation; these are called stop words, such as "在 (on)," "是 (is)," "一会 (a moment)," and so on. These words could affect the accuracy of the model and therefore should be removed. In this paper, we used the vocabulary of stop words developed by the Harbin Institute of Technology—Social Computing and Information Retrieval Research Center to remove stop words. As the focus of this paper was on the emotional analysis, we optimized the vocabulary of stop words by removing sentimental words, such as "愤然 (indignant)," "幸亏 (luckily)," "嘻 (hey)", etc.

2.3.2. Construction of the Word Vector List

The input used for the emotion classification model was a word vector matrix. We needed to convert each word in the micro-blog text into a multidimensional vector, and then convert the whole sentence into a word vector matrix. In this paper, we converted all previously processed texts into a word vector list. The training text and new text to be categorized were transformed into a word vector matrix by matching them with the word vector list. The method we used for this was word2vec [26], which projected every word in every sentence to a specified dimensional vector space.

There are two commonly used models in word2vec, which are skip-gram [27,28] and CBOW (Continuous Bag-of-Words) [27,28]. A large number of experiments have been done to compare these two models in terms of performance and accuracy [29] and the results show that the semantic accuracy rate of the skip-gram model is better than that of the CBOW model. Therefore, we used the Skip-gram model in our experiment to construct the text feature vector.

The skip-gram model can determine correlations between words for corpus training. These correlations are represented by the multidimensional feature vectors of each word. Additionally, these multidimensional feature vectors are calculated by taking full consideration of the context of semantic information. From the below formula, given a current word, w_i, this model tries to find words which have a contextual semantic relationship with the current word. The target of this model is to maximize the objective function, G:

$$G = \sum_{w_i \in C} \log P(Context(w_i)|w_i). \quad (1)$$

In this formula, w_i represents the current word and C represents the context window. $P(Context(w_i)|w_i)$ represents the probability of the context information in the current word.

When the training converges, words with similar semantic meanings are closer in the specified dimensional vector space. We exported the text feature vector of each word in the training corpus to generate word embedding. The structure of text feature vectors is shown as Figure 3.

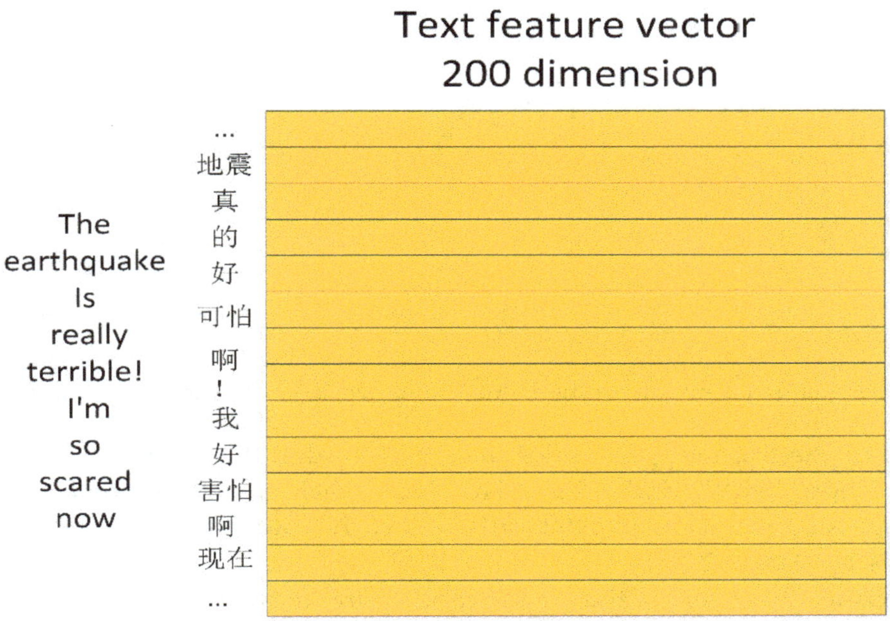

Figure 3. Structure of the text feature vector.

2.4. *Model Training*

The deep leaning model selected in this paper was the convolutional neural network. We read much related literature and found that different deep learning methods can be selected for emotion classification, such as the convolutional neural network (CNN), recurrent neural network (RNN), hierarchy attention network (HAN), etc. These models [30–32] all have their own characteristics and usage scenarios. According to the literature [33], CNN performs emotion classification well, especially in shorter sentences. RNN performs document-level emotion classification well [34]. A previous study [35] presented the performances of the CNN, RNN, and HAN in emotion classification. The results showed that when the training corpus is large enough, HAN has the highest accuracy, but CNN performs the best when the training corpus is not very large. The annotation of a large training corpus requires a lot of manpower and time. Additionally, it takes longer to train the HAN and RNN models than the CNN model. In this paper, the training corpus used was micro-blog texts, which are mainly short texts. Additionally, the amount of manually tagged training corpus data was more suitable for the CNN model. Therefore, we selected CNN as the method to extract the public emotion contained in social media. The training process of the model is shown below.

2.4.1. Word Segmentation, Removal of Stop Words, and Construction of the Feature Matrix

First, we segmented the training texts to obtain separate words. Then, we used a vocabulary of stop words to remove the meaningless words contained in those separate words. Finally, the remaining words were converted into word vectors through matching with a previously generated word vector list. Ultimately, every sentence was transformed into a feature matrix.

2.4.2. Training Convolutional Neural Network Model

The convolutional neural network (CNN) is a variant of the neural network. It was first successfully used for the recognition of images and videos. Later, some researchers introduced it into the field of natural language processing [36] and found that it had a good effect. The CNN model used in this paper consisted of an input layer, a convolutional layer, a pooling layer, a fully connected layer, and classification. The structure of the CNN is shown in Figure 4.

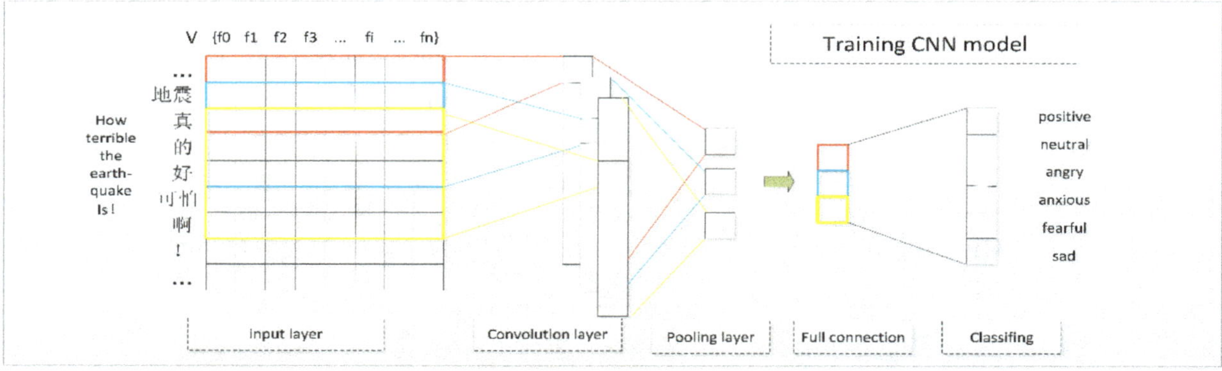

Figure 4. Structure of the convolutional neural network (CNN) model used in the paper.

In the course of the training process of the CNN model, the neurons in it are usually set to three dimensions: depth, width, and height. The size of each layer is depth × width × height [37]. For example, if there is a sentence with 140 words, and each word is set to be 200 dimensions, the size of the input layer is 1 × 140 × 200.

Next, we introduce the layers of the convolution neural network.

Input layer: The input layer of CNN is a matrix that consists of text feature vectors. This matrix is calculated using the skip-gram model, which was described in Section 2.2. The rows and columns (dimension) in this matrix were set before we put the matrix into the neural network model. Taking the Sina micro-blog text as an example, the number of characters in each sentence was less than 140. Therefore, we set the rows in the matrix to 140. If the number of words in a sentence was less than 140 characters, we used the empty character "space" to supplement the missing characters. Therefore, every sentence is expressed as follows:

$$S_{1:140} = S_1 \oplus S_2 \oplus S_3 \ldots \oplus S_{140.} \tag{2}$$

In this formula, S represents a character or "pad," and $\oplus$ is the concatenation operator.

Convolutional layer: The convolution layer is mainly used to extract features. It abstracts some fragmented elements into features which can be used to distinguish different categories. By convolution, many low-level features can be abstracted to higher level features. For example, the single word "打" or "call" has no emotional meaning. However, the higher-level feature "打call (praise)" can express an emotional attribute. The emotional attributes of these words can be acquired by the model through a large number of training corpus.

Given a matrix $\boldsymbol{u}$ that is from the input layer for convolution operation, the formula is as follows:

$$c_j = f(\boldsymbol{u} * \boldsymbol{k}_j + \boldsymbol{b}_j). \tag{3}$$

For the matrix $\mathbf{u} \in \mathbb{R}^{D \times L}$, D represents the embedding dimensionality, and L represents the sentence length. The parameter $\mathbf{k} \in \mathbb{R}^{D \times s}$ represents the j-th convolutional kernel, which is applied to a window of s words. The parameter $b_j \in \mathbb{R}$ represents a bias term. $f(\boldsymbol{u} * \boldsymbol{k}_j + \boldsymbol{b}_j)$ is a non-linear activation function.

Pooling layer: After the convolution operation, we can use the output features to directly classify emotions. However, in doing so, we will not only face the challenge of computational complexity, but also the problem of over-fitting, which will affect the classification accuracy. The pooling operation can solve these problems well. In addition, the pooling operation can also serve as a feature selector that can help to identify the most important features to improve the classification performance.

There are two methods that can be selected, namely max pooling and average pooling. We achieved better results with the max pooling method. This method selects global semantic features and attempts to capture the most important feature with the highest value for each feature map [37]. The output from the convolution operation c_j is used as the input of the pooling operation. The formula is as follows:

$$p_j = pooling(c_j) + b_j. \quad (4)$$

Fully Connected Layer: The neurons in this layer have full connections with all neurons in the previous layer. Meanwhile, the value of the full connected layer can be calculated through the neurons in the previous layer. In the calculation process, the dropout regularization method is usually used to avoid over-fitting.

Classifying: We can obtain the emotional labels of the original text through the softmax function. In other words, these calculated results represent the probability distribution of the emotional labels.

Based on the training corpus, we can identify the best parameters for the CNN model. Then, this trained model can be used to calculate the emotional categories of new texts.

2.5. Emotion Classification

We used the trained CNN model to analyze new texts. The emotions contained in these texts were divided into six categories: positive, neutral, angry, anxious, fearful, and sad. Among them, the positive emotion mainly included the public's satisfaction with disaster relief, the public's wishes for the disaster area, and the joy of surviving. The neutral emotion mainly included objective descriptions of the disaster. In the process of classification, new texts were first processed using word segmentation and the removal of stop words. Then, the previously trained word vector list was used to translate each word into a word vector. Furthermore, each new text was transformed into a word vector matrix. Finally, the word vector matrix was input into the trained CNN model. Through the calculation of the model, each new text was labeled into the different emotional categories. We classified all 39,344 pieces of texts into the six emotion categories based on this classification process.

2.6. Spatio-Temporal Analysis of the Public Emotions

The framework in this paper aims to assist with disaster mitigation by using the public emotional information contained in social media. In the process, emotional information was regarded as an attribute of geographic information. The powerful spatial analysis capability of a GIS was used to combine the emotional information with other geographic data to dig out more useful knowledge. For example, the population density distribution data can be added to carry out a spatio-temporal assessment of the affected population. The POI data (such as a sanctuary) can be considered to explore the spatio-temporal trajectory law of people in sudden disasters. In addition, emotional information can also help disaster reduction departments to screen out urgent public demands from a vast amount of information. The public demands that contain emotional information are also effective feedback for disaster reduction work. They can help to optimize decision-making to improve rescue efficiency.

3. Spatio-Temporal Analysis of Public Emotional Information

3.1. Spatio-Temporal Assessment of the Affected Population

It is very important to know the distribution of the affected population at the time when an earthquake happens. This can help to ensure an effective assessment of the disaster situation and rational deployment of rescue resources. In this section, we combined the population density distribution data related to the study area with the spatio-temporal information contained in social media to assist analysis. Among them, the population density distribution data was taken from the GHSL (Global Human Settlement Layer) (http://cidportal.jrc.ec.europa.eu/ftp/jrc-opendata/GHSL/GHS_POP_GPW4_GLOBE_R2015A/). The introduction of public emotional information can improve the accuracy of the assessment. It is generally believed that negative emotions indicate that the earthquake has a greater impact on the people. Further, according to the rules of time period division from the rescue department, we set up six time periods, which were: 0–4 hours, 4–12 hours, 12–24 hours, 24–48 hours, 48–72 hours, and 0–72 hours after the earthquake. Then, we used the overlay analysis of GIS software to process these data in each time period. The results of the analysis of the related data are shown in Figure 5.

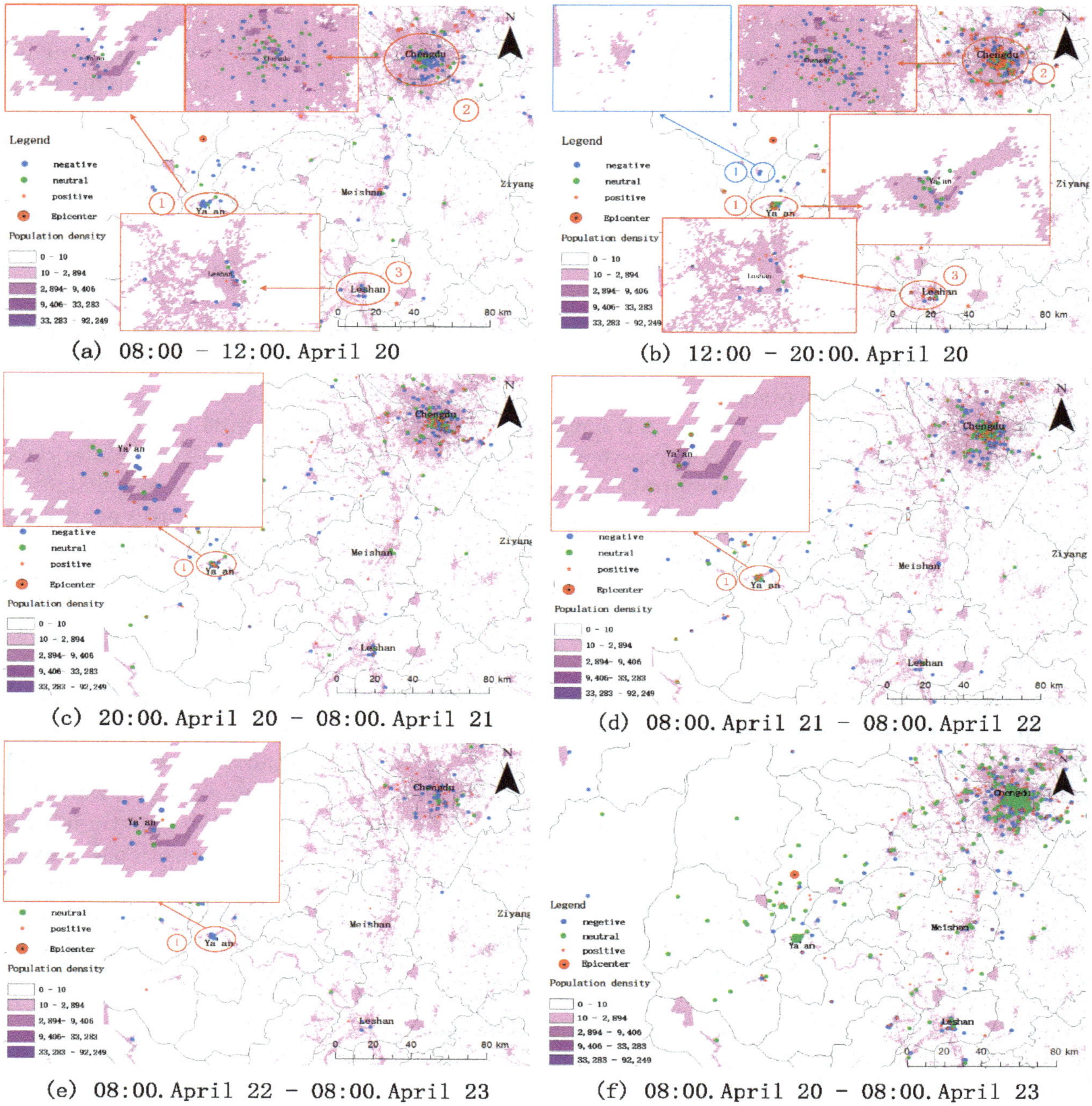

Figure 5. Emotional distribution characteristics of the affected population. The figure (**a**), (**b**), (**c**), (**d**) and (**e**) describe the distribution of emotions in different time periods within 72 hours after the disaster. The figure (**f**) shows the distribution of emotions over 72 hours. Among them, each of red circle 1, red circle 2, and red circle 3 in the figures represent the same area. The blue circle 1 in (**b**) shows that compared with (**a**), new negative emotions emerged in same area.

We know that: (1) The microblog data volume was larger in places with a high population density after the earthquake and negative emotions predominated. (2) Within four hours after the earthquake, as shown in Figure 5a, there were almost no positive emotions in the areas near to the epicenter. Areas far away from the epicenter, such as Leshan (red circle 3) and Chengdu (red circle 2), had fewer positive emotions. (3) From 4 hours to 12 hours after the earthquake, as shown in Figure 5b, compared with the previous distribution of emotional information, some new negative emotions emerged near the epicenter, such as the blue circle 1. This indicates that as time went on, some new disaster damage might have taken place in this region. We checked the corresponding text and found that these new emotion points were mostly anxiety. The reason people expressed anxiety was because the "Fan Min Road" was blocked by boulders, and people were worried that the rescue vehicles that did not know this information which would be delayed due to the incident. The emotions in other areas of this figure had also changed. For example, compared with the red circle 2 and red circle 3 in Figure 5a, the emotions in these areas of Figure 5b increased significantly. This indicates that the public's attention to earthquakes continued to increase in this time period. (4) We selected an area with a high population density near the epicenter to analyze how emotions changed over time in detail. The selected area was located in Yucheng District in Ya'an and it was marked with red circle 1 in Figure 5a–e. Figure 6 shows the changes in data volume in emotion categories in this area for different periods of time. We found that the positive emotions began to appear in the second period and then continued to increase. The reason was that as the rescue operation unfolded, people's gratitude for the rescuers increased. The number of negative emotions increased the most in the third period and then gradually reduced. Although this time period lasted only 12 hours, the number of negative emotions was the highest of all periods. Because this time period was the first night after the earthquake, most people were in urgent need of relief supplies, such as tents, clothes, etc. Therefore, anxiety was dominant. The number of neutral emotions expressed did not change very much. They mainly described the progress of earthquakes. (5) Figure 5f depicts the overall situation 72 hours after the earthquake. We found that the population density in Ya'an was not high, and the distribution of population was not uniform. However, the number of emotions expressed in this city was large and the distribution of them was fairly uniform, especially regarding negative emotions. This shows that the impact of the earthquake on Ya'an was the most serious. In addition, Chengdu is the capital of Sichuan Province and has the largest population density. During this earthquake, many negative emotions were expressed in this city. Although the distribution of these emotions was not uniform, more attention should be placed here to avoid unexpected accidents, such as people being hurt by rumors due to anxious emotions. The same approach can also be applied to other affected cities.

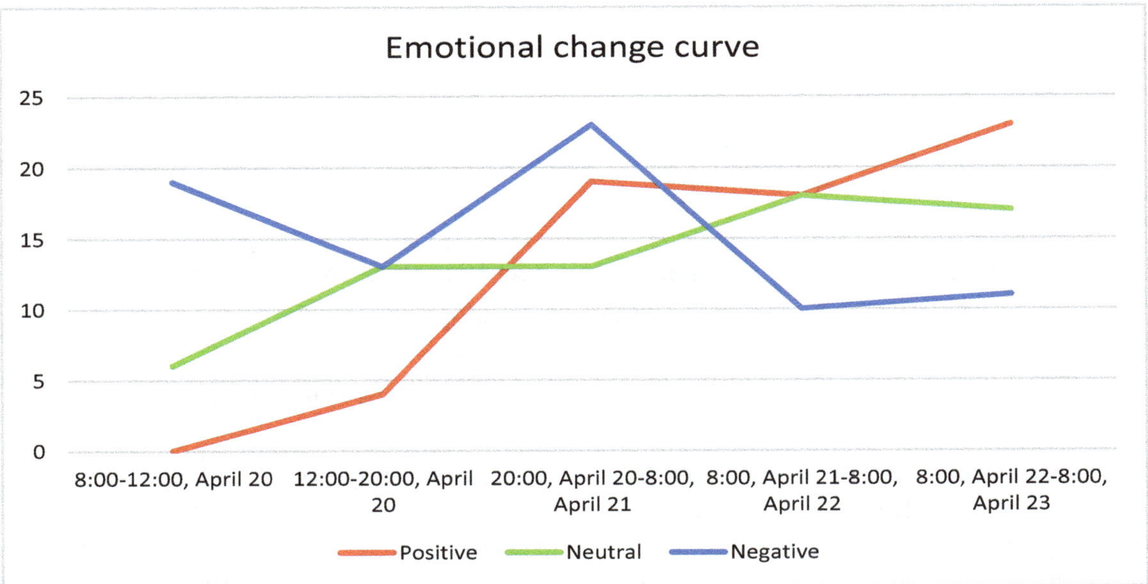

Figure 6. Changes in different emotion categories in data volume for different periods of time.

In this section, we conducted an overlay analysis containing public emotional information and population density distribution to assess the affected population. Spatio-temporal distribution characteristics of public emotional information can improve the accuracy of assessment and provides us with more valuable information. Although this emotional information was unevenly distributed, and even some areas with high population density had only a small amount of negative emotions, such as the area in the blue circle 1 in Figure 5b, we should still pay attention to them because emotions expressed by users of social media may also reflect emotions from the neighbors or communities around them, even if these neighbors or communities do not use social media [8].

3.2. Emotional Spatio-Temporal Trajectory Mining

Earthquakes, as a sudden disaster, cause tremendous damage in a short period of time, and is especially hazardous for of human life. Therefore, in most cities, there are many shelters for people to avoid these disasters. In this section, we explore how the spatio-temporal trajectories of human beings change when sudden disasters occur and whether these changes are related to the locations of shelters. Furthermore, in this process of change, we investigate which emotion categories are shown by human beings and how these emotion categories change. We used Chengdu as an example and determined the locations of the shelters in this city from "The Official Website of Chengdu Municipal People's Government (http://cdtf.gov.cn/chengdu/smfw/csyjbn.shtml)." Then, we translated these shelters into coordinates through the API of Baidu (http://api.map.baidu.com/lbsapi/getpoint/index.html) and vectorized the map of this region. Considering the sudden occurrence of earthquakes, we set up seven small time periods, which were 08:02 h to 08:12 h, 08:12 h to 08:22 h, 8:22 h to 08:40 h, 08:40 h to 09:30 h, 09:30 h to 10:30 h, 10:30 h to 11:30 h, and 11:30 h to 13:00 h, respectively (the earthquake occurred at 08:02 h), to analyze the changes of the crowd in these fine-grained time periods. Figure 7 shows the changes in the number of people over time. We can see that the population grew fastest in 08:40 h to 09:30 h. This may reflect that a large number of people had reached nearby shelters during this time period.

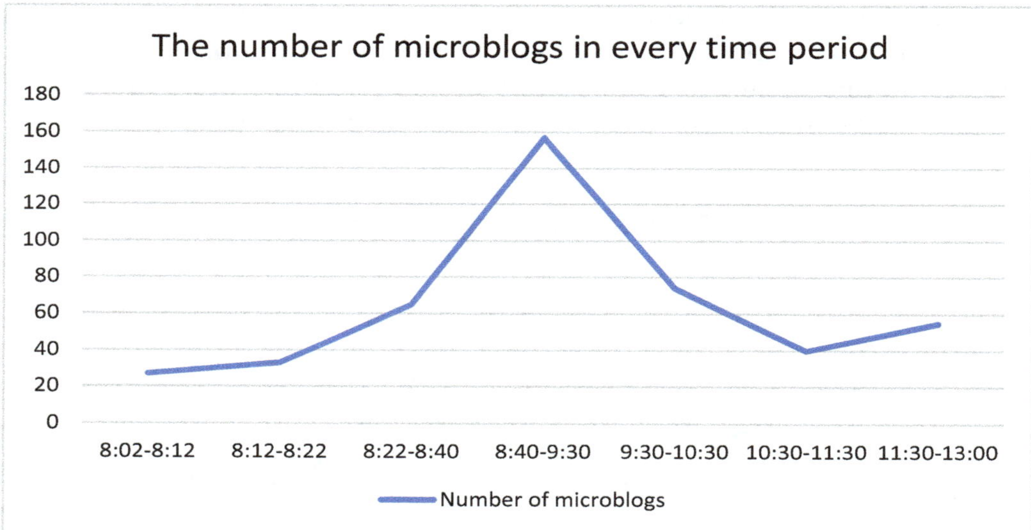

Figure 7. Changes of the crowd amount in each small time period.

We used the kernel density algorithm [38] to validate the change of crowd aggregation over time and explore the relationships between density centers of crowds and shelter locations, as shown in Figure 8. In Figure 8a, three clusters were formed within the first 10 minutes after the earthquake and we marked them as cluster 1, cluster 2, and cluster 3, respectively. Although the density of cluster 1 is small, we can see its core had been in the location of the shelter. This indicates that within 10 minutes of the earthquake, a small number of people had gathered in nearby shelters. Cluster 2 and cluster 3, especially cluster 3, had higher densities than cluster 1. However, the people in these areas had not yet gathered in the shelter. Ten minutes later, between 08:12 h to 08:22 h, the number of people increased as they further converged to shelters. We found that some shelters had gathered many people, such as cluster 1 and cluster 3, as shown in Figure 8b. As time went on, a large number of discrete and small clusters were fused into a larger clusters, as shown in Figure 8c–e. This indicates that during these periods, a large number of people had reached the shelters. Among them, the number of people between 08:40 h to 09:30 h was the largest, as shown in Figure 8d. Then, the number of people gathered in shelters began to decline. But the crowd hadn't dispersed at this time, as shown in Figure 8e. We can understand these changes through the corresponding value of crowd density. In Figure 8f,g, we can see that the big cluster began to disintegrate and was decomposed of small clusters. Then, these small clusters were gradually moving away from shelters. Perhaps it means that people's emotions were no longer so tense at this time.

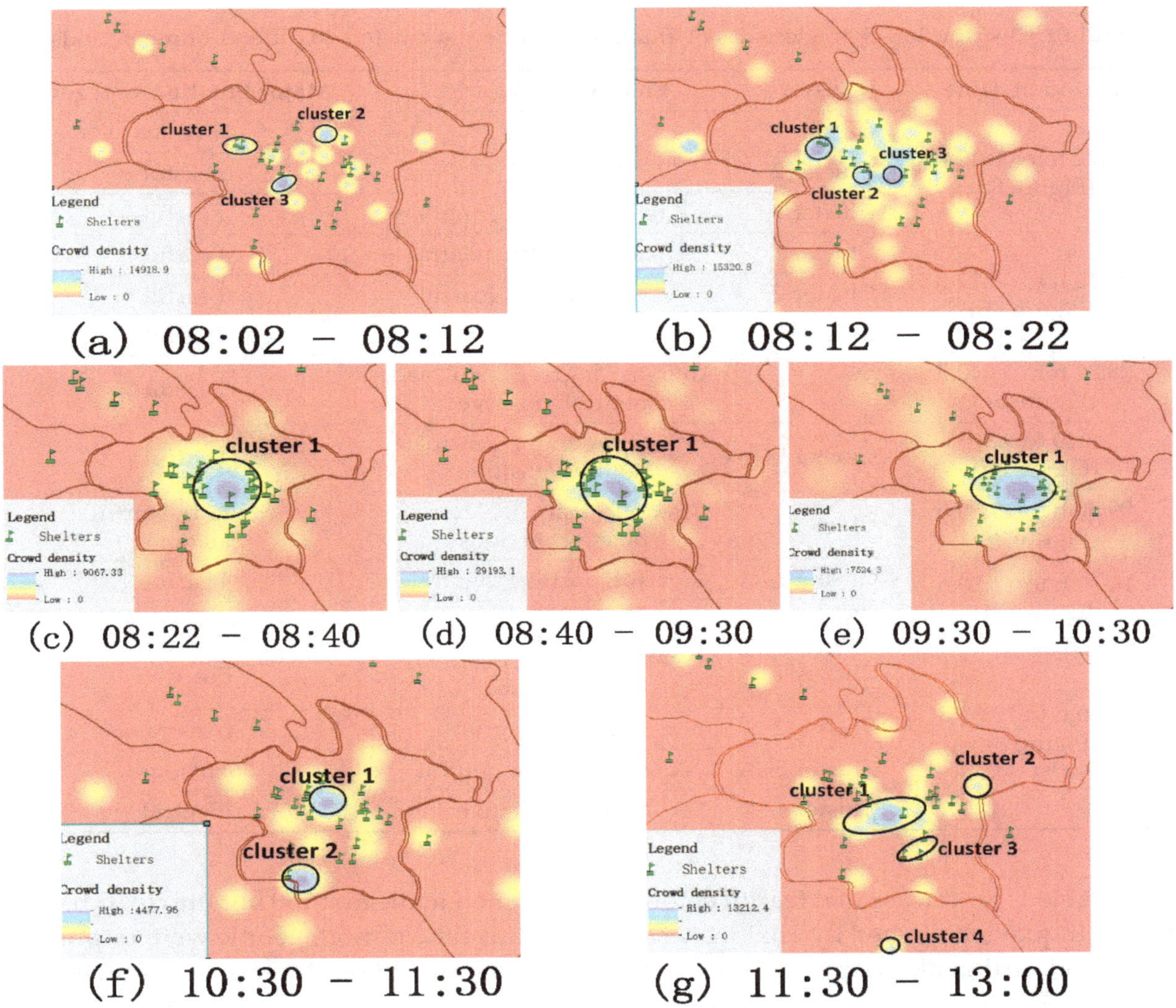

Figure 8. The change characteristics of public spatio-temporal trajectory. This sequence diagram describes how the crowd moved in different small time periods after the earthquake. Among them, the figure (**a**) shows the trajectories of public change in 10 minutes after the earthquake. Three clusters were formed in this period. The figure (**b**) shows the location relationship between each cluster and shelters in the second ten minutes. The figure (**c**), (**d**) and (**e**) shows that all small clusters formed a large cluster over time and it had the largest population between 08:40 and 09:00 as in figure (**d**). The figure (**f**) and (**g**) shows crowd was gradually dissipating and leaving the shelter. From the whole process of analysis, we determined that: (1) When the earthquake happened, people rushed to the

shelters in a very short time period. However, were these shelters reasonably laid out? We saw that some shelters did not contain many people, or even had no people. Therefore, the analysis results could be used as a reference for the rational layout of shelters. (2) The characteristics of crowd gathering and evacuation could be used as an effective reference to aid disaster reduction departments in dealing with future emergencies.

Further, we wanted to know which categories of emotion the public expressed during these periods and how these emotions changed because massive population movements in a short time may lead to some unnecessary accidents such as possible stampedes due to panic. Therefore, if the public emotions in this process can be monitored, it will help us take quick and effective measures to improve the efficiency of evacuation and prevent accidents. Table 1 presents the emotional characteristics in each period of time in Figure 8. Indicators in this table include clusters formed by kernel density clustering, the emotion categories and major emotion category contained in each cluster, etc.

Table 1. Distribution characteristics of emotions in different clusters in different time periods.

Period of Time	Cluster	Emotion Categories	Major Emotion Category
08:02 h to 08:12 h (Figure 8a)	Cluster 1	Anxious	Anxious
	Cluster 2	Fearful	Fearful
	Cluster 3	Anxious, fearful, angry	Fearful
08:12 h to 08:22 h (Figure 8b)	Cluster 1	Anxious, fearful, neutral	Fearful
	Cluster 2	Anxious, angry, fearful	Fearful
	Cluster 3	Angry, fearful, positive	Fearful
08:22 h to 08:40 h (Figure 8c)	Cluster 1	Anxious, angry, Fearful, sad, neutral, positive	Fearful
8:40 to 9:30 (Figure 8d)	Cluster 1	Anxious, angry, Fearful, neutral, positive	Fearful
09:30 h to 10:30 h (Figure 8e)	Cluster 1	Anxious, fearful, Neutral, positive anxious	Anxious, fearful anxious
10:30 h to 11:30 h (Figure 8f)	Cluster 1	Fearful, positive, neutral	Fearful
	Cluster 2	Neutral, sad	Neutral, sad
11:30 h to 12:30 h (Figure 8g)	Cluster 1	Fearful, neutral positive, anxious, sad	Positive
	Cluster 2	Neutral, angry	Angry
	Cluster 3	Angry, anxious, neutral	Anxious
	Cluster 4	Neutral, positive	Neutral

From Table 1, we can see that fearful emotions dominated in the first 150 minutes (8:02 h to 10:30 h) after the earthquake, followed by anxiousness. During this time period, people were unprepared for the unexpected earthquake and were afraid of losing their lives in the earthquake. Among them, in the first time period (8:02 h to 8:12 h), people expressed anxiousness in cluster 1, as shown in Figure 8a. Through the corresponding text content, we found that they did not know the details of the earthquake at this time (such as the location of the epicenter, the magnitude scale, etc.), so they were worried about the safety of their relatives, friends, and even others whom they did not know. Angry, neutral, and positive emotions began to appear in the second period of time (8:12 h to 8:22 h). However, the number of the positive and neutral emotions was relatively small, with one piece being positive and two pieces being neutral. Angry emotions in this time period mainly showed people's aversion to earthquakes. Sad emotions began to emerge in the third time period (08:22 h to 08:40 h) People who expressed sad emotions were mainly because this earthquake reminded them of a dreadful catastrophe that happened in Sichuan in 2008 (the earthquake that occurred in Wenchuan, Sichuan Province, on May 12, 2008, caused great damage). With more and more detailed information about earthquakes becoming available, people expressed more categories of emotion, and the reasons for these emotions had also changed. For example, in the fourth and fifth time periods, people expressed fearful and anxious emotions because they worried about there would be some aftershocks in the near future. This also shows that for a long time, people still gathered near the shelter, as shown in Figure 8c–e. We can see many people began to leave shelters in Figure 8f. Combining with emotions people expressed in this period, we found fearful emotions were no longer dominant. People expressed sad emotions in cluster 2 in the sixth time period because of grief for the victims in the worst-hit areas. In the seventh time period, the main emotion category in each cluster was different. People might have calmed down at this time. Even the main emotion of cluster 2 in Figure 8g was positive. People expressed their prayers for the disaster areas.

The fine-grained emotion analysis was not only a further explanation for human spatio-temporal trajectory mining, but also provides more details of the disaster for disaster reduction departments. On the one hand, it provides an understanding of the public's emergency awareness and movement law in the study area. On the other hand, based on characteristics of the emotional spatio-temporal trajectory, disaster reduction departments can provide timely management and guidance for "key nodes" in this process to avoid unexpected accidents. For example, we can provide effective guidance for areas where negative emotions are intense, i.e., in Figure 8a,b, to avoid possible stampedes due to panic.

3.3. Post-Disaster Emotional Change Analysis

The sudden catastrophe has a long-term impact on the public. By monitoring and analyzing the public fine-grained emotional information, we can mine a lot of important information from massive disaster-related data. This information can help us quickly understand the public's needs and feedback, even for some hard-to-find problems, such as mental health. This is very important for us to improve the efficiency of disaster emergency response and rescue. In this section, we took Ya'an as an example and used days as intervals to monitor public emotions for a longer period of time from the macro perspective. Meanwhile, we analyzed the causes of public emotion change using hot word extraction. This can help us quickly grasp the public's concerns from the mass emotional information. The tool we used to extract hot words was from the web (http://www.picdata.cn/).

Regarding positive emotions, as shown in Figure 9, on the second day (April 21) after the disaster, we brought the hot words "感动 (moved)" and 感激 (grateful)" into the corresponding text. We found that the public expressed their gratitude mainly to the professional rescue workers, such as the army. There were fewer volunteers at this time. However, more and more volunteers spontaneously joined the rescue effort over time, especially on the third and fourth days. Because in this time, "志愿者(volunteer)" appear more frequently. The civilian rescue gradually arrived in the disaster area from about the fourth day (April 23) after the disaster. The hot words "爱心 (love)"and "物资 (relief supplies)" indicated that people in the disaster-stricken areas were grateful for non-governmental spontaneous relief materials. Based on the changes in public positive emotions, we can understand the general process of rescue work.

Regarding anxiety, as shown in Figure 10, as time went on, the anxiety of the public gradually decreased. On the second day after the earthquake, there was more anxiety. The reason for this was because: (1) The roads were interrupted and some areas were isolated. We used the hot words "中断 (interrupt)" and "救援 (rescue)" in the original micro-blog texts to get detailed information. We found that "上里镇 (Shangli town)," "中里镇 (Zhongli town)," "下里镇 (Xiali town)," and "碧峰峡 (Bifengxia town)" were isolated from the outside world and needed urgent rescue after the earthquake. (2) Some areas were in urgent need of relief supplies, and these areas were suffering from bad weather. For example, through hot words, we found that there were some micro-blog texts saying that "Wangjia village in Longmen town was short of water, food, medicine, and tents". (3) Some people expressed anxiety because they could not contact their relatives and friends after the earthquake. From April 21 to April 26, we found the public's anxiety was mainly due to a lack of supplies and bad weather. Additionally, as time went on, the public's demand for relief supplies mainly concerned tents, especially on April 26th. The combination of micro-blog content and location information could have allowed a more precise rescue plan to be carried out.

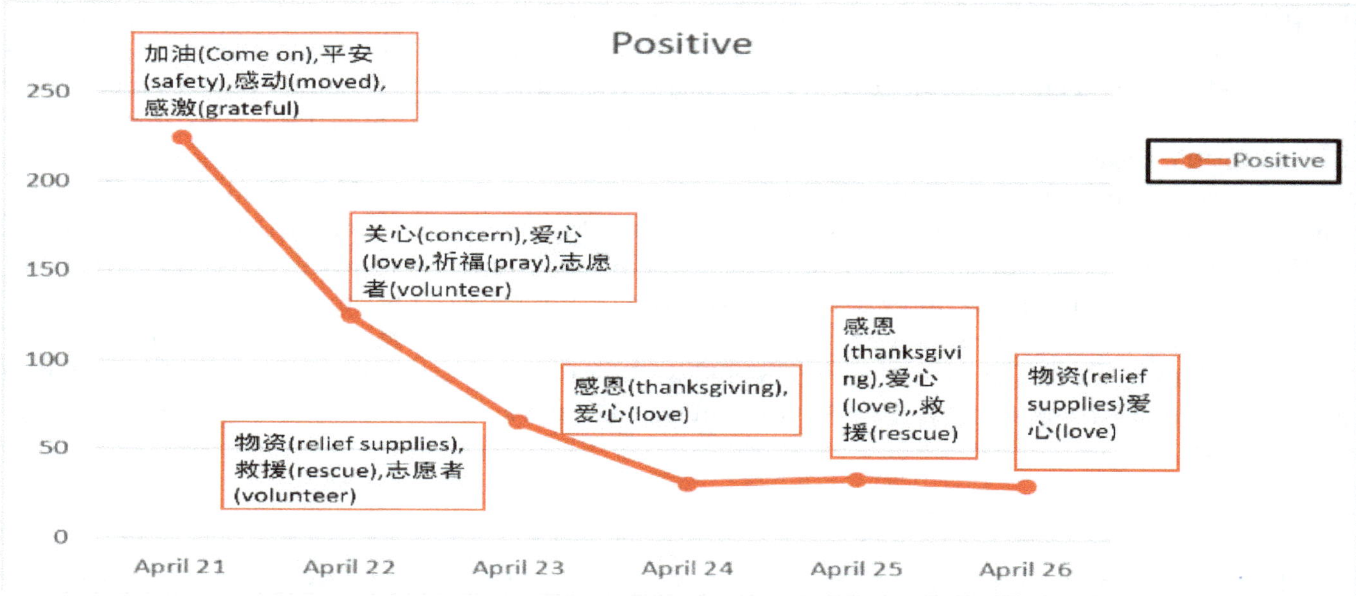

Figure 9. Sequence diagram of positive emotion (the words in the text box are the hot words related to this emotion in the corresponding time period).

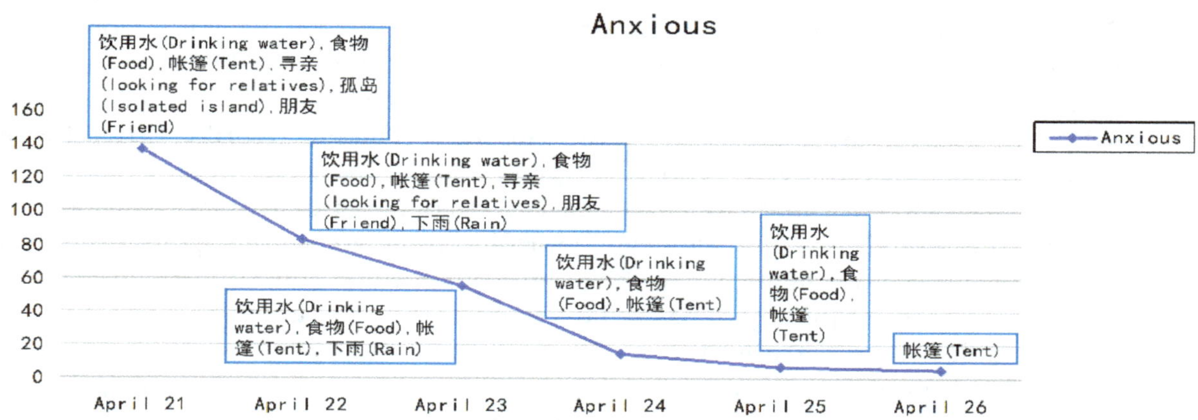

Figure 10. Sequence diagram of anxiety.

As shown in Figure 11, from April 21 to April 23, the public mainly expressed anger because of their aversion to the earthquake and because of some internet fraud. By combining the corresponding micro-blog content, we found that some of the Internet fraud was exposed by micro-blog users, such as "It's horrible. Some criminals cheated under the cover of the earthquake. Please pay attention to this telephone number: xxx.". These angry messages were used to help people guard against rumors (especially for anxious people). After April 23rd, public anger was mainly due to aversion against the disaster.

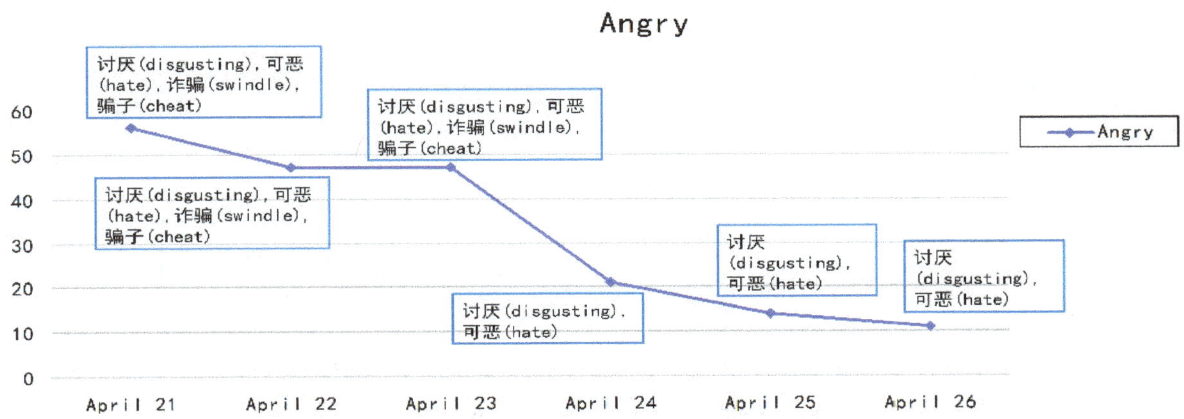

Figure 11. Sequence diagram of anger.

As shown in Figure 12, from April 21 to April 25, the sadness was due to the destruction of homes and the deaths of relatives or friends. Examples of corresponding micro-blog texts are: "Where is home? Where is the classroom? Yesterday? I was not an orphan yesterday" and "It's a scene of complete devastation and when can we rebuild our homeland?" On April 26, many mourning activities were carried out by official and non-governmental organizations, which was the reason why people expressed sadness. During the process of the earthquake rescue, disaster relief organizations could send psychological relief to places where sadness was intense, as determined by the locations of the corresponding micro-blog information.

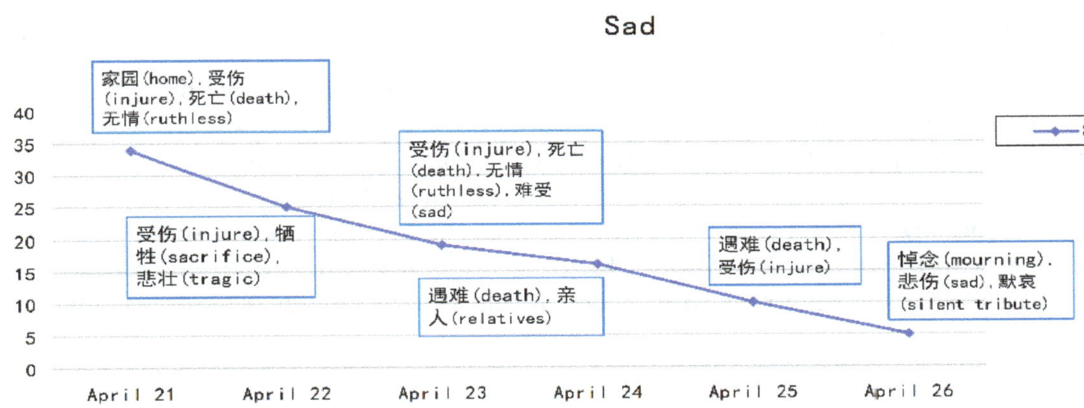

Figure 12. Sequence diagram of sadness.

In terms of fear, as shown in Figure 13, from April 22 to April 24, there were several aftershocks in Ya'an, which had a great impact on the public's life. Many hot words were observed, such as "余震 (aftershock)" and "可怕 (dreadful)". However, between April 24 and April 26, especially on April 26, there was a sudden increase in fear, and the hot words mainly included "心理咨询师 (psychological counselors)," "回忆 (recall)," and "惊吓 (startle)." We used these hot words to get the original micro-blog and saw some saying that: "A teacher in Zhongli Town reported that a girl was afraid of loud voices and kept eating. She said she would be afraid if she didn't eat. So this teacher hoped that the disaster reduction department could send a psychological counselor to help that girl" and "My brother said that as long as there was the thunder and lightning in Ya'an, he was very scared! Ask for help!" Therefore, the relief department should give some help to these people.

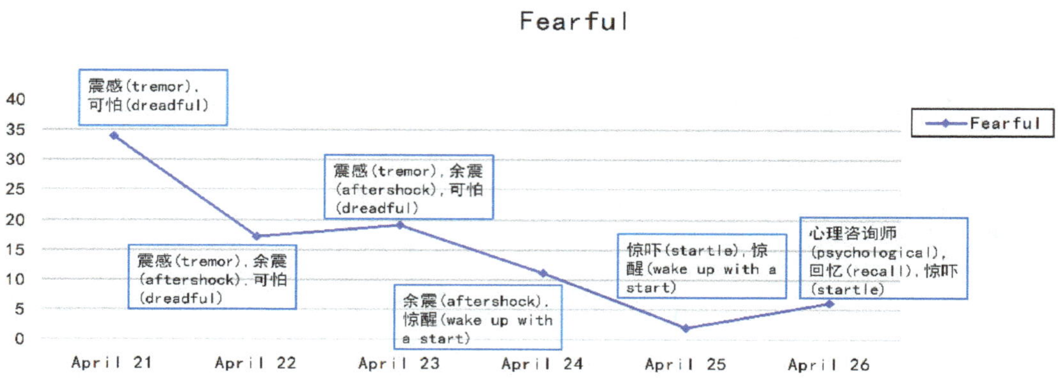

Figure 13. Sequence diagram of fear.

We can continue to explore other areas in the same way. This could help us to accurately understand the public's reactions to the progress of disaster mitigation. Further, the analysis results could help the rescue department to optimize rescue strategies and improve the efficiency of rescues.

4. Evaluation of the Experimental Indicators

4.1. Accuracy Evaluation of the Emotion Classification

4.1.1. Experimental Corpus

In the emotion classification experiment, we first manually annotated a corpus based on the six emotion categories. In this corpus, each emotional category contained 1000 text samples. Among these, 800 text samples were selected as the training corpus and 200 were selected as the testing corpus from each emotion category.

4.1.2. Experimental Environment

In order to improve the accuracy of the emotional classification, we translated special characters and symbols in the text into Chinese words. We integrated the word2vec framework (from Google [39]) and NLPIR-ICTCLAS (http://ictclas.nlpir.org) into our algorithmic framework to assist with the processing of this text. Finally, we built convolutional neural networks based on tensor flow [40], and optimized the model parameters to achieve the best results. In this process, we set the dimensions of the word vector as 200, the number of convolution kernels was 3 and their sizes were 3, 4, and 5. The size of max pooling was 4, the proportion dropout regularization was 0.3, and the stride was 1.

4.1.3. Experimental Results and Accuracy Comparison

We verified the accuracy of the algorithm based on the precision (P), recall (R), and comprehensive evaluation indexes (*F-1*). The formulas are shown below:

$$P = \frac{N\_Correct}{N\_Correct + N\_False} \tag{5}$$

$$R = \frac{N\_Correct}{N\_Category} \tag{6}$$

$$F - 1 = \frac{2 \times P \times R}{P + R}. \tag{7}$$

N_Correct represents the number of texts that were correctly classified into one category, *N_False* represents the number of texts that were misclassified into this category, and *N_Category* represents the number of texts that belonged to this category in the testing corpus.

Table 2 and Figure 14 show the accuracy of the CNN model in the fine-grained emotion classification. The comprehensive evaluation index scores for each category were all above 81%, which met the experimental requirements. In addition, in this paper, we also considered the use of slang, internet buzzwords, and special characters and symbols to enhance the performance of the model.

Table 2. Accuracy evaluation of positive emotion classification.

Emotional Category	Precision (P)	Recall (R)	Comprehensive Evaluation Index (*F-1*)
Positive	82.25%	80.00%	82.54%
Neutral	84.21%	87.91%	86.02%
Angry	91.57%	86.36%	88.89%
Sad	88.54%	85.00%	86.73%
Anxious	78.47%	85.27%	81.77%
Fearful	84.69%	85.57%	85.13%

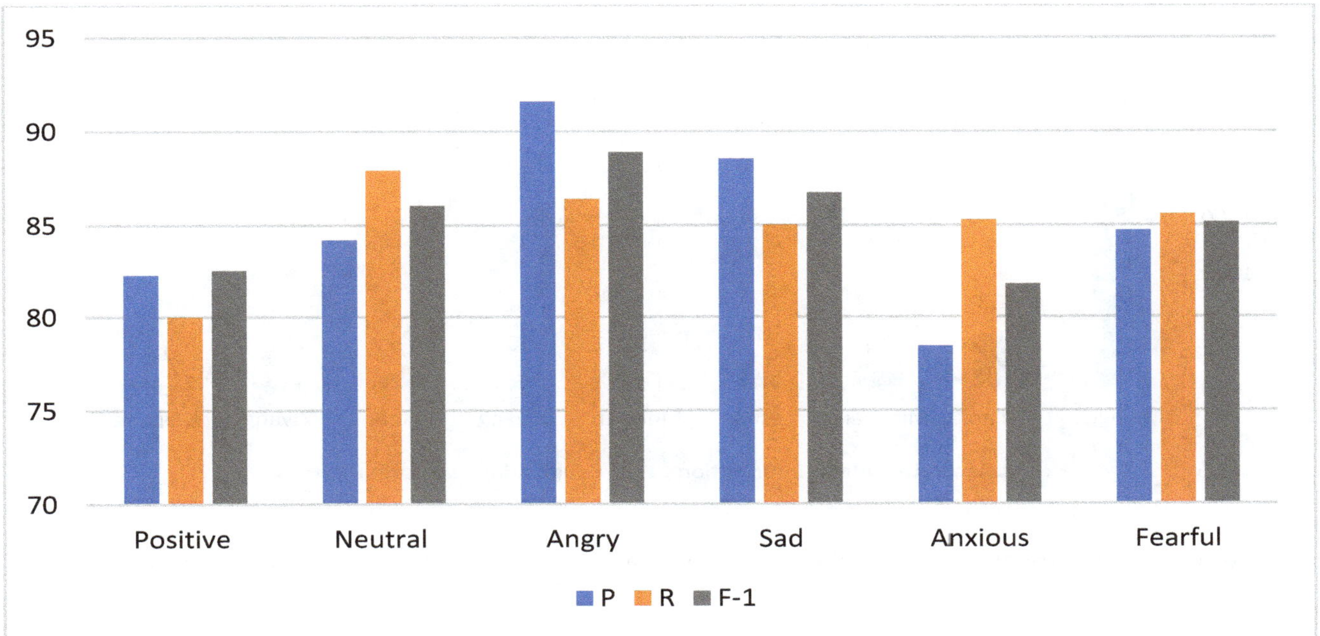

Figure 14. Classification accuracy of different emotions.

4.2. *Evaluation of Spatio-Temporal Analysis Experiments*

4.2.1. The Description of Data with Address Information

The number of texts in the data set of this paper was 39341. However, not all texts contained address information. We processed this address information based on the method described in Section 2.2. All address information can be divided into two categories in this paper, including rough address information and accurate address information. Among them, rough address information can only represent villages and towns, even districts and counties, such as "Lushan County, Ya'an City" and "Wuhou District, Chengdu City," etc. Accurate address information can represent streets and geographical entities, such as "Sishengci North Street," "Wangjiang Campus of Sichuan University," etc. Table 3 and Figure 15 depicts the proportion and number of data with different accuracy in different cities. Among them, the formula for calculating the proportion is as shown:

$$\text{Proportion} = \frac{\text{The number of specified texts in the city}}{\text{The number of all texts in the city}} \tag{8}$$

Table 3. Proportion of data with different accuracy in different cities.

City	The Proportion of Data with Accurate Address Information	The Proportion of Data with Rough Address Information	Total
Chengdu	8.08%	2.16%	10.24%
Ya'an	12.18%	13.91%	26.09%
Mianyang	9.45%	2.60%	12.05%
Leshan	9.81%	4.85%	14.66%
Meishan	13.30 %	7.44%	20.74%
Deyang	8.71%	5.79%	14.50%
Aba	11.29%	23.51%	34.80%
Ziyang	5.10%	3.10%	8.20%
Neijiang	4.79%	2.33%	7.12%

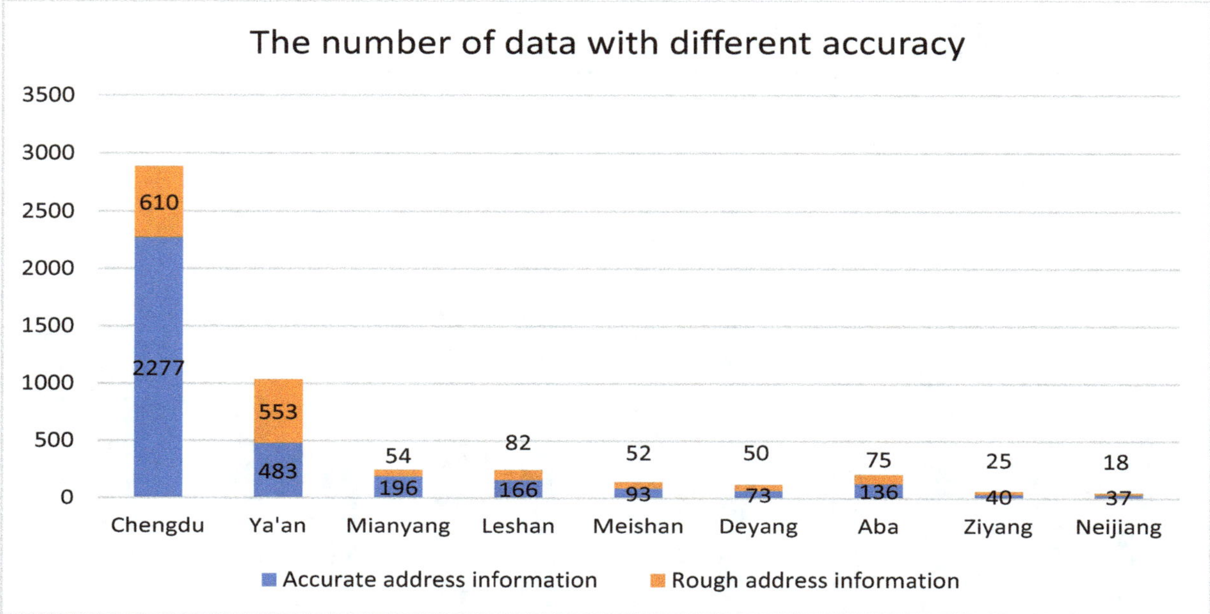

Figure 15. Comparison of the number of pieces of address information with different accuracy in each city.

4.2.2. Evaluation of the Experimental Process and Results

In Section 3.1, we used the population density distribution data provided by GHSL (Global Human Settlement Layer) to assist in assessing the affected population. The scale of maps we used was relatively small. Therefore, we considered that all the data with both accurate address information and rough address information can be used. Among them, the data in Aba and Ya'an better reflected the real situation expressed by social media in the region because although the social media data volume in these two cities is small, the data with address information accounted for a larger proportion; they reached 34.8% and 26.09%, respectively. Considering population density and epicenter location (the epicenter of the earthquake occurred in Ya'an), in Section 3.1, we mainly focused on the disaster situation in Ya'an and several cities close to Ya'an.

In Section 3.2, we explored how the spatio-temporal trajectories of human beings changed when sudden disasters occur and used geographic data (shelters) with accurate location information. Therefore, we considered that the social media data with accurate address information can be used for analysis. In addition, earthquakes are sudden disasters and cause tremendous damage in a short period of time. Therefore, in this paper, we set up seven small periods of time to explore the public movement after the disaster and combined this with public emotions to mine more details regarding the disaster. This required more data with accurate address information. We can see that Chengdu meets the requirements most from Figure 15 and this produced satisfactory analysis results in Section 3.2. Of course, the same method can also be used to analyze the disaster in Ya'an; however, time granularity would be coarse.

In Section 3.3, we mainly monitored the emotional information in each city for a long time. Therefore, we just needed to know which city the social media data belongs to. Although a lot of social media data had no location information, they all had their own labels. This was explained in Section 2.1. Considering that Ya'an was the worst hit by the earthquake, we selected Ya'an as the research object.

5. Conclusions

When a disaster occurs, social media can provide a large amount of important disaster-related geographic information to the disaster reduction departments in near real-time. In this paper, we regarded the fine-grained public emotional information extracted from social media as an attribute of geographic information to assist in disaster mitigation. In the process of extracting emotional information, we fully analyzed the characteristics of Chinese social media and selected a suitable algorithm (convolution neural network model). Meanwhile, a large number of special characters and symbols with emotional characteristics contained in social media were also considered to improve the accuracy of classification. The methods in this paper achieved satisfactory results.

In order to verify the effectiveness of the method in this paper in disaster mitigation, we used the 7.0 earthquake that occurred on April 20, 2013, in Ya'an City, Sichuan Province, China, as a case study. We classified the social media texts related to areas affected by the earthquake into six different emotion categories. Then, with the help of GIS software and other traditional geographic information data (population density distribution data and shelter data), we explored the role of public emotional information that is helpful for disaster reduction. The results showed that fine-grained public emotions can provide more powerful data support for disaster reduction departments to optimize rescue strategies and improve rescue efficiency.

Although social media plays an important role in assisting disaster mitigation, it also has some limitations. (1) Social media data is unevenly distributed. The economically developed and populous areas tend to have more users of the Sina micro-blog. In the research area of this paper, Chengdu had the most Sina micro-blog data and these data were more concentrated in the urban area, but it was not the worst-hit city. Therefore, in future research, more abundant data that include other sources is also needed to supplement social media data, such as image data, vehicle-borne GPS data, etc. (2) Not all social media users are willing to share their location information. In the dataset used in this paper, the proportion of text with location information was very small. This limits the use of some spatio-temporal analysis methods. However, we found that there are many geographically named entities in texts and many of them can respect the user's location. Therefore, an effective method is needed to automatically extract these geographically named entities to supplement the deficiencies of geographic location information in social media.

In addition, the use of social media for disaster mitigation is far from enough. With the development of data mining technology, more disaster-related information contained in social media can be extracted, such as different categories of disaster loss information, etc. With the help of the powerful spatio-temporal analysis ability of GISs, this useful information can play a greater role.

Author Contributions: T.Y., J.X. and G.L. conceived and designed the paper; T.Y. and J.X. wrote the paper; T.Y. and N.M. designed and implemented the algorithmic framework; T.Y. and Z.L. realized the visualization; C.T. and J.Z. collected data and processed them.

Acknowledgments: We thank Edward T.-H. Chu, Associate Professor at Yunlin University of Science and Technology for his advice and collaborative work on the emotion classification under the framework of cooperation project. We also thank Zhenyu Lin and Qinglan Zhang, Master students at Henan Polytechnic University for their advice on visualization.

References

1. Goodchild, M.F. Citizens as sensors: Web 2.0 and the volunteering of geographic information. *GeoFocus. Rev. Int. Ciencia Tecnol. Inf. Geogr.* **2007**, *7*, 8–10.
2. Zhou, Y.; Yang, L.; Van de Walle, B.; Han, C. Classification of microblogs for support emergency responses: Case study Yushu earthquake in China. In Proceedings of the 2013 46th Hawaii International Conference on System Sciences, Wailea, Maui, HI, USA, 7–10 January 2013; pp. 1553–1562.
3. Qu, Y.; Huang, C.; Zhang, P.; Zhang, J. Microblogging after a major disaster in China: A case study of the 2010 Yushu earthquake. In Proceedings of the ACM 2011 Conference on Computer Supported Cooperative Work, Hangzhou, China, 19–23 March 2010; pp. 25–34.
4. Chae, J.; Thom, D.; Jang, Y.; Kim, S.; Ertl, T.; Ebert, D.S. Special Section on Visual Analytics: Public behavior response analysis in disaster events utilizing visual analytics of microblog data. *Comput. Graph.* **2014**, *38*, 51–60. [CrossRef]
5. Li, Z.; Wang, C.; Emrich, C.T.; Guo, D. A novel approach to leveraging social media for rapid flood mapping: A case study of the 2015 South Carolina floods. *Cartogr. Geogr. Inf. Sci.* **2017**, 1–14. [CrossRef]
6. *Social Media in Disasters and Emergencies*; The Drum: Washington, DC, USA, 3 March 2013.
7. Ekman, P. Basic Emotions. In *Handbook of Cognition & Emotion*; Dalgleish, T., Power, M.J., Eds.; John Wiley & Sons: Hoboken, NY, USA, 1999.
8. Gruebner, O.; Lowe, S.R.; Sykora, M.; Shankardass, K.; Subramanian, S.V.; Galea, S. A novel surveillance approach for disaster mental health. *PLoS ONE* **2017**, *12*, e0181233. [CrossRef] [PubMed]
9. Lindell, M.K.; Prater, C.S.; Perry, R.W.; Nicholson, W.C. *Fundamentals of Emergency Management*; Emond Montgomery Publications: Toronto, ON, Canada, 2006.
10. Camras, L.A. An Event—Emotion or Event—Expression Hypothesis? A Comment on the Commentaries on Bennett, Bendersky, and Lewis (2002). *Infancy* **2010**, *6*, 431–433. [CrossRef]
11. Goltz, J.D.; Russell, L.A.; Bourque, L.B. Initial behavioural response to a rapid onset disaster. *Int. J. Mass Emerg. Disasters* **1992**, *10*, 43–69.
12. Yi, J.; Nasukawa, T.; Bunescu, R.; Niblack, W. Sentiment Analyzer: Extracting Sentiments about a Given Topic using Natural Language Processing Techniques. In Proceedings of the IEEE International Conference on Data Mining, Melbourne, FL, USA, 22 November 2003; pp. 427–434.
13. Xu, R.; Wong, K.F.; Xia, Y. Coarse-Fine Opinion Mining—WIA in NTCIR-7 MOAT Task. In Proceedings of the NTCIR 2008, Tokyo, Japan, 16–19 December 2008.
14. Turney, P.D. Thumbs up or thumbs down? Semantic orientation applied to unsupervised classification of reviews. In Proceedings of the Annual Meeting of the Association for Computational Linguistics, Philadelphia, Pennsylvania, 6 July 2012; pp. 417–424.
15. Xiao, Z.; Li, X.; Wang, L.; Yang, Q.; Du, J.; Sangaiah, A.K. Using convolution control block for Chinese sentiment analysis. *J. Parallel Distrib. Comput.* **2017**, *116*, 18–26. [CrossRef]
16. Narayanan, V.; Arora, I.; Bhatia, A. Fast and accurate sentiment classification using an enhanced Naive Bayes model. In *Proceedings of the International Conference on Intelligent Data Engineering and Automated Learning*; Springer: Berlin/Heidelberg, Germany, 2013; pp. 194–201.
17. Pang, T.B.; Pang, B.; Lee, L. Thumbs up? Sentiment Classification using Machine Learning. *Empir. Methods Nat. Lang. Process.* **2002**, *10*, 79–86.
18. Jozefowicz, R.; Vinyals, O.; Schuster, M.; Shazeer, N.; Wu, Y. Exploring the Limits of Language Modeling. *arXiv* **2016**, arXiv:1602.02410.
19. Bengio, Y.; Ducharme, R.; Vincent, P.; Jauvin, C. A neural probabilistic language model. *J. Mach. Learn. Res.* **2003**, *3*, 1137–1155.
20. Norris, F.H.; Friedman, M.J.; Watson, P.J.; Byrne, C.M.; Diaz, E.; Kaniasty, K. 60,000 Disaster Victims Speak: Part I. An Empirical Review of the Empirical Literature, 1981–2001. *Psychiatry-Interpers. Biol. Process.* **2002**, *65*, 207–239. [CrossRef]
21. Goldmann, E.; Galea, S. Mental health consequences of disasters. *Ann. Rev. Public Health* **2014**, *35*, 169. [CrossRef] [PubMed]
22. Neria, Y.; Shultz, J.M. Mental health effects of Hurricane Sandy: Characteristics, potential aftermath, and response. *J. Am. Med. Assoc.* **2012**, *308*, 2571–2572. [CrossRef] [PubMed]
23. Coyle, D.; Meier, P. New technologies in emergencies and conflicts: The role of information and social

networks. *Washington D* **2009**. Available online: https://www.popline.org/node/209135 (accessed on 12 January 2019).
24. Tausczik, Y.R.; Pennebaker, J.W. The Psychological Meaning of Words: LIWC and Computerized Text Analysis Methods. *J. Lang. Soc. Psychol.* **2009**, *29*, 24–54. [CrossRef]
25. Oh, O.; Kwon, K.H.; Rao, H.R. An Exploration of Social Media in Extreme Events: Rumor Theory and Twitter during the Haiti Earthquake 2010. In Proceedings of the International Conference on Information Systems, Icis 2010, Saint Louis, MO, USA, 12–15 December 2010; p. 231.
26. Mikolov, T.; Sutskever, I.; Chen, K.; Corrado, G.; Dean, J. Distributed representations of words and phrases and their compositionality. *Adv. Neural Inf. Process. Syst.* **2013**, *26*, 3111–3119.
27. Mikolov, T.; Chen, K.; Corrado, G.; Dean, J. Efficient estimation of word representations in vector space. *arXiv* **2013**, arXiv:1301.3781.
28. Mikolov, T.; Kopecky, J.; Burget, L.; Glembek, O.; Cernocky, J. Neural network based language models for highly inflective languages. In Proceedings of the IEEE International Conference on Acoustics, Speech and Signal Processing, Taipei, Taiwan, 19–24 April 2009; pp. 4725–4728.
29. Xiong, F.; Deng, Y.; Tang, X. The Architecture of Word2vec and Its Applications. *J. Nanjing Norm. Univ.* **2015**, *1*, 43–48.
30. Kim, Y. *Convolutional Neural Networks for Sentence Classification*; Association for Computational Linguistics: Doha, Qatar, 2014.
31. Zhang, D.; Wang, D. Relation Classification: CNN or RNN? 2016. Available online: https://link.springer.com/chapter/10.1007/978-3-319-50496-4_60 (accessed on 12 January 2019).
32. Pappas, N.; Popescu-Belis, A. Multilingual Hierarchical Attention Networks for Document Classification. *arXiv* **2017**, arXiv:1707.00896.
33. Yin, W.; Kann, K.; Yu, M.; Schtze, H. Comparative Study of CNN and RNN for Natural Language Processing. *arXiv* **2017**, arXiv:1702.01923.
34. Tang, D.; Qin, B.; Liu, T. Document modeling with gated recurrent neural network for sentiment classification. In Proceedings of the 2015 Conference on Empirical Methods in Natural Language Processing, Lisbon, Portugal, 17–21 September 2015; pp. 1422–1432.
35. Report on Text Classification Using CNN, RNN & HAN. Available online: https://medium.com/jatana/report-on-text-classification-using-cnn-rnn-han-f0e887214d5f (accessed on 12 January 2019).
36. Collobert, R.; Weston, J. A unified architecture for natural language processing:deep neural networks with multitask learning. In Proceedings of the International Conference on Machine Learning, Helsinki, Finland, 5–9 July 2008; pp. 160–167.
37. Wang, Y.; Feng, S.; Wang, D.; Yu, G.; Zhang, Y. Multi-label Chinese Microblog Emotion Classification via Convolutional Neural Network. In Proceedings of the Web Technologies and Applications: 18th Asia-Pacific Web Conference, APWeb 2016, Suzhou, China, 23–25 September 2016.
38. Silverman, B.W. *Density Estimation for Statistics and Data Analysis*; Chapman and Hall: San Francisco, CA, USA, 1986; pp. 296–297.
39. Word2Vec. Available online: https://code.google.com/archive/p/word2vec/ (accessed on 12 January 2019).
40. Abadi, M.; Agarwal, A.; Barham, P.; Brevdo, E.; Chen, Z.; Citro, C.; Corrado, G.S.; Davis, A.; Dean, J.; Devin, M. TensorFlow: Large-Scale Machine Learning on Heterogeneous Distributed Systems. *arXiv* **2016**, arXiv:1603.04467.

Permissions

All chapters in this book were first published in MDPI; hereby published with permission under the Creative Commons Attribution License or equivalent. Every chapter published in this book has been scrutinized by our experts. Their significance has been extensively debated. The topics covered herein carry significant findings which will fuel the growth of the discipline. They may even be implemented as practical applications or may be referred to as a beginning point for another development.

The contributors of this book come from diverse backgrounds, making this book a truly international effort. This book will bring forth new frontiers with its revolutionizing research information and detailed analysis of the nascent developments around the world.

We would like to thank all the contributing authors for lending their expertise to make the book truly unique. They have played a crucial role in the development of this book. Without their invaluable contributions this book wouldn't have been possible. They have made vital efforts to compile up to date information on the varied aspects of this subject to make this book a valuable addition to the collection of many professionals and students.

This book was conceptualized with the vision of imparting up-to-date information and advanced data in this field. To ensure the same, a matchless editorial board was set up. Every individual on the board went through rigorous rounds of assessment to prove their worth. After which they invested a large part of their time researching and compiling the most relevant data for our readers.

The editorial board has been involved in producing this book since its inception. They have spent rigorous hours researching and exploring the diverse topics which have resulted in the successful publishing of this book. They have passed on their knowledge of decades through this book. To expedite this challenging task, the publisher supported the team at every step. A small team of assistant editors was also appointed to further simplify the editing procedure and attain best results for the readers.

Apart from the editorial board, the designing team has also invested a significant amount of their time in understanding the subject and creating the most relevant covers. They scrutinized every image to scout for the most suitable representation of the subject and create an appropriate cover for the book.

The publishing team has been an ardent support to the editorial, designing and production team. Their endless efforts to recruit the best for this project, has resulted in the accomplishment of this book. They are a veteran in the field of academics and their pool of knowledge is as vast as their experience in printing. Their expertise and guidance has proved useful at every step. Their uncompromising quality standards have made this book an exceptional effort. Their encouragement from time to time has been an inspiration for everyone.

The publisher and the editorial board hope that this book will prove to be a valuable piece of knowledge for researchers, students, practitioners and scholars across the globe.

List of Contributors

Xiaochen Kang, Chun Dong and Shenghua Xu
Research Center of Government Geographic Information System, Chinese Academy of Surveying and Mapping, Beijing 100830, China

Jiping Liu
Research Center of Government Geographic Information System, Chinese Academy of Surveying and Mapping, Beijing 100830, China
Institute of Geographical Sciences, Henan Academy of Sciences, Zhengzhou 450052, China

Martin Breunig, Patrick Erik Bradley, Markus Jahn, Paul Kuper, Nima Mazroob and Norbert Rösch
Geodetic Institute, Karlsruhe Institute of Technology, 76128 Karlsruhe, Germany

Mulhim Al-Doori
College of Information Technology, University of Fujairah, Fujairah, UAE

Emmanuel Stefanakis
Department of Geomatics Engineering, Schulich School of Engineering, University of Calgary, Calgary, AB T2N 1N4, Canada

Mojgan Jadidi
Department of Earth & Space Science & Engineering, Lassonde School of Engineering, York University, Toronto, ON M3J 1P3, Canada

José Lucas Safanelli, Raul Roberto Poppiel, Luis Fernando Chimelo Ruiz, Fellipe Alcantara de Oliveira Mello, Rodnei Rizzo and José A. M. Demattê
Department of Soil Science, Luiz de Queiroz College of Agriculture, University of São Paulo, Pádua Dias Av., 11, Piracicaba, São Paulo Postal Code 13416-900, Brazil

Benito Roberto Bonfatti
State University of Minas Gerais, 700 Colorado Street, Passos, Minas Gerais Code 37902-092, Brazil

Hangbin Wu and Zeran Xu
College of Surveying and Geo-informatics, Tongji University, Shanghai 200092, China

Guangjun Wu
1RenData (ShangHai) Technology Co., Ltd., Shanghai 200092, China

Junghee Jo
Busan National University of Education, Busan 46241, Korea

Kang-Woo Lee
Electronics and Telecommunications Research Institute (ETRI), Daejeon 34129, Korea

Shu Wang, Xueying Zhang, Peng Ye, Mi Du, Yanxu Lu and Haonan Xue
Key Laboratory of Virtual Geographic Environment (Nanjing Normal University), Ministry of Education, Nanjing 210023, China
State Key Laboratory Cultivation Base of Geographical Environment Evolution (Jiangsu Province), Nanjing 210023, China
Jiangsu Center for Collaborative Innovation in Geographical Information Resource Development and Application, Nanjing 210023, China

Shuang Li
School of Earth Sciences, Zhejiang University, Hangzhou 310027, China

Junfeng Kang
School of Earth Sciences, Zhejiang University, Hangzhou 310027, China
School of Architectural and Surverying & Mapping Engineering, Jiangxi University of Science and Technology, Ganzhou 341000, China

Lei Fang and Xiangrong Wang
Department of Environmental Science and Engineering, Fudan University, Shanghai 200438, China

Juozas Gaigalas, Liping Di and Ziheng Sun
Center for Spatial Information Science and Systems, George Mason University, Fairfax, VA 22030, USA

Can Zhuang, Kai Ma and Mingqiang Guo
School of Information Engineering, China University of Geosciences, Wuhan 430074, China

Liang Wu and Zhong Xie
National Engineering Research Center for GIS, Wuhan 430074, China

Tong Zhang, Jianlong Wang, Chenrong Cui, Yicong Li and Wei He
State Key Laboratory of Information Engineering in Surveying, Mapping and Remote Sensing, Wuhan University, Wuhan 430079, China

Yonghua Lu
Shenzhen Investigation and Research Institute Co., Ltd., Shenzhen 518026, China

Qinghua Qiao
Chinese Academy of Surveying & Mapping, Beijing 100830, China

Kang Zhao and Hong Fan
State Key Laboratory for Information Engineering in Surveying, Mapping, and Remote Sensing, Wuhan University, 129 Luoyu Road, Wuhan 430079, China

Baoxuan Jin
Information Center, Department of Land and Resources of Yunnan Province, Kunming 650504, China

Weiwei Song, Sunyu Zhou and Yuanyi Jiang
Department of Geoinformation Science, Kunming University of Science and Technology, Kunming 650504, China

Jibo Xie and Guoqing Li
Institute of Remote Sensing and Digital Earth, Chinese Academy of Sciences, Beijing 100094, China

Tengfei Yang, Chuanzhao Tian and Jing Zhao
Institute of Remote Sensing and Digital Earth, Chinese Academy of Sciences, Beijing 100094, China
University of Chinese Academy of Sciences, Beijing 100094, China

Naixia Mou and Zhenyu Li
College of Geomatics, Shandong University of Science and Technology, Qingdao 266590, China

Index

T

U

V

W